VOLUME SEVEN

STRATIGRAPHY & TIMESCALES

Integrated Quaternary Stratigraphy

VOLUME SEVEN

STRATIGRAPHY & TIMESCALES

Integrated Quaternary Stratigraphy

Edited by

MICHAEL MONTENARI

Geography, Geology and the Environment
Keele University
Newcastle (Staffs.)
United Kingdom

ELSEVIER

ACADEMIC PRESS

An imprint of Elsevier

Academic Press is an imprint of Elsevier
50 Hampshire Street, 5th Floor, Cambridge, MA 02139, United States
525 B Street, Suite 1650, San Diego, CA 92101, United States
The Boulevard, Langford Lane, Kidlington, Oxford OX5 1GB, United Kingdom
125 London Wall, London, EC2Y 5AS, United Kingdom

First edition 2022

ISBN: 978-0-323-98913-8
ISSN: 2468-5178

For information on all Academic Press publications
visit our website at https://www.elsevier.com/books-and-journals

Publisher: Zoe Kruze
Acquisitions Editor: Jason Mitchell
Developmental Editor: Naiza Ermin Mendoza
Production Project Manager: Sudharshini Renganathan
Cover Designer: Miles Hitchen

Typeset by STRAIVE, India

Contents

Contributors

Upasana S Banerji
National Centre for Earth Science Studies, Thiruvananthapuram, Kerala, India

Isabelle Billy
Laboratoire Environnements et Paléoenvironnements Océaniques et Continentaux (EPOC)—UMR CNRS 5805, Université de Bordeaux, Pessac, France

Noel Thomas Boaz
Emory & Henry College School of Health and Science, Marion; International Institute for Human and Evolutionary Research, Integrative Centers for Science and Medicine, Martinsville, VA, United States

Karine Charlier
Laboratoire Environnements et Paléoenvironnements Océaniques et Continentaux (EPOC)—UMR CNRS 5805, Université de Bordeaux, Pessac, France

Jimmy Daynac
Laboratoire Environnements et Paléoenvironnements Océaniques et Continentaux (EPOC)—UMR CNRS 5805, Université de Bordeaux, Pessac, France

Maxime Debret
Département de Géosciences et Environnement—Morphodynamique Continentale et Côtière, Sciences et Techniques, Université de Rouen, Mont-Saint-Aignan, France

Chandra Prakash Dubey
National Centre for Earth Science Studies, Thiruvananthapuram, Kerala, India

Moftah Hasan El-Shawaihdi
Earth Science Department, Faculty of Sciences, University of Benghazi, Benghazi, Libya

Frédérique Eynaud
Laboratoire Environnements et Paléoenvironnements Océaniques et Continentaux (EPOC)—UMR CNRS 5805, Université de Bordeaux, Pessac, France

Thierry Garlan
Service hydrographique et océanographique de la Marine (SHOM), Brest, France

Jacques Giraudeau
Laboratoire Environnements et Paléoenvironnements Océaniques et Continentaux (EPOC)—UMR CNRS 5805, Université de Bordeaux, Pessac, France

Wiktoria Gornig
Department of Paleozoology, University of Wrocław, Wrocław, Poland

Kumar Batuk Joshi
National Centre for Earth Science Studies, Thiruvananthapuram, Kerala, India

Aleksandra Kropczyk
Department of Paleozoology, University of Wrocław, Wrocław, Poland

Tebogo V. Makhubela
Department of Geology, University of Johannesburg, Johannesburg, South Africa

Bruno Malaize
Laboratoire Environnements et Paléoenvironnements Océaniques et Continentaux (EPOC)—UMR CNRS 5805, Université de Bordeaux, Pessac, France

Elodie Marches
Service hydrographique et océanographique de la Marine (SHOM), Brest, France

Adrian Marciszak
Department of Paleozoology, University of Wrocław, Wrocław, Poland

Silindokuhle S. Mavuso
Department of Geology, Rhodes University, Eastern Cape, South Africa

Christoph Mayr
Institute of Geography, Friedrich-Alexander-University Erlangen-Nürnberg, Erlangen; Department of Earth and Environmental Sciences, Paleontology and Geobiology; GeoBio-Center, Ludwig-Maximilians-University München, München, Germany

Ahmed Muftah
Earth Science Department, Faculty of Sciences, University of Benghazi, Benghazi, Libya

Laxmi Pandey
CSIR-National Institute of Oceanography, Goa, India

Bogdan Ridush
Department of Physical Geography, Geomorphology and Paleogeography, Yuriy Fedkovych Chernivtsi National University, Chernivtsi, Ukraine

Linda Rossignol-Malaize
Laboratoire Environnements et Paléoenvironnements Océaniques et Continentaux (EPOC)—UMR CNRS 5805, Université de Bordeaux, Pessac, France

Gertrud E. Rössner
Staatliche Naturwissenschaftliche Sammlungen Bayerns, Bayerische Staatssammlung für Paläontologie und Geologie; Department für Geo- und Umweltwissenschaften, Palaeontologie und Geobiologie, Ludwig Maximilians Universität München, München, Germany

Marjolaine E.J. Sabine-Lamoureux
Laboratoire Environnements et Paléoenvironnements Océaniques et Continentaux (EPOC)—UMR CNRS 5805, Université de Bordeaux, Pessac, France

Fathi Salloum
Earth Science Department, Faculty of Sciences, University of Benghazi, Benghazi, Libya

Philipp Stojakowits
State Authority for Mining, Energy and Geology, Hannover, Germany

Sébastien Zaragosi
Laboratoire Environnements et Paléoenvironnements Océaniques et Continentaux (EPOC)—UMR CNRS 5805, Université de Bordeaux, Pessac, France

Preface

The Quaternary Period represents the past 2.58 Myr of Earth's history. It is a period characterized by severe climatic and environmental changes as well as significant developments and shifts within the global fauna and flora. In this book, a series of contributions are presented that describe and discuss various aspects of these changes, with the final chapter even reaching out beyond the Quaternary into the Neogene. Chapter 1 by **Upasana S. Banerji, Kumar Batuk Joshi, Laxmi Pandey, and Chandra Prakash Dubey** on "An outline of geochemical proxies used on marine sediments deposited during the Quaternary Period" describes that sediments have been a ubiquitous archive for paleoclimate reconstruction, whereas marine sediments are considered to be serene and mostly undisturbed from anthropogenic encroachments. Unlike continental records, deep-sea sediments have elucidated the evidence of at least 50 glacial and interglacial stages during the Quaternary Period and has thus proved their applicability in Quaternary climate reconstruction. The geochemical variations in marine sediments are often used as a proxy to decode past productivity, redox, weathering, and provenance changes as a function of past climate and oceanographic perturbations. The geochemical behavior of the elements in the sedimentary environment primarily relies on ionic potential along with redox potential and pH, which leads to the mobility and enrichment of selected elements, thus acting as potential evidence for ambient temporal changes. This chapter aims at providing an overview of frequently used geochemical proxies and their applicability. Furthermore, this chapter also tries to provide the significance of statistical and machine learning approaches to geochemical datasets for understanding the climatic processes that led to changes in geochemical variability. Chapter 2 by **Marjolaine E.J. Sabine, Frédérique Eynaud, Sébastien Zaragosi, Jacques Giraudeau, Maxime Debret, Linda Rossignol-Malaize, Karine Charlier, Isabelle Billy, Bruno Malaize, Jimmy Daynac, Elodie Marches, and Thierry Garlan** titled "Stratigraphy in the Greenland/Iceland/Norwegian (GIN) seas: a multiproxy approach on Pleistocene sediments" highlights that a multiproxy sedimentological study was conducted on five sediment cores retrieved between 67°N and 79°N in the Greenland/Iceland/Norwegian seas in order to infer a common chronostratigraphic model for the selected cores. This model is based on the use of a series of geochemical, physical, and micropaleontological proxies, which are routinely measured in the sedimentological investigations of marine sediment cores: the major and minor element content derived from X-ray fluorescence (XRF) core scanner analyses; the lightness and color of the sediments derived from spectrophotometric analyses; the magnetic susceptibility of the sediments; and the distribution of planktonic foraminiferal assemblages combined with coccolith stratigraphy (acme zones) to provide preliminary stratigraphic tie points. All these proxies are independently correlated between the five sediment cores. Our results demonstrate that high-resolution studies using these standard paleoceanographic tools are powerful for establishing a core-to-core correlation. A chronostratigraphic framework is proposed based on the correlation of the planktonic $\delta^{18}O$ isotopic record obtained from core M17KC03 with the LR04 benthic $\delta^{18}O$ reference stack. A comparison between the M17KC03 stratigraphy and other dated cores from the subpolar North Atlantic confirms the robustness of the approach. Our study suggests that XRF core scanner-derived elemental ratios (especially those related to Ca), the lightness L*, and the coccolith-based biostratigraphy provide robust stratigraphic tie points at the scale of the whole Greenland/Iceland/Norwegian seas. The elemental ratio Sr/sum, magnetic susceptibility, and planktonic foraminiferal records have a more local stratigraphic significance in the GIN seas. Over the last 1100 ka BP, our datasets are indicative of: (1) an excursion of Ca content and planktonic foraminiferal abundances during MIS 11 and (2) an overall decrease toward the present of both the detrital content (with the exception of values characterizing MIS 12) and the magnetic susceptibility of sediments deposited during Glacials. Chapter 3 by **Philipp Stojakowits and Christoph Mayr** on "Quaternary Palynostratigraphy of Germany with special emphasis on the Late Pleistocene" summarizes and reviews the current state of knowledge on Quaternary palynostratigraphy in Germany. Our focus is especially on the Pleistocene and a comparison between northern and southern Germany, while we consider the Holocene only marginally, since many review articles already exist about this epoch. In the region under consideration, several warm and cold stages were reported for the Early Pleistocene. The thermomers are characterized by the partially frequent occurrences of relict Neogene elements. Concerning the last interglacial of the Early Pleistocene (i.e., the Osterholz Interglacial), for instance, *Eucommia* and *Azolla filiculoides* are important for stratigraphical purposes. The Middle Pleistocene encompasses the Cromerian Complex with five interglacials detected so far in northern Germany.

Subsequently, the Elsterian and Saalian glacial cycles occurred intercalated by the Holsteinian interglacial. This warm stage was detected at many sites all over Germany with a characteristic vegetational development, including the presence of *Fagus* and *Pterocarya*. In contrast, just a few records exist from the warm stages within the Cromerian Complex and the Saalian and their vegetational succession is poorly known. The interglacial preceding the Holocene, i.e., the Eemian interglacial, in turn, is the palynostratigraphically best defined Pleistocene warm stage. The subsequent alternation between interstadial and stadial conditions during the last glacial (i.e., Würmian or Weichselian) led to the formation and decline of boreal forests and shrub tundra, respectively. However, these vegetation changes are well-understood only up to early Marine Isotope Stage (MIS) 3. The further course of the Middle Würmian and Weichselian vegetation development is still insufficiently reconstructed, since most pollen records provide only short time intervals and do not include the full range of stadials and interstadials found, for example, in Greenland ice cores. There are also few continuous pollen records for the Last Glacial Maximum. In contrast, the vegetation history of the Late Glacial and subsequent Holocene is again well-known and, in some places, is very precisely dated based on varve chronologies. We conclude that despite a long palynological tradition in Germany, some periods are underrepresented or hardly studied for various reasons and a correlation between north and south is often not provided. The focus of future studies should therefore be on these periods and on a better correlation of pollen stratigraphy using state-of-the-art dating technologies. Chapter 4 by **Adrian Marciszak, Wiktoria Gornig, Aleksandra Kropczyk, and Gertrud Rößner** titled "The latest European record of *Chasmaporthetes lunensis lunensis* (Del Campana, 1914) from Schernfeld (Bavaria, Germany) in terms of the changes in the European carnivore paleoguilds" explains that *Chasmaporthetes lunensis* lacks the robust and hyper-robust dentition of the bone-cracking hyenas. Known as the "hunting hyena," it was characterized by a gracile build and was well-adapted to running and chasing down prey. The animal was widespread throughout the Pliocene and the Early Pleistocene, ca. 4.5–1.5 Myr. This hyena has been an important component of the fauna as a cursorial carnivore. By the earliest Pleistocene, the relatively long-lived and seemingly fairly stable guild of the European Pliocene was changing and canids from the genus *Lycaon* and *Canis* replaced *Chasmaporthetes*. Its last known European appearance was at the

German site Schernfeld, where five isolated teeth documented the occurrence of one, large individual. Chapter 5 by **Bogdan Ridush** titled "The Quaternary vertebrate fauna of cave deposits of the Podillia-Bukovynian Karst-Speleological Area (Western Ukraine)" is highlighting that The Podillia-Bukovinian Karst-Speleological Area is rich in numerous caves of karst, suffosion, and weathering origin. The gypsum karst is the most widespread in the area. It is famous for the World's largest maze-caves in gypsum. Some of these caves contain Quaternary sediments with vertebrate fauna that can be used for biostratigraphy. The first faunal remains were found as early as 1886. Up to now, at least 30 cave sites with Quaternary vertebrate fauna are known. Most of the sites are in horizontal caves, which results in specific bone- and sediment accumulations. The formation of most of the bone accumulations is related to the inhabitation of caves by the *troglophilic* species, especially carnivores, and only a few inhabited by humans. The oldest among the cave sites is Chortkiv, dated to the late Odesian Fauna (Kryzhanivka Stage, MIS 41-62). The best-studied cave sections with cave fauna are Bukovynka, Kryshtaleva (Kryvchanska), Pryima I, Chortovi Skeli (Lviv VIII). Several caves are promising for further research, such as Martynivka, Tovtry, and others. A very small part of the osteological material was published in detail. Only 15 radiocarbon dates are known for the cave sites in the region. Many finds in the caves were made in the 1950s-1960s, sometimes just by accident. Part of the bone collections were lost. The material stored in museums should be revised. There still is a large potential for the future research of the Quaternary cave deposits of the Podillia-Bukovynian Karst-Speleological Area (Western Ukraine). Chapter 6 by **Tebogo V. Makhubela and Silindokuhle S. Mavuso** on "An overview of the geology of paleoanthropological and archaeological sites of South Africa" highlights that paleoanthropological and archaeological remains in South Africa are found in various sites formed in diverse environments of caves, rock shelters, and open-air settings. These sites contain deposits mainly dating back to the Quaternary but also extending back into the Late Pliocene. The Cradle of Humankind (Cradle) is the main locality of the Fossil Hominid Sites of South Africa's UNESCO World Heritage Site. It has 15 sites that have preserved fossils belonging to several hominin species that lived during the Late Pliocene and Pleistocene times. Sites in the Cradle were developed in continuously evolving cave systems due to lithological differences and tectonic processes that promote and control the dissolution of the Late Archean–Early Proterozoic dolostone host bedrock. Thus, the deposits in the caves have complex and intermittent stratigraphy, difficult to resolve even with advanced dating techniques. Outside the Cradle, there are innumerable sites with deposits that mostly date later than the Early Pleistocene. Thus, these sites have preserved paleoanthropological and archaeological remains associated with archaic humans and *Homo sapiens*. Here, we provide an overview of the stratigraphy and timescales of 6 sites in the Cradle and 20 sites outside the Cradle to highlight the differences between the sites, their importance, and the need for further research at those sites. Chapter 7 by **Moftah Hasan El-Shawaihdi, Noel Thomas Boaz, Ahmed Muftah, and Fathi Salloum** reaches out beyond the Quaternary and establishes a "New lithostratigraphic nomenclature of Neogene rock units in As-Sahabi area, Sirt Basin, northeast Libya" by reviewing the lithostratigraphic nomenclature of Neogene formal and informal rock units in the As-Sahabi area of northeast Libya. Formally named lithostratigraphic units are modified, whereas informal units are redesignated and given new names. These modifications are proposed on the basis of regional and local correlations, the new identification of an unconformity, and the designation of a new type of locality. The lowermost unit is a limestone of the Miocene (Langhian–Tortonian) age. This limestone was previously designated as Formation "M," but the term Formation "M" is to be abandoned and this limestone should be named the Benghazi Formation of the Ar Rajmah Group. This limestone is capped by a regional unconformity, which is overlain by a 10–40-m-thick unit of Miocene (Messinian) sand, calcareous sand, gypsiferous clay, silty clay, and sandy clay. This unit was previously designated as Formation "P" and the lower part of Member "T" of the Sahabi Formation, but these designations (Formation "P" and Member "T" of the Sahabi Formation) are to be abandoned and this unit is named as the Sabkhat Al Qunnayyin Member of the Sahabi Formation. The 10–40-m-thick unit of Miocene (Messinian) sand, calcareous sand, gypsiferous clay, silty clay, and sandy clay (Sabkhat Al Qunnayyin Member) is overlain by the following nongypsiferous units (from base to top) of the Miocene (Messinian) age: a 20-m-thick unit of white to yellow sand (with balls and lenses of clay); a 10-m-thick unit of white sand, green to gray clay, pink to purple clay, and cross-bedded sandstone; and a 4-m-thick unit of green clay and sand (with two beds of brown sandstone). Collectively, these strata form a 34-m-thick interval of nongypsiferous sand and clay that was previously designated as Members "U1," "UD," and "U2," but these designations

(Members "U1," "UD," and "U2") are to be abandoned and these nongypsiferous strata are collectively named the Qarat Makada Member of the Sahabi Formation. This 34-m-thick interval of nongypsiferous sand and clay (Qarat Makada Member) is capped by a regional unconformity, which is overlain by a 10–40-m-thick unit of Pliocene white to green sand, pebbly sand, and sandy clay. This unit was previously designated as Member "V" of the Sahabi Formation, but this designation is to be replaced by the name Qarat Weddah Formation. The 10–40-m-thick unit of Pliocene white to green sand, pebbly sand, and sandy clay (Qarat Weddah Formation) is overlain by a 1.5–2-m-thick unit of carbonate soil/calcrete. This carbonate soil/calcrete was previously designated as Member "Z" of the Sahabi Formation, but this designation (Member "Z") is to be replaced by the name Shagig Formation. These new formal names should be used from now onward, and all informal names should be discarded. This new lithostratigraphic nomenclature of the Neogene Sahabi units is designated according to the North American Commission on Stratigraphic Nomenclature and the International Stratigraphic Guide.

Michael Montenari
Keele University
United Kingdom

CHAPTER ONE

An outline of geochemical proxies used on marine sediments deposited during the Quaternary Period

Upasana S. Banerji[a,*], Kumar Batuk Joshi[a], Laxmi Pandey[b], and Chandra Prakash Dubey[a]
[a]National Centre for Earth Science Studies, Thiruvananthapuram, Kerala, India
[b]CSIR-National Institute of Oceanography, Goa, India
*Corresponding author: e-mail address: upasana.s.banerji@gmail.com

Contents

Abstract

The sediments have been a ubiquitous archive for paleoclimate reconstruction while the marine sediments are considered to be serene and mostly undisturbed from anthropogenic encroachments. Unlike continental records, the deep-sea sediments elucidated the evidence of at least 50 glacial and interglacial stages during the Quaternary period and thus proved its applicability in quaternary climate reconstruction. The geochemical variations in the marine sediments are often used as a proxy to decode the past productivity, redox, weathering and provenance changes as a function of past climate and oceanographic perturbations. The geochemical behavior of the elements in the

Stratigraphy & Timescales, Volume 7
ISSN 2468-5178
https://doi.org/10.1016/bs.sats.2022.09.002

sedimentary environment primarily relies on ionic potential along with redox potential and pH which leads to mobility and enrichment of selected elements and thus acting as potential evidence for ambient temporal changes. The present chapter aims to provide an overview of frequently used geochemical proxies and their applicability. Further, the chapter also tries to provide the significance of statistical and machine learning approaches to the geochemical datasets in understanding the climatic processes that led to changes in the geochemical variability.

1. Introduction

The historical climate data has ratified the prevalence of global warming, especially during the last century. But due to the restricted extension of the climate data, it is still ambivalent whether such a warming trend is a cyclic process or we are on the brink of a devastating future. The assessment of the present and future climatic trends demands better and higher resolution investigation of the past climatic trends with a specific focus on the Quaternary Period. The Quaternary Period broadly divided into Pleistocene (~2 Ma to 11.8 ka) and Holocene (11.8 ka to Present), is not just a coda to 4500 Myr Earth's history but it is the geological time that witnessed frequent and rapid global climate perturbations leading to noticeable sea-level changes. The earlier phase of Quaternary Period (2–0.9 Myr) exhibited a cyclicity of 41 kyr for the ice sheet contraction and expansion. On the contrary, the last 0.9 Myr depicted a 100 kyr periodicity (Williams et al., 1998). Such disparity in the ice sheet spread within the Quaternary Period demands the need to reconstruct a high-resolution climate perturbation. Additionally, it is also necessary to address the cyclicity, mechanism and feedback processes attributed to global climate system. The climate perturbations and cyclic mechanism has been addressed through various geochemical, mineralogical, isotopic and biotic proxies on the natural archives from terrestrial (tree rings, lake sediments, peats, speleothems) and marine (corals, coastal and marine sediments, coastal peats) realms (Banerji et al., 2020; Joshi et al., 2021a) supported by several geochronological techniques using radioisotopes (Banerji et al., 2022b; Joshi et al., 2021b).

The ubiquitous presence of sediment in both marine and terrestrial environment is a crucial and frequently used archive for the past climate reconstructions as the layered accretion of the sediments under lacustrine, coastal and marine environment catalogues the ambient conditions for its deposition (Bradley, 2015). However, the signatures of the past climate variability from the sediment deposition under terrestrial environment is ambivalent due to poor preservation caused by physio-chemical weathering, erosion, human interference (Calvert and Pedersen, 2007) or site inaccessibility. Under such circumstances, the marine sediments have been proved to be an imperative archive to decrypt the past climate and oceanographic conditions.

Often the constituents of marine sediments are broadly divided into cosmogenous, lithogenous, biogenous and hydrogenous based on their origin (Goldberg, 1961). The cosmogenous and lithogenous together forms part of the detrital component. Cosmogenous constituent comprises of silicate chondrules and cosmic spherules with particles of FeNi and is beyond the scope of the present study. The constituents derived from the rock weathering on land and/or seafloor and volcanic eruptions represents lithogenous. The biogenous fraction is represented by planktonic and benthic organism along with biogenic apatite and the constituents form by the inorganic precipitation of the sea water is identified as hydrogenous component of the marine sediment (Berger, 1976; Goldberg, 1961; Piper and Heath, 1989). The present chapter primarily focuses on the lithogenous, biogenous and hydrogenous fraction of the marine sediments.

Usually, the elemental and isotopic variations in the marine sediment have been gleaned to decode the climate changes over terrestrial system by judiciously selecting the sediment cores from the marine environment which in turn also provided cues on paleoprovenance, paleosealevel and paleoweathering of the selected region. Nevertheless, the durability of climate records profoundly relies on the constituents based on which the climate has been reconstructed. Considering the geochemical aspect, it can be broadly divided in to labile that can easily suffer changes or breakdown and refractory which resists against any change or alteration. With the passage of time, the refractory component gets enriched while the labile components get depleted or reduced. Eventually, the geochemical component significantly varies from the time of formation/deposition to the time of its estimation for paleoclimate reconstruction. Thus, multiproxy approaches including geochemical, isotopic and biotic components are simultaneously implemented to delineate the past climatic perturbations (Higginson, 2009) The present chapter aims to discuss the frequently used geochemical proxies for the reconstruction of past climate variations especially during Quaternary period.

2. Approach

During the last century, the systematic application of the proxies was attempted to reconstruct the past climate and oceanographic conditions using marine sediments. The proxies for marine sediments can be categorized in to six broad divisions, viz., microfossils, stable isotopes, radiogenic isotopes, biogenic compounds, sedimentology, and elements (Gornitz, 2006). The present chapter aims to provide an overview of the elemental concentration and their ratios used as proxies in the reconstruction of the paleoclimatic and paleoceanographic conditions using marine sediments. Further, the chapter also provides a comprehensive view of the statistical and machine learning approaches made on the geochemical datasets to address the paleoclimate and paleoceanographic reconstruction.

3. Major and trace elemental proxies

The paleohydrological conditions, paleotectonics settings, past water compositions paleoweathering, past redox environment and sedimentary provenance have been assessed through the geochemical composition of the sediments and sedimentary rocks (Claes et al., 2019; Sun et al., 2012). A variety of factors like the source rock composition, weathering intensity in the source region, depositional environment and post-depositional changes plays a vital role in controlling the chemical and mineralogical contents of the sediments/sedimentary rocks (Banerji et al., 2022a; McLennan et al., 1993; Nesbitt and Markovics, 1980; Roser and Korsch, 1988). The distribution of the elements is heterogenous in the Earth ranging from core, mantle to crust and the elements are broadly divided in to major (>0.4%), minor (0.1–0.4%) and trace (<0.1%) elements (Lee, 2019). The chemical behavior of each of the elements is controlled by its stability of the outer electrons rendering it high electronegativity (tend to attract electrons due to nearly filled electrons in the outmost shell) or low electronegativity (tend to donate electron due to partially filled outer shell) thereby forming anions (Cl^-, OH^-, SO_4^{2-}, etc.) or cations (Na^+, Ca^+, K^+, etc.) respectively. Further, the Ionic Potential (IP) represented by charge to ionic radius of the elements plays a crucial role to decipher the mobility of the selected elements (Fig. 1). Ions with low IP (<3) fails to break the water molecule bonds (O—H bond) and thus remains in the solution as hydrated cations (such as Na^+, K^+, Rb^+, Sr^+, Cs^+) and also referred as large ion lithophile elements

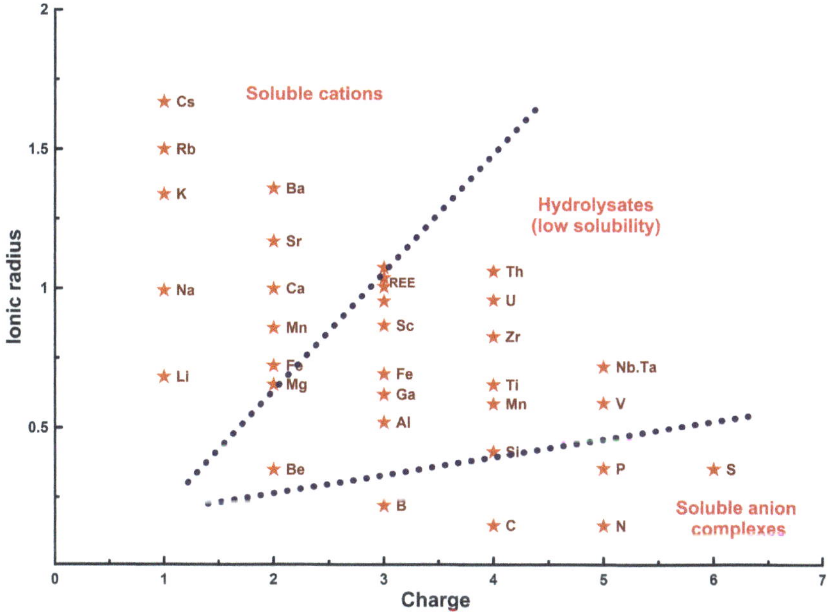

Fig. 1 The plot depicts the ionic radius and charge (valence) for selected geochemically significant elements. The cations with low IP are soluble while the ions bonds with OH^- with intermediate IP and the soluble cation complexes are formed with high IP (Bjørlykke, 2010).

(LILE) while elements with comparatively high IP (such as Ti^{4+}, Zr^{4+}, Nb^{5+}) referred as High Field Strength Elements (HFSE) (Bjørlykke, 2010; Lee, 2019). The elements with IP (nearly equivalent to the strength of O—H bond in H_2O, i.e., 3–12) usually bonds with OH^- group and has low solubility thereby forming hydrolysates such as Al $(OH)_3$, Fe $(OH)_3$, etc. Thus, during weathering, generally low IP elements remains in solution along while the hydrolysates elements such as Al^{+3}, Fe^{+3}, Mn^{4+}, Ti^{4+}, etc. gets enriched as residues on land (Bjørlykke, 2010). Ions with IP > 12 are much stronger than O—H bonds yielding soluble anions (SO_4^{2-}, PO_4^{3-}) and releasing H^+ ion in the solution (Bjørlykke, 2010). In addition to IP, there are other factors such as redox potential (Eh), distribution coefficient and pH, that also plays crucial role in elemental variation in the sedimentary environment. Moreover, the changes in the pH and Eh are the prime factors that encompasses the mechanism of chemical precipitation or dissolution of the selected elements provided the environment should not be influenced by extensive evaporation (Bjørlykke, 2010; Garrels, 1965; Lindsay, 1979).

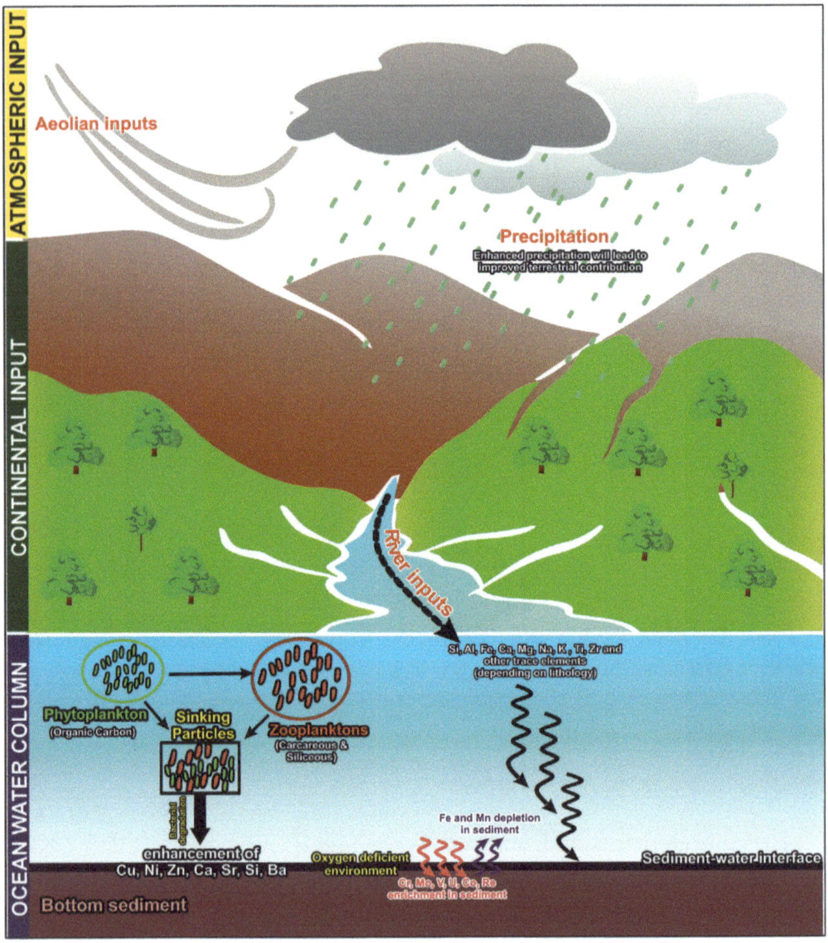

Fig. 2 Schematic of plausible sources and/or processes resulting in the deposition of the elements in the bottom sediments and thus representing geochemical indicator of detrital influence, in-situ productivity, redox conditions and weathering intensity.

The following section discusses the applicability of the elemental contents and their ratios depicting detrital source weathering, in-situ productivity and paleo redox conditions in deciphering the past environmental changes in the region. A schematic of the is also shown in Fig. 2.

3.1 Detrital source and grainsize indicators

The formation of terrigenous detritus along with the dissolved cations and anions is controlled by the climate and tectonics, predominantly the interaction between the continental crust upliftment and the physical and/or chemical weathering processes (Sageman and Lyons, 2003). The mineralogy of source region, weathering intensity and the reactivity of the products during its transportation produced through weathering controls the chemical composition of the detrital fraction of the sediment (Sageman and Lyons, 2003). The climatic parameters significantly controls the transportation of the weathered products which is either mediated by riverine or wind driven flux while for marine environment, currents also plays a crucial role (Bertrand et al., 1996; Sageman and Lyons, 2003).

The silica (SiO_2), aluminum dioxide (Al_2O_3) and titanium dioxide (TiO_2) are the major component of aluminosilicate phase (Agnihotri et al., 2003b). Among these, the SiO_2 normally acts as a general building block for the aluminosilicates. Further, the SiO_2 also separately occurs as quartz mineral that remains associated with silt and sand fractions of sediments (Calvert and Pedersen, 2007; Feniak, 1944). In addition to SiO_2, Al_2O_3 and TiO_2 acts as resistant fractions towards chemical leaching and continues to prevail its properties even under several erosional and depositional cycles (Calvert and Pedersen, 2007). The Al_2O_3 and TiO_2 acts as conservative elements which not only retains its properties under weathering but it can also resist redox reactions (Nesbitt and Markovics, 1997) hence considered as a terrestrial proxy (Agnihotri et al., 2003b; Kemp and Izumi, 2014) to infer the changes in the hydrological conditions. Moreover, Ti can exist in aluminosilicates (in place of some major elements) or as a discrete Ti oxide and silicate mineral phases. The formation of Ti oxides during weathering leads its enrichment in laterites and bauxite. The Ti is transported along with silt, fine-sand and coarser quartz grains in the form of rutile (TiO_2), sphene ($CaTiSiO_5$) and ilmenite ($FeTiO_3$) (Calvert and Pedersen, 2007; Dolcater et al., 1970; Spears and Kanaris-Sotiriou, 1976) while trivial amount of Ti can also reside in clay wherein it can substitute aluminum, iron and manganese and plausibly silicon in the aluminosilicates such as biotite, amphiboles, pyroxenes and selected clay minerals (Correns, 1969; Wedepohl et al., 1969). The TiO_2 varies distinctly in different rocks than SiO_2 and Al_2O_3 and it is hosted in Ti-bearing minerals which are transported with silt and sand fraction of the sediment (Spears and Kanaris-Sotiriou, 1976). Among these selected major elements, the Al_2O_3 is widely accepted as lithogenous reference element and is endorsed for the elemental normalization purpose such as the SiO_2/Al_2O_3 provides cues on the quartz to aluminosilicate minerals which in turn also helps in inferring the grain size variability as quartz and Al_2O_3 remain associated with coarser and clay fractions respectively (Calvert and Pedersen, 2007). Similarly, the Ti/Al or TiO_2/Al_2O_3 ratio can also be used as a grain-size indicator as the heavy minerals are the main Ti carrier in varied sediments (Correns, 1969; Spears and Kanaris-Sotiriou, 1976). Apart from these major elements, the zirconium (Zr) primarily exhibited by the zircon mineral remains chemically resistant during weathering, gets transported with fine to medium sand quartz grains (Feniak, 1944) and survives recycling (Joshi et al., 2022a; Mason and Jacobs, 1998; Sudom and Arnaud, 1971) and thus also serves as grainsize proxy as Zr content and/or in the form of Zr/Al_2O_3 within the sediments (Ganeshram, 1998).

A number of studies have utilized the variation in the detrital flux to predict paleomonsson. One of the recent studies demonstrated an enhanced detrital flux and coarser grainsize in the Vellayani lake sediment, Kerala, southern India due to the improved hydrological conditions owing to enhanced Indian summer monsoon (Banerji et al., 2021b). In another study, a lowering of Al content in the sediment from Benital lake, higher central Himalaya was suggested to be associated with cooler climatic events and corroborated well with the bond cycle of north Atlantic ocean (Bhushan et al., 2018).

3.2 Paleoweathering and weathering intensities

The chemical weathering plays a vital role in global hydrogeochemical cycle of elements (Warrier and Shankar, 2009; Zhao and Zheng, 2015) and leads to partitioning of the elements between continental rocks and water (Taylor and McLennan, 1985). Sediment weathering can be mediated by several factors like microbes, climate, nature of source rocks, and relief (Joshi, 2014; Madhavaraju et al., 2016; McLennan et al., 1993; Taylor and McLennan, 1985; Xi et al., 2014). The alkaline and alkaline-earth metals are highly mobile in humid conditions while the arid conditions, lower moisture content, as well as temperature in colder regions, inhibits weathering. However, in exceptional cases, lower temperatures have been known to cause exceptionally significant weathering (Hall et al., 2002; Marra et al., 2017; Nesbitt and Young, 1982; Schneider et al., 2016).

The immobile major elements, HFSEs, transition trace elements, REEs, or elemental ratios are often utilized to study the nature of the source weathering (Condie, 1993; Greber and Dauphas, 2019; Lupker et al., 2013; Taylor and McLennan, 1985). In certain cases, the geochemistry of sediments gets altered by weathering and can be used to understand the palaeo-weathering conditions (McLennan, 1993; McLennan et al., 1993). The chemical weathering preferentially leaches the mobile elements

(Ca, Na, K, and Sr) which in turn enriched the immobile (Al, Ti, Th, Sc) and REEs in the sediment (Jin et al., 2006; Li et al., 2016; Resmi et al., 2020). The parent rock weathering depletes Na and Ca in the subsequently formed sediments as these elements are preferentially removed. On the other hand, the concentration of K, Mg, Rb and other immobile elements increases as the former get adsorbed by clay and the latter are concentrated in the residue. Thus, the elemental ratios such as Na/Al, Na/K, Ca/K and Rb/Sr has been extensively used to evaluate the chemical weathering (Lü et al., 2016; Yang et al., 2004; Zhang et al., 2014). Furthermore, the Mg/Al and Ti/Al ratios have also been used as proxies for weathering while Ti/Ca has been used as proxy for the enhanced clastic flux (Warrier and Shankar, 2009; Clift et al., 2014).

Various weathering indices such as plagioclase index of alteration (PIA), chemical index of alteration (CIA), index of chemical variability (ICV), chemical index of weathering (CIW), etc. have been implemented to study weathering intensity (Table 1). Out of these CIA is an extensively used weathering index (Fedo et al., 1996; Nesbitt and Young, 1982). Weathering oflabile cations (e.g., Ca^{2+}, Na^{+}, and K^{+}) relative to stable residual constituents (Al^{3+}, Ti^{4+}) are denoted by higher CIA values (Fig. 3)while low values indicate minimal chemical degradation (Sandeep et al., 2022). In certain cases, post-deposition addition of K may compromise the CIA values (Fedo et al., 1995; Garzanti et al., 2014; Joshi et al., 2021c). In order overcome post-depositional K-addition, CIW and PIA are calculated wherein values closer to 100 indicate intense plagioclase breakdown.

3.3 Paleoproductivity proxies

The primary productivity in the ocean is the foremost driving mechanism in the marine carbon cycle that regulates the global climate system. The past productivity variations has been deciphered based on paleontological and geochemical variations (Sanchez, 2019). In palaeoceanographic studies usually the export productivity is being reconstructed which includes the accumulation of organic and inorganic contents in the marine sediment and thus relies on export and burial efficiencies.

The opal, organic carbon and calcium carbonate content have served as the paleoproductivity indicators and had played a major role in portioning of CO_2 between the atmosphere and deep-sea environment under the oceanic conditions (Ragueneau et al., 2000). But they tend to suffer variable degree of recycling within the water column which may lead to fading of their

Table 1 The table represents the frequently used weathering indices and its formulae.

Weathering index	Formulae	Remarks	References
Chemical Index of Alteration (CIA)	$CIA = \{Al_2O_3/(Al_2O_3 + K_2O + CaO^* + Na_2O)\} \times 100$	Estimates the silicate weathering with values for igneous rock <50; Shale: 70–75; residual Clays: close to 100	Nesbitt and Young (1982)
Plagioclase Index of Alteration (PIA)	$PIA = \{(Al_2O_3 - K_2O)/(Al_2O_3 + CaO^* + Na_2O - K_2O)\} \times 100]$	Plagioclase weathering can be deciphered in addition to silicate weathering	Fedo et al. (1995)
Chemical Weathering Intensity (CWI)	$CWI = [(CaO + MgO + Na_2O)/Al_2O_3 \times 100]$	Additional support to CIA wherein higher CWI values represents higher soluble and mobile element input	Sun et al. (2010)
Chemical Index of Weathering (CIW)	$CIW = [Al_2O_3/(Al_2O_3 + CaO + Na_2O)]$	Improved approach to calculate the degree of chemical weathering; avoid; avoids K_2O in the equation as it may get leached or can get accumulated in the residue during weathering	Harnois (1988)
Index of compositional variability (ICV)	$ICV = [(CaO + K_2O + Na_2O + Fe_2O_3 + MgO + MnO + TiO_2)/Al_2O_3]$	Rock-forming minerals >0.84; clay minerals <0.84	Cox et al. (1995)

respective signatures (Dymond et al., 1992). Instead of direct indicators the geochemical proxies in the form of substitutes for siliceous, organic carbon and calcium carbonate productivity have been identified and adopted for successful reconstruction of the paleoproductivity. The calcareous productivity is frequently identified in the form of foraminifera, bivalves, gastropods etc. can be supported by $CaCO_3$ and CaO/Al_2O_3 (Achyuthan et al., 2014; Bhushan et al., 2001; Wei et al., 2004).

The total organic carbon (TOC) content in the sediments from marine, coastal and lacustrine sediments have provided extensive cues on the in-situ productivity variations (Agnihotri et al., 2002; Banerji et al., 2017; Bhushan et al., 2018; Chandana et al., 2017). However, the oxidation of TOC leading to its degradation and such process is instigated by the accumulation of iron oxides and manganese (Fe-Mn oxyhydroxides) thereby affecting several other trace elements as Fe-Mn oxyhydroxides has strong adsorption capacity that can decline the trace element diffusion rates from the sediment (McKenzie, 1980; Tribovillard et al., 2006). The degradation of TOC might mask the variation of the in-situ productivity and under such circumstances copper (Cu) and nickel (Ni) forms organometallic complexes and gets delivered to the sediment thereby representing the organic carbon productivity in the sediments (Tribovillard et al., 2006). Moreover, the phytoplankton continues incorporating barium (Ba) through metabolic uptake or adsorption during their life span but the decay of phytoplankton results in the release of Ba that might get precipitated in the form of barite and thus yielding cues on the in-situ productivity (Daihres et al., 1980; Dymond et al., 1992). In addition, the phosphorous (P) in the sediment is mediated by the phytoplankton as the P is released in the form of PO_4^{3-} into the sediment-water interface during the bacterial decay of organic matter from oxic to anoxic conditions and thus also acts as a proxy to paleoproductivity (Kidder et al., 2003; Sannigrahi and Ingall, 2005).

3.4 Redox sensitive elements and their mobilization

In oxic-environment (O_2 content in water is 8–2 mL/L) (Tyson and Pearson, 1991), the aerobic organism consumes the dissolved oxygen from the interstitial and ambient waters in order to breakdown the organic matter while the continued degradation of organic matter causes oxygen depletion. Under such conditions the organism tends to exploit the secondary oxidants (nitrate, Mn- oxides and Fe-oxides and oxyhydroxides and sulfate) which in turn causes the development of oxygen deficient (suboxic, anoxic or euxinic) conditions within the sediment and/or sediment-water interface (Tribovillard et al., 2006). Thus, based on the oxygen [O_2] content, four types of redox depositional conditions can prevail i.e., oxic (>2 mL/L), dysoxic (2–0.2 mL/L), suboxic (0.2–0 mL/L), anoxic (0 mL/L) and euxinic (zero along with free H_2S) (Algeo and Li, 2020; Tyson and Pearson, 1991). The oxygen deficient condition triggers the remobilization and enrichment of the elements from sediment to porewater or from water to sediment-water interface (Banerji et al., 2016; Tribovillard et al., 2006). Such diffusion and enrichment patterns of the elemental content has been significantly used to decipher paleoredox which persisted during the sediment depositional phase (Fig. 4).

The iron (Fe) has been explored extensively to improve the understanding of the paleoredox conditions of the sediment depositional environment. Generally, the reduction and dissolution of Fe^{3+} in the suboxic sediments leads to the release of Fe into the water column wherein the Fe^{2+} within the water column if migrates to euxinic environment in the water column it tend to react with dissolved sulfide in order to form Fe sulfide minerals that sink to underlying sediments (Scheiderich et al., 2010). Likewise, in case of Manganese (Mn), under oxygen deficient environment the reduction of Mn takes place and soluble cation (Mn^{2+} or MnCl) are formed (Calvert and Pedersen, 1993). As the Mn within the sediments cannot be taken up by the organic matter, they tend to diffuse upwards until they witness oxic conditions, where they may dissolve to water column or gets precipitated as Mn oxides. However, under certain circumstances if the dissolved Mn encounters carbonate they may get trapped in authigenic Mn-carbonates or as oxides (Hild and Brumsack, 1998). Further, the ratios of Fe/Mn have also served as redox proxy as both are sensitive to varied oxidative and reductive conditions (Davison, 1993). The oxidation of Fe^{+2} is easier than Mn^{+2} while the hydroxide of Mn tend to deoxidize more easily than Fe and thus, the enhanced values for the Fe/Mn generally occurs as a result of reduced conditions (Wersin et al., 1991; Zhu et al., 2008).

The chromium (Cr) prevails as Cr^{+4} under oxic environment which reduces to Cr^{+3} forming aqua hydroxyl and hydroxyl cations under anoxic conditions that gets adsorbs to Fe and Mn oxyhydroxides or complexes well with humic and fluvic acids (Algeo and Maynard, 2004; Breit and Wanty, 1991). However, the uptake of Cr^{+3} by authigenic Fe-sulfides is limited due to its structural and electronic incompatibilities with pyrite and Cr is seldom known to form insoluble sulfides (Huerta-Diaz and Morse, 1992; Morse and Luther, 1999). Moreover, the Cr content in the sediment could be a land driven clastic fraction such as clay minerals, chromite, Fe-Mg minerals (Cr substitute for Mg) thus reduces its applicability as a redox sensitive proxy (Brumsack, 1989; Francois, 1988; Hild and Brumsack, 1998).

Within the oxygenated environment, Co generally exist as dissolved cation (Co^{+2}) or else is complexed with humic/fluvic acid (Saito et al., 2002; Whitfield, 2001). Under anoxic conditions, the Co forms insoluble sulfides (CoS) which is taken up by solid solution by authigenic Fe-sulfides but due to slow kinetics, CoS incorporation into authigenic sulfides is limited (Huerta-Diaz and Morse, 1992). The Co content in the sediment can also be associated with the clastic material and thus restricts its application as redox sensitive elements (Tribovillard et al., 2006). The Ni in the sediment, despite played an important proxy for the in-situ productivity, but it has also displayed a role in paleoredox conditions. The Ni remains as a soluble cation or adsorbed on to humic/fluvic acids (Calvert and Pedersen, 1993) but in the absence of Mn oxides and sulfide under the oxygen deficient environment, the recycling of Ni from sediments to the water column prevails (Tribovillard et al., 2006). However, during the post depositional oxidation, Ni gets easily oxidized and diffuses into the water column and hence limits its applicability as redox proxy (Martinez-Ruiz et al., 2015). Nevertheless, higher values of Ni/Co indicated reducing conditions as both can be found in pyrite fraction in the sediments and diagenetic pyrite generally displays Ni/Co > 1 (Jones and Manning, 1994).

The vanadium (V) is closely linked with the Mn redox cycle and gets easily adsorbed onto Fe-Mn oxyhydroxides (Calvert and Piper, 1984; Wehrli and Stumm, 1989). Under the mild oxygen deficient conditions in the presence of humic/fluvic acids vanadate ion already prevailing in the oxic environment is reduced to vanadyl (VO^{2+}), related hydroxyl species $VO(OH)_3^-$ and insoluble hydroxides $VO(OH)_2$ (Wilson and Weber, 1979) thereby precipitating within the sediment and enhancing the concentration of vanadium. Moreover, the ratios such as V/(V + Ni) and V/Cr have also been used to depict the redox environment (Hatch and Leventhal, 1992). The V/(V + Ni) with values >0.84, 0.54–0.82 and 0.46–0.60 indicates euxinic, anoxic and dysoxic conditions respectively. While in case of V/Cr values <2 represents oxic conditions but 2–4.25 and > 4.25 represents dysoxic and suboxic to anoxic environment (Rimmer, 2004).

The increased Mo concentrations are often observed in MnO_2-rich sediments as the Mo gets adsorbed onto the Mn oxides (Bertine and Turekian, 1973; Kato, 1995). The disintegration of organic matter through sulfate reducing bacteria under anoxic conditions, leads to release of molybdate and the Mo^{6+} is reduced to Mo^{5+} or Mo^{4+} species (Calvert and Pedersen, 1993) while a rapid uptake of Mo by the authigenic sulfides produced from free H_2S can occur during the prevalence of euxinic environment (Crusius et al., 1996, Morse and Luther, 1999).

In the oxic conditions the uranium (U) prevails as dissolved U^{6+} which gets reduced and forms U^{+4} under oxygen deficient environment thereby forming autogenous U minerals and hence enhancing the concentration of U in the sediment (Calvert and Pedersen, 1993; Jones and Manning, 1994).

Unlike V, Mo and U, the redox behavior of Rhenium (Re) is comparatively less complicated as it is not related to Fe or Mn cycling (Colodner et al., 1993; Crusius et al., 1996). The detrital contribution of Re is trivial and is believed to exist as Re^{+7} in aquatic environment which gets removed from water and get enriched in the sediments under anoxic conditions making it a crucial authigenic redox sensitive proxy (Anbar et al., 1992). Moreover, the ratios between Re and Mo (Re/Mo) has also been utilized to indicate sulfidic water column. Under oxic conditions the Re/Mo indicates crustal ratios (0.3 ng/μg) while in absence of Mn oxide under suboxic conditions the Re/Mo might increase as Re enriched compared to Mo. On the contrary, the Re and Mo concentrations are significantly high under anoxic conditions leading decrease in the Re/Mo ratios (Nameroff et al., 2004).

Overall, the Fe and Mn gets more soluble under oxygen deficient conditions (anoxic) while Cr despite being enriched under poor oxygen conditions but may be influenced by terrestrial inputs. The trace elements such as U, V, Mo are sensitive towards the redox conditions and gets enriched in the sediment providing a robust cue on redox conditions.

3.5 Sediment provenance

The sediment provenance refers to the source or parent rock from which the sediment is being derived and it depends on several geological disciplines (Haughton et al., 1991). For decades, mineralogy, bulk sediment/rock geochemistry, Sr-Nd isotopes, heavy mineral dating and chemistry, etc. have been utilized to estimate the petrogenesis (Joshi et al., 2022b), source of rocks and sediments (Banerji et al., 2021a, 2022a; Chaudhuri et al., 2020; Joshi et al., 2021a; Singh et al., 2008; Slabunov et al., 2017). The geochemical signatures preserved in the sediments/rocks are controlled by several geologic factors such as source rock type, climate and topography, weathering (chemical, biological and physical), mixing from variable sources grain size and hydrodynamic processes, transportation, and post-depositional changes (Garzanti et al., 2010; Hall et al., 2002; Joshi, 2014; Pang et al., 2018; Taylor and McLennan, 1985).

In the recent past, a number of studies have utilized the geochemical behavior of elements and elemental ratios for understanding provenance (Agnihotri et al., 2003a; Calvert and Pedersen, 2007; Feniak, 1944; Shaji et al., 2022; Spears and Kanaris-Sotiriou, 1976). The rock composition of the source region can be recognized using major and trace element concentrations (McLennan et al., 1993). For instance, quartz being resistant to weathering, can survive several erosional and depositional cycles and the concentration of SiO_2 in sediments and sedimentary rocks can be attributed to a nearby felsic or intermediate source (Blatt et al., 1972). Similarly, the other major oxides like CaO, Na_2O and K_2O are leachates of feldspar and volcanic rock weathering while MgO is generally derived from sheet silicates, mafic minerals and clays (Banerji et al., 2021b; Nesbitt and Young, 1984; Pettijohn et al., 1987). Their mobility fails to provide evidence of chemical weathering and sediment provenance but it does provide indirect evidence of the rock source.

The Al_2O_3/TiO_2 and TiO_2/Zr ratios have been implemented to infer felsic, intermediate, and mafic igneous source rock compositions (Hayashi et al., 1997; Hossain et al., 2017). The TiO_2 being associated with the heavy mineral can be transported with the coarser particles, therefore, the Al_2O_3 or Al normalized TiO_2 or Ti (Ti/Al) and Zr (Zr/Al) have been implemented to track the source of riverine and wind-blown dust in marine sediments (Lückge et al., 2001; Martinez-Ruiz et al., 2015; Scheuvens et al., 2013). Additionally, a number of studies have utilized A-CN-K and A-CNK-FM ternary diagrams to address source rock compositions (Awasthi et al., 2014; Banerji et al., 2021a; Li et al., 2018; Shaji et al., 2022) which when combined with other parameters can point towards tentative source areas (Fig. 3).

The enrichment and depletion of transition metals, HFSE and rare earth elements (REEs) have acted as reliable tools for source discrimination (Asiedu et al., 2017; Asiedu et al., 2019; Roddaz et al., 2007). During crystallization and anatexis, the incompatible elements gets enriched in felsic rocks while the mafic rocks get enriched in the compatible elements (Cullers, 2000; Feng and Kerrich, 1990). Elevated concentrations of Cr

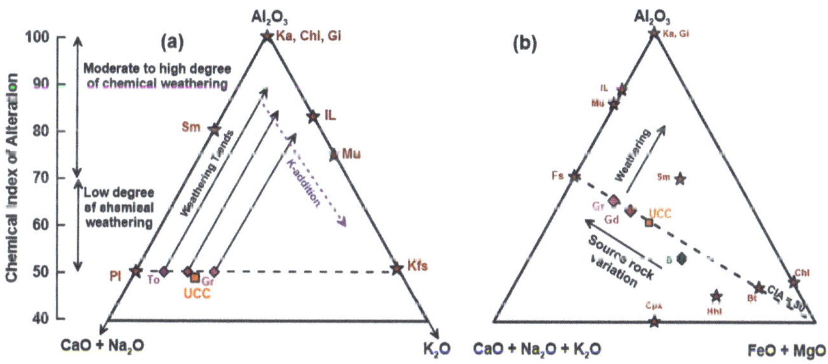

Fig. 3 The ternary plot representing the typical weathering trend in (A) A-CN-K and (B) A-CNK-FM plot. (Note: The plot is not as per the scale.)

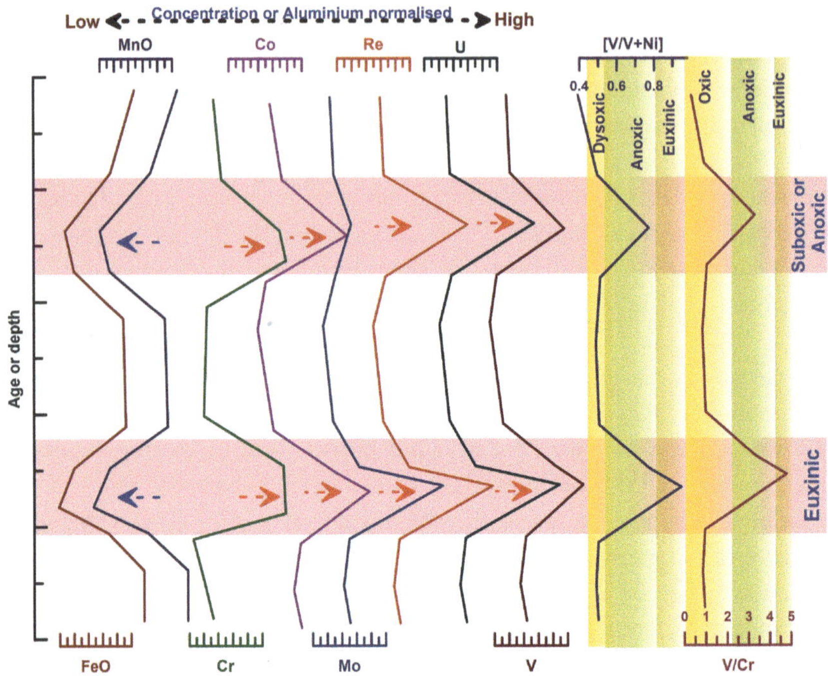

Fig. 4 Typical pattern of the redox sensitive elements under oxygen deficient conditions. (Please note that other variable such as local lithology, level of oxygen deficient conditions, other geochemical processes, etc., might play role in the variation of the redox sensitive elements.)

(>150 ppm) and Ni (>100 ppm) have been considered to be indicative of ultramafic rock as the major sediment source (Garver et al., 1996). Further, incompatible to compatible element ratios and the respective biplots (Th/Cr versus La/Cr, Th/Sc versus La/Sc, La/Th vs. Hf, Cr/V vs. Y/Ni and Co/Th vs. La/Sc) are utilized to delineate the sediment provenance (Bankole et al., 2020; Joshi et al., 2021c). The Th/Sc and Th/Co ratios have been successfully used to track the source terrains wherein Th/Sc values of <0.8 are typical of mafic rocks, while the upper crust's typical ratio is 1 (Taylor and McLennan, 1985).

In addition to incompatible and compatible element ratios, REE patterns, the LREE/HREE ratios, europium (Eu) anomaly, the Eu/Eu* versus Gd_N/Yb_N variation have also been frequently used to differentiate sources of terrigenous sediments (Cullers, 2000; El-Bialy, 2013; Hossain, 2020; Mclennan, 1989; McLennan et al., 1990; McLennan et al., 1995; Slack and Stevens, 1994). Generally, high LREE/HREE are obtained for the sediments derived from the felsic source, while sediments derived from the mafic source are characterized by low LREE/HREE ratios (Cullers et al., 1997). The sediments with negative Eu anomalies ($Eu/Eu^* < 1$) indicates the sediment source to be of silicic igneous rocks and presence or absence of trivial Eu anomalies ($Eu/Eu^* = 1$) is a representative of basic igneous rocks.

4. Mathematical and statistical analysis of climate variables

A significant change in biological, chemical and physical processes near to the surface of the earth is highly impacted by the climate. Such processes promote the thermal variations of the atmosphere and environmental changes at both regional and global scales. These changes can be analyzed using mathematical and statistical approaches based on several classical as well as machine learning based algorithms to postulate the temporal variations of the climate system. However, several geochemical and isotopic proxies provide a quantitative relationship between data and climate but the interpretations sometimes remains perplexed due to various processes acting at the same time (Song et al., 2022). All of these climate variables when used independently or together can aid in decoding the past climate of a particular time period or a region.

Several mathematical and statistical approaches like detrended correspondence analysis (DCA), redundancy analysis (RDA) or canonical correspondence analysis (CCA), continuous wavelet transform (CWT), gaussian response analysis (GRA), rate of change analysis (RCA) using standardized Euclidean distance (SED), principal component analysis (PCA) is generally used to assess the quantitative relationship between the proxies and the climate variables for reconstruction purposes. The other classical approaches include univariate methods such as leaf-margin analysis (LMA) (Geissler et al., 2017) and leaf-area analysis (LAA), multivariate method like climate-leaf analysis multivariate program (CLAMP), and digital leaf physiognomy (DLP). The DLP has been found to be one of the most considerable methods for reconstructing past climate as it overcomes the precision issues related with CLAMP.

Over the past two decades, it has been found that the involvement of machine learning (ML), with mainly Artificial Neural Network (ANN), in climate science has been very slow and more investment has been seen in physical models, like global climate models (GCMs), owing to their importance to the concept of human induced global warming. However, the complexity in the climate systems and limited knowledge of all the physical processes leads to ambiguous results including the Equilibrium Climate Sensitivity (ECS). Unfortunately, these traditional methods restrain in handling large and multi-dimensional dataset. Also, the models built including the DLP method and other sets of classical approaches still lack precision, mainly for mean annual precipitation and other climate proxies. Therefore, the need for improving the accuracy and robustness of predictions and their advancements motivated the use of Artificial intelligence (AI) based ML algorithms. These algorithms use ensemble models of regression tree [bagging, random forest (RF), rotation forest, boosted regression tree (BRT), gradient boosted regression tree (GBRT) and extremely randomized trees (ETREES)] and ANN, using both the traditional and MI approaches (Abbot and Marohasy, 2017; Bochenek and Ustrnul, 2022; Déchelle-Marquet, 2019; Huang et al., 2020; Huntingford et al., 2019; Juggins et al., 2015; Liang et al., 2021; Salonen et al., 2012, 2019; Salonen et al., 2014; Salonen et al., 2016; Wei et al., 2021; Zhang et al., 2022). These algorithms are capable of modeling nonlinear data with complex interactions and have been used successfully to address various problems in Earth Sciences. The ML based algorithms are trained (input to model) to derive information from the data in both supervised and unsupervised environments and can be applied to the available climate data and observations for future climate prediction.

However, quantifying the model variables is crucial for establishing the Quaternary climate models and providing a benchmark for predicting future climate. Thus, considering the completeness of the observational records, and the comparability of existing quantitative Quaternary reconstruction using geochemical and isotope data sets, mean annual temperature, mean annual precipitation, relative humidity, and mean annual wind speed can also be utilized for the reconstruction of paleoclimate. Before modeling, all these datasets need to be firstly selected for interpolation (methods like ANUSPLIN) to construct climate data surfaces and improve the smoothness and accuracy (Hancock and Hutchinson, 2006).

4.1 Climate reconstruction model

Classical multivariate linear calibration methods are generally used for quantitative Quaternary climate reconstructions (Camuera et al., 2022). Many researchers have shed light on the nature of the drivers of climate change across space and time using probability density function (PDF), parametric curve fitting technique, weighted averaging (WA) and later modified as weighted averaging partial least square (WA-PLS), modern analogue technique (MAT) for the proxy's samples (Chevalier, 2022; Overpeck et al., 1985; Peng et al., 2022; Salonen et al., 2016; Salonen et al., 2019; Ter Braak and Juggins, 1993). These methods are based on unique assumptions or limitations for climate reconstructions. Recent publications on reducing the data set while generating models (Roy et al., 2021a, 2021b), periodic geodynamic events on a global scale using CWT and coherence (Joshi et al., 2022a) suggest that similar techniques can be applied for understanding the paleoclimate and its future predictions.

Few of the techniques such as MAT or WA-PLS are primarily designed to associate modern proxy observations with their "most likely" or "mean" climate values. On the other hand, CWT-based modeling and CREST-Climate Reconstruction Software (Chevalier, 2022) can estimate and weighs all the climate values that are compatible with the observed proxy data. CWT-based Monte Carlo Simulation and the application of CREST are used to estimate the possible outcome of uncertain events. These yield a probabilistic quantification of all the climate values that are well-suited to the utilized data. Other models provide ambiguous model interpretation in terms of simpler, less informative "most likely" or "best" climate estimates. It is very uncertain to get the "best estimate" climate model and such an approach might be optimal when proxy assemblage is analyzed in complete isolation. Fig. 5 shows conceptual illustration of the differences between several modeling approaches based on the full spread of data distribution with the probabilities distributed along the climate variables (Chevalier, 2022). Herein, three different modeling approaches are elaborated (a) statistical-based modeling demonstrates the best estimate of all climate variables with small statistical error compared to probability-based data distribution, (b) standard modeling illustrates the best estimates with some statistical uncertainties using two different sample data to show mutually independent construction, and (c) uncertainty based modeling approach which highlights the joint reconstruction of two different samples to exhibit the best estimate or most likely climate model.

Often, additional information obtained by the presence of independent–local or regional–information may not always seem to be consistent with the best estimate reconstructions (Fig. 5). Thus, sometimes developed models from the joint reconstructions based on all the available proxy information deviates from best climate estimates (Fig. 5A–C). Therefore, we need a more robust modeling algorithm to estimate the full range of climate uncertainties

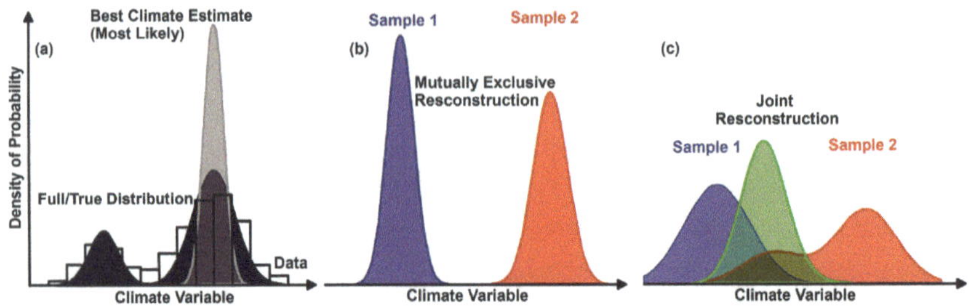

Fig. 5 A conceptual illustration of the differences between various modeling approaches based on the full spread of data distribution with the probabilities distributed along the climate variables. (A) statistical based modeling demonstrates the best estimate of climate variables with small statistical error (in light gray) as compared to probability-based data distribution (in dark gray), (B) standard modeling shows the best estimates with some statistical uncertainties using two different sample data show mutually independent construction, and (C) uncertainty based modeling approach highlight the joint reconstruction of two different samples to exhibit the best estimate or most likely climate model (in green color) using the complete uncertainty distribution of the given sample data sets (Chevalier, 2022).

associated with a proxy sample. Such assumptions of modeling approaches are perilous to transport reconstructions and even climate simulations together in a cohesive way. The statistical approach along with mathematical modeling can be utilized to unravel climate variable relationships that have remained unnoticed and would be important for predicting future successional trends under changing climates.

Apart from the classical approaches discussed so far, recent advancement has been seen in the AI based ML methods in which ANN has been successfully used in paleo-reconstruction and future projections of climatic and meteorological archives such as temperature, rainfall, wind speed, solar radiation, ECS computations, etc. (Abbot and Marohasy, 2017; Wei et al., 2021). Wei et al. (2021) developed five ML (tree based and SVM) based models for mean annual temperature and precipitation which was compared with the multilinear regression (MLR) models. Depending on the training data set, a single regression tree can produce versatile results with low prediction power. However, predicted results can be improved by combining several ensemble-tree methods (Breiman, 2001). It was observed that the ML models predicted an improved paleoclimatic reconstruction in comparison to those predicted from MLR and ANN models. Although, ANN could model nonlinear and complex relation among the data but still compared to other ML based methods, ANN model was found to be less reliable in terms of precision while predicting mean annual precipitation. In calibration of aquatic microfossil proxies, regression tree-ensemble methods overcome the performance of parametric transfer functions (PLS, WA, WA-PLS) approaches while handling large datasets (Salonen et al., 2016).

The Bidirectional long-short term memory (Bi-LSTM) method which belongs to ML category, takes input both in forward and backward direction based on the serial correlation of time series (Huang et al., 2020). Compared to classical multivariate methods (PCA, CCA), it does not consider linear and stable relation between the underlying proxy network and the targeted climate field. Zhang et al. (2022) claimed that while mapping past climate summer temperatures for northern, hemisphere and north Atlantic European, nonlinear Bi-LSTM neural network method proved as one of the potential and reliable approaches compared to linear PCA and CCA approaches. Support vector machine (SVM), one of the supervised ML algorithms, maps features vectors to high dimensional space via kernel functions and aims at building optimal linear discriminating function that could fit the training model (Wei et al., 2021). Support vector regression (SVR) an extended version of SVM ignores the issue of overfitting and robustly can handle large set of high dimensional climatic variables. Wei et al. (2021) used radial basis function (RBF) as a kernel in SVM to develop models for paleoclimatic reconstructions from leaf traits. A probable flowchart is given in Fig. 6 for conventional and ML based algorithms that can be used for any type of Quaternary climate reconstruction.

4.2 Dominant climatic variable and model reliability

There is a huge dataset available that involves data retrieved from the natural proxies and online satellite data. Therefore, the vital task to identify the most dominant climatic variable from the available datasets for model initiation. There are several statistical approaches like DCA (detrending by segments, non linear rescaling of axes, and the down weighting of proxies), Canonical coordination analysis, and Monte Carlo permutation tests to determine which method is the most appropriate for determining the dominant climatic variable, with gradient length (the range of variation) as the criterion (Ter Braak and Prentice, 2004). The reliability of the quantitative reconstructions of the Quaternary climate model can be assessed using the performance statistics of the WA PLS based transfer function model under cross validation. The model averaging method can be used to reduce uncertainty associated with the model and further to improve predictive accuracy (Marmion et al., 2009). This averaging method has been successfully applied to numerous

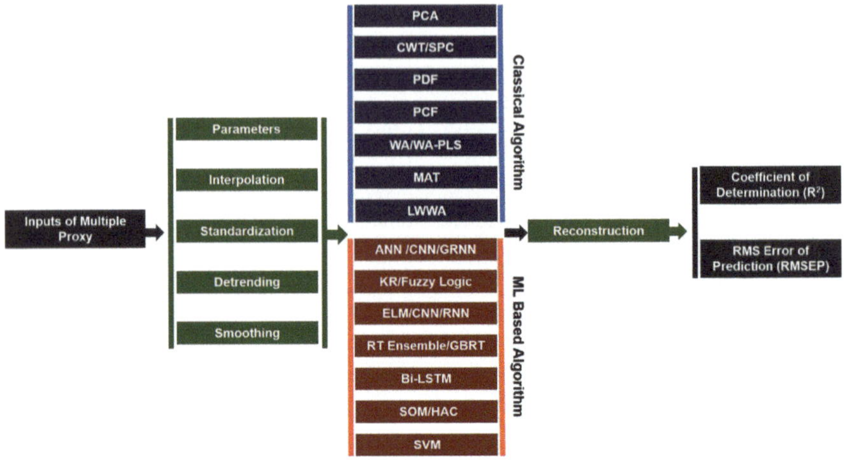

Fig. 6 Flow chart diagram of Quaternary climate reconstruction using possible classical and ML-based modeling approaches.

studies for model reconstruction (Benestad, 2004; Goswami and O'Connor, 2007; Nilsson et al., 2000; Wei et al., 2021).

Recent studies have claimed that among the ML based algorithms, RF (Random Forest) is one of the most reliable data analysis algorithms that computes two indices in regression (Lesmeister, 2017; Salonen et al., 2019). The evaluation proxies for the classical and ML based models include the maximum bias or the largest mean of prediction residuals found for any of the 10 equal length segments of the calibration data climate gradient, coefficient of determination (R^2), and the root means square error of prediction (RMSEP) expressed as:

$$RMSEP = \sqrt{\frac{1}{n}\sum_{i=1}^{n}\left(y_{observed} - y_{predicted}\right)^2} \quad (1)$$

$$R^2 = 1 - \frac{\sum_{i=1}^{n}\left(y_{observed} - y_{predicted}\right)^2}{\sum_{i=1}^{n}\left(y_{observed} - y_{mean}\right)^2} \quad (2)$$

where n is the number of observations in Eqs. (1) and (2). The RMSEP is the average deviation between the predicted and the observed values. The smaller value of RMSEP, indicates the minimum error between the observation and the prediction made. The R^2 which lies between 0 and 1 is the measurement of the deviation expressed by the prediction. It means that approximation of R^2 closer to 1 indicates the minimum error between the observed and predicted value. The model can be more précised and reliable if the unnecessary climatic variables are dropped from the model (Wei et al., 2021). The rule for removing redundant variables is based on the (1) dominant variable measured by the RF method that is once the variable is eliminated, the model is trained using the remaining variables and process is repeated until no further variables remain for elimination, and (2) estimation of R^2 that is if the R^2 does not decrease on dropping the variable then in that case it is removed from the model else retained (Wei et al., 2021). Based on Eqs. (1) and (2), the accuracy of different algorithms is tested and optimal set of input is determined. Fig. 7 shows the Cross-validation (CV) results with RMSEP estimated using different classical (MAT, WA, WA-PLS) and ML (ELM, NNET, RF, BRT and ETREES) based approaches for the July mean temperature climate variable in North America. It is obvious to note from Fig. 7 that ML based technique like RF, BRT and ETREES show their

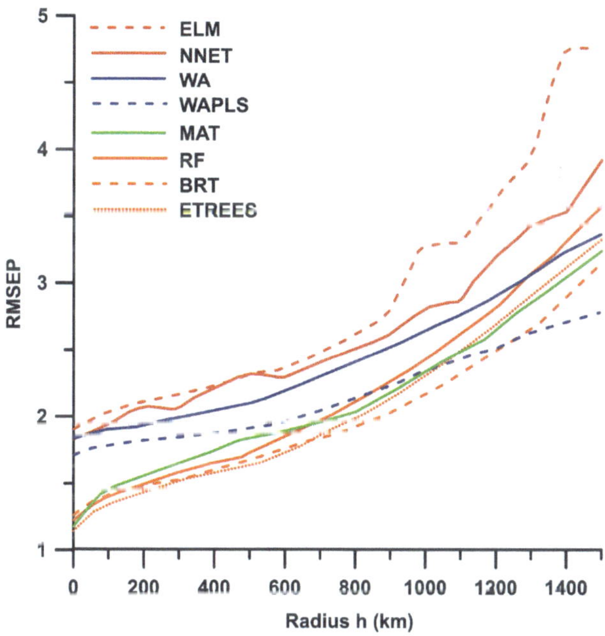

Fig. 7 Cross-validation (CV) results with root-mean-square error of prediction (RMSEP) estimated using different classical (MAT, WA, WA-PLS) and ML based approaches (ELM, NNET, RF, BRT and ETREES) for the July mean temperature climate variable in North America (Salonen et al., 2019).

capabilities to produce best reconstruction climate model as compared to others classical approaches.

4.3 Factors affecting model and source of error in the model

There are several factors which can affect the model accuracy viz. collecting the true proxy's samples, sample analysis and data processing (in-situ measurements), different scales of data calibration, modeling assumptions and mathematical errors, location. The other probable factors include the errors and bias associated to the ambiguity of character definitions, use of discontinuous variables, weak correlations between climate and some character states (Peppe et al., 2011; Wei et al., 2021). These sources of error tend to impact the quality of reconstructed climate model where few of them are not in human controls and thus, lead to model inaccuracy.

Apart from the uncertainties associated with the sparse data coverage and those in archives owing to the non-climatic variability, there are also uncertainties present in the finally reconstructed model due to the statistical methods. However, different climate systems control surface as well as atmospheric processes that lead to spatial and temporal patterns of a high degree of variability (Song et al., 2022). The other factor may be the amount of variance demonstrated by the dominant climatic variables due to the redundancy in wide distribution of proxy's samples (Wu et al., 2020). Ultimately, different mathematical algorithms used to develop the climate reconstruction models exhibit a wide range of accuracy among themselves. For example, Locally-Weighted Weighted Averaging (LWWA) performs much better on large, heterogeneous training datasets as compared to traditional WA or WA-PLS approach (Song et al., 2022). It utilizes the advantage of increased environmental and biological coverage provided by extensive training sets without the disadvantage of an increased prediction error (Birks et al., 2012). Finally, all developed models of climate reconstruction should be always superior to the available previous models in terms of performance and applicability.

5. Summary and remarks

In recent years, a growing attention is given on the past climate reconstruction with specific focus on the marine sediments as an archive, as it often remains undisturbed and less interrupted by anthropogenic encroachments. In this chapter, we have tried to highlight the utility of elemental variations in the marine sediments as it has provided a holistic understanding of the paleoclimate and paleoceanography thorough the proxies depicting terrestrial influence, *in-situ* productivity and associated redox conditions, weathering intensities and paleoprovenance variations.

Overall, the IP, Eh, pH and the distribution coefficient has been the major factor that significantly controls the mobility of each of the elements. Based on the elemental mobility and their enrichments, several geochemical proxies have been delineated that had provided comprehensive understanding of the temporal variations in the climate and other associated variables such as weathering, provenances, etc.

Detrital proxies are generally considered as those elements that remains immobile and impervious towards weathering and redox changes such as the Si, Al, Ti, Zr. Moreover, most of them are also associated with the textural variations such as Ti, Zr and Si were associated with coarser fraction of the sediments while Al has been associated with the clay fraction. In general, Al has been also used for the normalization of the elemental proxies which is assumed to take care of the textural variations in the sediments. The mobile elements such as Ca, K, Na, Sr, etc. has provided cues on the weathering changes in the region, while certain indices such as CIA, PIA, ICV, WIA, etc. have helped in quantification of the weathering changes. The paleoproductivity proxies especially the export proxies represented by TOC and $CaCO_3$ supported by other major and trace elemental variations has provided colossal evidence of past productivity changes and their association with the climate and oceanographic conditions. Despite its enormous applicability, the TOC can be either prone to degradation or can be influenced by the terrestrially derived TOC. The Al normalized trace elements such as Cu, Ni and Sr and Ba while major element CaO and Si has substantially proved as a reliable productivity indicators. The enhanced productivity can lead to oxygen deficient conditions which in turn causes migration and enrichment of selected redox sensitive elements from sediment to the pore waters. The oxygen deficient conditions can be of dysoxic, sub oxic, anoxic and euxinic depending on the oxygen content in the water column and the sediment-water interface has been the most effected region of elemental depletion and enrichments.

The statistical approach attests to the reliability of the geochemical proxies while ML can help in understanding the present processes and anticipating future trends. The ML methods have outperformed the classical univariate and multivariate methods by improving the precision and reducing the pre dictive error of estimating paleoclimate. The ML-based methods are flexible enough to model the strong nonlinear relation between leaf traits and climate, which is considered to be more précised and reasonable compared to those standard regression approaches. The accuracy of ML-based reconstructed climate model can be improved by using the continuous and highly reproducible variables that generally link to climate. When unnecessary variables were dropped, the prediction errors of all ML models were reduced in terms of R^2 and RMSEP and they offer a possible understanding of past climate change. Thus, overall, the use of AI based ML approaches can improve the robustness of paleoclimatic reconstructions by allying the effects of spatial autocorrelation and temporally varying cross-correlation.

Where ML methods offer certain advantages at the same time, they also pose certain limitations or challenges. That is when it comes to data labelling, it needs to be well organized for ML but if the data is not of high quality and well labelled then it is generally not recommendable to adopt ML approaches. Lack of reproducibility is also a complex and growing issue being faced in the application of ML to paleoclimate. No doubt ML models may be accurate enough in paleoclimate reconstructions but there are more likely chances that may not validate its solutions. ML algorithms require huge set of high-quality data to train their models, lack of which may significantly reduce the model's accuracy and thereby the predictions.

The combination of geochemical datasets with statistical and ML techniques can provide enormous information not only to address the past climate scenarios but also help in understanding the present and projecting the future climatic trends. Understanding the present climate patterns will further enhance the ability to address the linkage of local and regional climate system with the global climate dynamics. The cues on the present climate perturbation can help policy makers to identify the vulnerabilities of climate change which in turn can aid them in proceeding with the appropriate mitigation steps for societal benefits.

Acknowledgments

This research was supported by National Centre for Earth Science Studies (NCESS), Ministry of Earth Science, Government of India. The authors are grateful to the Director, NCESS for his encouragement and permission to conduct the present work. USB is thankful to Dr. D. Padmalal for his constant support and encouragement.

References

Abbot, J., Marohasy, J., 2017. The application of machine learning for evaluating anthropogenic versus natural climate change. GeoResJ 14, 36–46.

Achyuthan, H., Nagasundaram, M., Gourlan, A.T., Eastoe, C., Ahmad, S.M., Padmakumari, V.M., 2014. Mid-Holocene Indian summer monsoon variability off the Andaman Islands, Bay of Bengal. Quat. Int. 349, 232–244.

Agnihotri, R., Dutta, K., Bhushan, R., Somayajulu, B.L.K., 2002. Evidence for solar forcing on the Indian monsoon during the last millennium. Earth Planet. Sci. Lett. 198, 521–527.

Agnihotri, R., Bhattacharya, S.K., Sarin, M.M., Somayajulu, B.L.K., 2003a. Changes in surface productivity and subsurface denitrification during the Holocene: a multiproxy study from the eastern Arabian Sea. Holocene 13, 701–713.

Agnihotri, R., Sarin, M.M., Somayajulu, B.L.K., Jull, A.J.T., Burr, G.S., 2003b. Late-quaternary biogenic productivity and organic carbon deposition in the eastern Arabian Sea. Palaeogeogr. Palaeoclimatol. Palaeoecol. 197, 43–60.

Algeo, T.J., Li, C., 2020. Redox classification and calibration of redox thresholds in sedimentary systems. Geochim. Cosmochim. Acta 287, 8–26.

Algeo, T.J., Maynard, J.B., 2004. Trace-element behavior and redox facies in core shales of Upper Pennsylvanian Kansas-type cyclothems. Chem. Geol. 206, 289–318.

Anbar, A.D., Creaser, R.A., Papanastassiou, D.A., Wasserburg, G.J., 1992. Rhenium in seawater: confirmation of generally conservative behavior. Geochim. Cosmochim. Acta 56, 4099–4103.

Asiedu, D.K., Asong, S., Atta-Peters, D., Sakyi, P.A., Su, B.X., Dampare, S.B., Anani, C.Y., 2017. Geochemical and Nd-isotopic compositions of juvenile-type Paleoproterozoic Birimian sedimentary rocks from southeastern West African Craton (Ghana): constraints on provenance and tectonic setting. Precambrian Res. 300, 40–52.

Asiedu, D.K., Agoe, M., Amponsah, P.O., Nude, P.M., Anani, C.Y., 2019. Geochemical constraints on provenance and source area weathering of metasedimentary rocks from the Paleoproterozoic (~2.1 Ga) Wa-Lawra Belt, southeastern margin of the West African Craton. Geodin. Acta 31, 27–39.

Awasthi, N., Ray, J.S., Singh, A.K.A.K., Band, S.T., Rai, V.K., 2014. Provenance of the Late Quaternary sediments in the Andaman Sea: implications for monsoon variability and ocen circulation. Geochem. Geophys. Geosystems 15, 3890–3906.

Banerji, U.S., Bhushan, R., Maurya, D.M., 2016. Sedimentary records of paleoredox conditions at relict mudflat of Vasoj, Southern Saurashtra coast, Gujarat. Earth Sci. India 9, 114–125.

Banerji, U.S., Bhushan, R., Jull, A.J.T., 2017. Mid–late Holocene monsoonal records from the partially active mudflat of Diu Island, southern Saurashtra, Gujarat, western India. Quat. Int. 443, 200–210.

Banerji, U.S., Arulbalaji, P., Padmalal, D., 2020. Holocene climate variability and Indian Summer Monsoon: an overview. Holocene 30, 744–773.

Banerji, U.S., Bhushan, R., Joshi, K.B., Shaji, J., Jull, A.J.T., 2021a. Hydroclimate variability during the last two millennia from the mudflats of Diu Island, Western India. Geol. J., 3584–3604.

Banerji, U.S., Shaji, J., Arulbalaji, P., Maya, K., Vishnu Mohan, S., Dabhi, A.J., Shivam, A., Bhushan, R., Padmalal, D., 2021b. Mid-late Holocene evolutionary history and climate reconstruction of Vellayani lake, south India. Quat. Int. 599–600, 72–94.

Banerji, U.S., Dubey, C.P., Goswami, V., Joshi, K.B., 2022a. Geochemical indicators in provenance estimation. In: John, A.-A., Kailasa, P., Verma, S.K. (Eds.), Geochemical Treasures and Petrogenetic Processes. Springer Nature.

Banerji, U.S., Goswami, V., Joshi, K.B., 2022b. Quaternary dating and instrumental development: an overview. J. Asian Earth Sci. X 7, 100091.

Bankole, O.M., El Albani, A., Meunier, A., Poujol, M., Bekker, A., 2020. Elemental geochemistry and Nd isotope constraints on the provenance of the basal siliciclastic succession of the middle Paleoproterozoic Francevillian Group. Gabon. Precambrian Res. 348, 105874.

Benestad, R.E., 2004. Empirical-statistical downscaling in climate modeling. EOS Trans. Am. Geophys. Union 85, 417–422.

Berger, W.H., 1976. Biogenous deep-sea sediments: production, preservation and interpretation. Chem. Oceanogr. 5, 265–372.

Bertine, K.K., Turekian, K.K., 1973. Molybdenum in marine deposits. Geochim. Cosmochim. Acta 37, 1415–1434.

Bertrand, P., Shimmield, G., Martinez, P., Grousset, F., Jorissen, F., Paterne, M., Pujol, C., Bouloubassi, I., Buat Menard, P., Peypouquet, J.P., Beaufort, L., Sicre, M.A., Lallier-Verges, E., Foster, J.M., Ternois, Y., 1996. The glacial ocean productivity hypothesis: the importance of regional temporal and spatial studies. Mar. Geol. 130, 1–9.

Bhushan, R., Dutta, K., Somayajulu, B.L.K., 2001. Concentrations and burial fluxes of organic and inorganic carbon on the eastern margins of the Arabian Sea. Mar. Geol. 178, 95–113.

Bhushan, R., Sati, S.P., Rana, N., Shukla, A.D., Mazumdar, A.S., Juyal, N., 2018. High-resolution millennial and centennial scale Holocene monsoon variability in the Higher Central Himalayas. Palaeogeogr. Palaeoclimatol. Palaeoecol. 489, 95–104.

Birks, H.J.B., Lotter, A.F., Juggins, S., Smol, J.P., 2012. Tracking Environmental Change Using Lake Sediments: Data Handling and Numerical Techniques. Springer Science & Business Media.

Bjørlykke, K., 2010. Sedimentary geochemistry. In: Bjørlykke, K. (Ed.), Petroleum Geoscience: From Sedimentary Environments to Rock Physics. Springer-Verlag, Berlin Heidelberg, pp. 1–508, https://doi.org/10.1007/978-3-642-02332-3.

Blatt, H., Middleton, G.V., Murray, R.C., 1972. Origin of Sedimentary Rocks. Prentice-Hall, NJ.

Bochenek, B., Ustrnul, Z., 2022. Machine learning in weather prediction and climate analyses—applications and perspectives. Atmosphere (Basel). 13, 180.

Bradley, R.S., 2015. Paleoclimatology: Reconstructing Climates of the Quaternary, third ed. Oxford.

Breiman, L., 2001. Random Forests. vol. 94720 Statistics Department, Univ. of California, Berkeley, CA, p. 33.

Breit, G.N., Wanty, R.B., 1991. Vanadium accumulation in carbonaceous rocks: a review of geochemical controls during deposition and diagenesis. Chem. Geol. 91, 83–97.

Brumsack, H.-J., 1989. Geochemistry of recent TOC-rich sediments from the Gulf of California and the Black Sea. Geol. Rundsch. 78, 851–882.

Calvert, S.E., Pedersen, T.F., 1993. Geochemistry of recent oxic and anoxic marine sediments: implications for the geological record. Mar. Geol. 113, 67–88.

Calvert, S.E., Pedersen, T.F., 2007. Elemental proxies for palaeoclimatic and palaeoceanographic variability in marine sediments: interpretation and application. Dev. Mar. Geol. 1, 567–644.

Calvert, S.E., Piper, D.Z., 1984. Geochemistry of ferromanganese nodules from DOMES site a, Northern Equatorial Pacific: multiple diagenetic metal sources in the deep sea. Geochim. Cosmochim. Acta 48, 1913–1928.

Camuera, J., Ramos-Román, M.J., Jiménez-Moreno, G., García-Alix, A., Ilvonen, L., Ruha, L., Gil-Romera, G., González-Sampériz, P., Seppä, H., 2022. Past 200 kyr hydroclimate variability in the western Mediterranean and its connection to the African Humid Periods. Sci. Rep. 12, 1–13.

Chandana, K.R., Bhushan, R., Jull, A.J.T., 2017. Evidence of poor bottom water ventilation during LGM in the equatorial Indian Ocean. Front. Earth Sci. 5, 84.

Chaudhuri, A., Banerjee, S., Prabhakar, N., Das, A., 2020. The use of heavy mineral chemistry in reconstructing provenance: a case study from Mesozoic sandstones of Kutch Basin, India. Geol. J. 55, 7807–7817.

Chevalier, M., 2022. crestr: an R package to perform probabilistic climate reconstructions from palaeoecological datasets. Clim. Past 18, 821–844.

Claes, H., Huysmans, M., Soete, J., Dirix, K., Vassilieva, E., Erthal, M.M., Vandewijngaerde, W., Hamaekers, H., Aratman, C., Özkul, M., 2019. Elemental geochemistry to complement stable isotope data of fossil travertine: importance of digestion method and statistics. Sediment. Geol. 386, 118–131.

Clift, P.D., Wan, S., Blusztajn, J., 2014. Reconstructing chemical weathering, physical erosion and monsoon intensity since 25Ma in the northern South China Sea: a review of competing proxies. Earth-Science Rev. 130, 86–102. https://doi.org/10.1016/j.earscirev.2014.01.002.

Colodner, D., Sachs, J., Ravizza, G., Turekian, K., Edmond, J., Boyle, E., 1993. The geochemical cycle of rhenium: a reconnaissance. Earth Planet. Sci. Lett. 117, 205–221.

Condie, K.C., 1993. Chemical composition and evolution of the upper continental crust: contrasting results from surface samples and shales. Chem. Geol. 104, 1–37.

Correns, C.W., 1969. Titanium. In: Wedepohl, K.H., Correns, C.W., Shaw, D.M., Turekian, K.K., Zemann, J. (Eds.), Handbook of Geochemistry. Springer Berlin, Berlin, p. 22.

Cox, R., Lowe, D.R., Cullers, R.L., 1995. The influence of sediment recycling and basement composition on evolution of mudrock chemis- try in the southwestern United States. Geochim. Cosmochim. Acta 59 (14), 2919–2940. https://doi.org/10.1016/0016-7037(95) 00185-9.

Crusius, J., Calvert, S., Pedersen, T., Sage, D., 1996. Rhenium and molybdenum enrichments in sediments as indicators of oxic, suboxic and sulfidic conditions of deposition. Earth Planet. Sci. Lett. 145, 65–78.

Cullers, R.L., 2000. The geochemistry of shales, siltstones and sandstones of Pennsylvanian Permian age, Colorado, USA: implications for provenance and metamorphic studies. Lithos 51, 181–203.

Cullers, R.L., Bock, B., Guidotti, C., 1997. Elemental distributions and neodymium isotopic compositions of Silurian metasediments, western Maine, USA: redistribution of the rare earth elements. Geochim. Cosmochim. Acta 61, 1847–1861.

Daihres, F., Chesselet, R., Jedwab, J., 1980. Discrete suspended particles of barite and barium cycle in the ocean. Earth Planet. Sci. Lett. 49, 528–550.

Davison, W., 1993. Iron and manganese in lakes. Earth Sci. Rev. 34, 119–163.

Déchelle-Marquet, M., 2019. Paleoclimate Reconstruction from Climate Proxiesby Neural Methods. Degree Project, Department of Sustainable Development, Environmental Science and Engineering, Stockholm, Sweden.

Dolcater, D.L., Syers, J.K., Jackson, M.L., 1970. Titanium as free oxide and substituted forms in kaolinites and other soil minerals. Clay Clay Miner. 18, 71–79.

Dymond, J., Suess, E., Lyle, M., 1992. Barium in deep-sea sediment: a geochemical proxy for paleoproductivity. Paleoceanography 7, 163–181.

El-Bialy, M.Z., 2013. Geochemistry of the neoproterozoic metasediments of Malhaq and Um Zariq formations, Kid metamorphic complex, Sinai, Egypt: implications for source-area weathering, provenance, recycling, and depositional tectonic setting. Lithos 175–176, 68–85.

Fedo, C.M., Nesbitt, H.W., Young, G.M., 1995. Unravelling the effects of potassium metasomatism in sedimentary rocks and paleosols, with implications for paleoweathering conditions and provenance. Geology 23, 921–924.

Fedo, C.M., Eriksson, K.A., Krogstad, E.J., 1996. Geochemistry of shales from the Archean (~3.0 Ga) Buhwa Greenstone Belt, Zimbabwe: implications for provenance and source-area weathering. Geochim. Cosmochim. Acta 60, 1751–1763.

Feng, R., Kerrich, R., 1990. Geochemistry of fine-grained clastic sediments in the Archean Abitibi greenstone belt, Canada: implications for provenance and tectonic setting. Geochim. Cosmochim. Acta 54, 1061–1081.

Feniak, M.W., 1944. Grain sizes and shapes of various minerals in igneous rocks. Am. Mineral. J. Earth Planet. Mater. 29, 415–421.

Francois, R., 1988. A study on the regulation of the concentrations of some trace metals (Rb, Sr, Zn, Pb, Cu, V, Cr, Ni, Mn and Mo) in Saanich Inlet Sediments, British Columbia, Canada. Mar. Geol. 83, 285–308.

Ganeshram, R.S., 1998. Factors controlling the burial of organic carbon in laminated and bioturbated sediments off NW Mexico. Mineral. Mag. 62A, 1723–1734.

Garrels, R.M., 1965. Solutions, Minerals, and Equilibrium. Freeman, p. 363.

Garver, J.I., Royce, P.R., Smick, T.A., 1996. Chromium and nickel in shale of the Taconic foreland; a case study for the provenance of fine-grained sediments with an ultramafic source. J. Sediment. Res. 66, 100–106.

Garzanti, E., Andò, S., France-Lanord, C., Vezzoli, G., Censi, P., Galy, V., Najman, Y., 2010. Mineralogical and chemical variability of fluvial sediments. 1. Bedload sand (Ganga-Brahmaputra, Bangladesh). Earth Planet. Sci. Lett. 299, 368–381.

Garzanti, E., Vermeesch, P., Padoan, M., Resentini, A., Vezzoli, G., Andò, S., 2014. Provenance of passive-margin sand (Southern Africa). J. Geol. 122 (1), 17–42.

Geissler, W.H., Jokat, W., Jegen, M., Baba, K., 2017. Thickness of the oceanic crust, the lithosphere, and the mantle transition zone in the vicinity of the Tristan da Cunha hot spot estimated from ocean-bottom and ocean-island seismometer receiver functions. Tectonophysics 716, 33–51.

Goldberg, E.D., 1961. Chemical and mineralogical aspects of deep-sea sediments. Phys. Chem. Earth 4, 281–302. https://doi.org/10.1016/0079-1946(61)90009-X.

Gornitz, V., 2006. Encyclopedia of paleoclimatology and ancient environments. In: Encyclopedia of Paleoclimatology and Ancient Environments. Springer.

Goswami, M., O'Connor, K.M., 2007. Real-time flow forecasting in the absence of quantitative precipitation forecasts: a multi-model approach. J. Hydrol. 334, 125–140.

Greber, N.D., Dauphas, N., 2019. The chemistry of fine-grained terrigenous sediments reveals a chemically evolved Paleoarchean emerged crust. Geochim. Cosmochim. Acta 255, 247–264.

Hall, K., Thorn, C.E., Matsuoka, N., Prick, A., 2002. Weathering in cold regions: some thoughts and perspectives. Prog. Phys. Geogr. 26, 577–603.

Hancock, P.A., Hutchinson, M.F., 2006. Spatial interpolation of large climate data sets using bivariate thin plate smoothing splines. Environ. Model Softw. 21, 1684–1694.

Harnois, L., 1988. The CIW index: a new chemical index of weathering. Sediment. Geol. 55 (3), 319–322.

Hatch, J.R., Leventhal, J.S., 1992. Relationship between inferred redox potential of the depositional environment and geochemistry of the Upper Pennsylvanian (Missourian) Stark Shale Member of the Dennis Limestone, Wabaunsee County, Kansas, U.S.A. Chem. Geol. 99, 65–82.

Haughton, P.D.W., Todd, S.P., Morton, A.C., 1991. Sedimentary provenance studies. Geol. Soc. Lond. Spec. Publ. 57, 1–11.

Hayashi, K.-I., Fujisawa, H., Holland, H.D., Ohmoto, H., 1997. Geochemistry of ~1.9 Ga sedimentary rocks from northeastern Labrador, Canada. Geochim. Cosmochim. Acta 61, 4115–4137.

Higginson, M.J., 2009. Geochemical proxies (non-isotopic). In: Encyclopedia of Paleoclimatology and Ancient Environments. Springer.

Hild, E., Brumsack, H.-J., 1998. Major and minor element geochemistry of Lower Aptian sediments from the NW German Basin (core Hoheneggelsen KB 40). Cretac. Res. 19, 615–633.

Hossain, H.M.Z., 2020. Major, trace, and REE geochemistry of the Meghna River sediments, Bangladesh: constraints on weathering and provenance. Geol. J. 55, 3321–3343.

Hossain, H.M.Z., Kawahata, H., Roser, B.P., Sampei, Y., Manaka, T., Otani, S., 2017. Geochemical characteristics of modern river sediments in Myanmar and Thailand: implications for provenance and weathering. Geochemistry 77, 443–458.

Huang, Y., Yang, L., Fu, Z., 2020. Reconstructing coupled time series in climate systems using three kinds of machine-learning methods. Earth Syst. Dynam. 11, 835–853.

Huerta-Diaz, M.A., Morse, J.W., 1992. Pyritization of trace metals in anoxic marine sediments. Geochim. Cosmochim. Acta 56, 2681–2702.
Huntingford, C., Jeffers, E.S., Bonsall, M.B., Christensen, H.M., Lees, T., Yang, H., 2019. Machine learning and artificial intelligence to aid climate change research and preparedness. Environ. Res. Lett. 14, 124007.
Jin, Z., Li, F., Cao, J., Wang, S., Yu, J., 2006. Geochemistry of Daihai Lake sediments, Inner Mongolia, north China: implications for provenance, sedimentary sorting, and catchment weathering. Geomorphology 80, 147–163.
Jones, B., Manning, D.A.C., 1994. Comparison of geochemical indices used for the interpretation of palaeoredox conditions in ancient mudstones. Chem. Geol. 111, 111–129.
Joshi, K.B., 2014. Microbes: mini iron factories. Indian J. Microbiol. 54, 483–485.
Joshi, K.B., Banerji, U.S., Dubey, C.P., Oliveira, E.P., 2021a. Heavy minerals in provenance studies: an overview. Arab. J. Geosci. 14, 1–16.
Joshi, K.B., Goswami, V., Banerji, U.S., Shankar, R., 2021b. Recent Developments in Instrumentation and its application in absolute geochronology: historical perspective and Overview. J. Asian Earth Sci. 211, 104690.
Joshi, K.B., Ray, S., Ahmad, T., Manavalan, S., Aradhi, K.K., 2021c. Geochemistry of meta-sediments from Neoproterozoic Shimla and Chail Groups of outer lesser Himalaya: implications for provenance, tectonic setting, and paleo-weathering conditions. Geol. J. 56, 4451–4478.
Joshi, K.B., Banerji, U.S., Dubey, C.P., Oliveira, E.P., 2022a. Detrital zircons in crustal evolution: a perspective from the Indian subcontinent. Lithosphere 2022, 3099822.
Joshi, K.B., Singh, S.K., Halla, J., Ahmad, T., Rai, V.K., 2022b. Neodymium Isotope Constraints on the Origin of TTGs and High-K Granitoids in the Bundelkhand Craton, Central India: implications for Archaean Crustal Evolution. Lithosphere 2022.
Juggins, S., Simpson, G.L., Telford, R.J., 2015. Taxon selection using statistical learning techniques to improve transfer function prediction. Holocene 25, 130–136.
Kato, Y., 1995. Remobilization of Transition Elements in Pore Water of Continental Slope Sediments. Flux West. Pacific.
Kemp, D.B., Izumi, K., 2014. Multiproxy geochemical analysis of a Panthalassic margin record of the early Toarcian oceanic anoxic event (Toyora area, Japan). Palaeogeogr. Palaeoclimatol. Palaeoecol. 414, 332–341.
Kidder, D.L., Krishnaswamy, R., Mapes, R.H., 2003. Elemental mobility in phosphatic shales during concretion growth and implications for provenance analysis. Chem. Geol. 198, 335–353.
Lee, C.-T., 2019. Geochemical classification of elements. In: White, W.M., Casey, W.H., Marty, B., Yurimoto, H. (Eds.), Encyclopedia of Geochemistry-A Comprehensive Reference Source on the Chemitry of the Earth. Springer, pp. 137–138.
Lesmeister, C., 2017. Mastering Machine Learning with R. Packt Publishing Ltd.
Li, Z., Wang, N., Cheng, H., Li, Y., 2016. Early–middle Holocene hydroclimate changes in the Asian monsoon margin of northwest China inferred from Huahai terminal lake records. J. Paleolimnol. 55, 289–302.
Li, Y., Clift, P.D., Böning, P., Blusztajn, J., Murray, R.W., Ireland, T., Pahnke, K., Helm, N.C., Giosan, L., 2018. Continuous Holocene input of river sediment to the Indus Submarine Canyon. Mar. Geol. 406, 159–176.
Liang, S., Wang, D., Wu, J., Wang, R., Wang, R., 2021. Method of bidirectional LSTM modelling for the atmospheric temperature. Intell. Autom. SOFT Comput. 30, 701–714.
Lindsay, W.L., 1979. Chemical Equilibria in Soils. John Wiley and Sons Ltd.
Lü, X., Versteegh, G.J.M., Song, J., Li, X., Yuan, H., Li, N., 2016. Geochemistry of Middle Holocene sediments from south Yellow Sea: implications to provenance and climate change. J. Earth Sci. 27, 751–762. https://doi.org/10.1007/s12583-015-0577-0.

Lückge, A., Doose-Rolinski, H., Khan, A.A., Schulz, H., Von Rad, U., 2001. Monsoonal variability in the northeastern Arabian Sea during the past 5000 years: geochemical evidence from laminated sediments. Palaeogeogr. Palaeoclimatol. Palaeoecol. 167, 273–286.
Lupker, M., France-Lanord, C., Galy, V., Lavé, J.Ô., Kudrass, H., 2013. Increasing chemical weathering in the Himalayan system since the Last Glacial Maximum. Earth Planet. Sci. Lett. 365, 243–252.
Madhavaraju, J., Tom, M., Lee, Y.I.L., Balaram, V., Ramasamy, S., Carranza-Edwards, A., Ramachandran, A., 2016. Provenance and tectonic settings of sands from Puerto Peñasco, Desemboque and Bahia Kino beaches, Gulf of California, Sonora, México. J. S. Am. Earth Sci. 71, 262–275.
Marmion, M., Parviainen, M., Luoto, M., Heikkinen, R.K., Thuiller, W., 2009. Evaluation of consensus methods in predictive species distribution modelling. Divers. Distrib. 15, 59–69.
Marra, K.R., Madden, M.E.E., Soreghan, G.S., Hall, B.L., 2017. Chemical weathering trends in fine-grained ephemeral stream sediments of the McMurdo Dry Valleys, Antarctica. Geomorphology 281, 13–30.
Martinez-Ruiz, F., Kastner, M., Gallego-Torres, D., Rodrigo-Gámiz, M., Nieto-Moreno, V., Ortega-Huertas, M., 2015. Paleoclimate and paleoceanography over the past 20,000yr in the Mediterranean Sea Basins as indicated by sediment elemental proxies. Quat. Sci. Rev. 107, 25–46.
Mason, J.A., Jacobs, P.M., 1998. Chemical and particle-size evidence for addition of fine dust to soils of the midwestern United States. Geology 26, 1135–1138.
McKenzie, R.M., 1980. The adsorption of lead and other heavy metals on oxides of manganese and iron. Soil Res. 18, 61–73.
McLennan, S.M., 1989. Rare earth elements in sedimentary rocks: influence of provenance and sedimentary processes (Chapter 7). In: Reviews in Mineralogy and Geochemistry. Walter de Gruyter.
McLennan, S.M., 1993. Weathering and global denudation. J. Geol. 101, 295–303.
McLennan, S.M., Taylor, S.R., McCulloch, M.T., Maynard, J.B., 1990. Geochemical and NdSr isotopic composition of deep-sea turbidites: crustal evolution and plate tectonic associations. Geochim. Cosmochim. Acta 54, 2015–2050.
McLennan, S.M., Hemming, S., McDaniel, D.K., Hanson, G.N., 1993. Geochemical approaches to sedimentation, provenance, and tectonics. Spec. Pap. Geol. Soc. Am. 284, 21–40.
McLennan, S.M., Hemming, S.R., Taylor, S.R., Eriksson, K.A., 1995. Early Proterozoic crustal evolution: geochemical and NdPb isotopic evidence from metasedimentary rocks, southwestern North America. Geochim. Cosmochim. Acta 59, 1153–1177.
Morse, J.W., Luther, G.W., 1999. Chemical influences on trace metal-sulfide interactions in anoxic sediments. Geochim. Cosmochim. Acta 63, 3373–3378.
Nameroff, T.J., Calvert, S.E., Murray, J.W., 2004. Glacial-interglacial variability in the eastern tropical North Pacificoxygen minimum zone recorded by redox-sensitive trace metals. Paleoceanography 19, PA1010.
Nesbitt, H.W., Markovics, G., 1997. Weathering of granodioritic crust, long-term storage of elements in weathering profiles, and petrogenesis of siliciclastic sediments. Geochim. Cosmochim. Acta 61, 1653–1670.
Nesbitt, H.W., Young, G.M., 1982. Early Proterozoic climates and plate motions inferred from major element chemistry of lutites. Nature 299, 715–717.
Nesbitt, H.W., Young, G.M., 1984. Prediction of some weathering trends of plutonic and volcanic rocks based on thermodynamic and kinetic considerations. Geochim. Cosmochim. Acta 48, 1523–1534.
Nesbitt, H.W., Markovics, G., Price, R.C., 1980. Chemical processes affecting alkalis and alkaline earths during continental weathering. Geochim. Cosmochim. Acta 44, 1659–1666.

Nilsson, L., Groening, O., Emmenegger, C., Kuettel, O., Schaller, E., Schlapbach, L., Kind, H., Bonard, J.M., Kern, K., 2000. Scanning field emission from patterned carbon nanotube films. Appl. Phys. Lett. 76, 2071–2073.

Overpeck, J.T., Webb, T., Prentice, I.C., 1985. Quantitative interpretation of fossil pollen spectra: dissimilarity coefficients and the method of modern analogs. Quat. Res. 23, 87–108.

Pang, H., Pan, B., Garzanti, E., Gao, H., Zhao, X., Chen, D., 2018. Mineralogy and geochemistry of modern Yellow River sediments: implications for weathering and provenance. Chem. Geol. 488, 76–86.

Peng, J., Zhao, H., Dong, Z., Zhang, Z., Yang, H., Wang, X., 2022. Numerical methodologies and tools for efficient and flexible unmixing of single-sample grain-size distributions: application to late Quaternary aeolian sediments from the desert-loess transition zone of the Tengger Desert. Sediment. Geol. 438, 106211.

Peppe, D.J., Royer, D.L., Cariglino, B., Oliver, S.Y., Newman, S., Leight, E., Enikolopov, G., Fernandez-Burgos, M., Herrera, F., Adams, J.M., 2011. Sensitivity of leaf size and shape to climate: global patterns and paleoclimatic applications. New Phytol. 190, 724–739.

Pettijohn, F.J., Potter, P.E., Siever, R., 1987. Introduction and source materials. In: Sand and Sandstone. Springer, pp. 1–21.

Piper, D.Z., Heath, G.R., 1989. Hydrogenous Sediment. The Eastern Pacific Ocean and Hawaii. Geoscience world.

Ragueneau, O., Tréguer, P., Leynaert, A., Anderson, R.F., Brzezinski, M.A., DeMaster, D.J., Dugdale, R.C., Dymond, J., Fischer, G., François, R., Heinze, C., Maier-Reimer, E., Martin-Jézéquel, V., Nelson, D.M., Quéguiner, B., 2000. A review of the Si cycle in the modern ocean: recent progress and missing gaps in the application of biogenic opal as a paleoproductivity proxy. Glob. Planet. Chang. 26, 317–365.

Resmi, M.R., Achyuthan, H., Babeesh, C., 2020. Holocene evolution of the Palar river, Southern India: tracking history of migration, provenance, weathering and tectonics. Quat. Int. 575–576, 358–374.

Rimmer, S.M., 2004. Geochemical paleoredox indicators in Devonian-Mississippian black shales, Central Appalachian Basin (USA). Chem. Geol. 206, 373–391.

Roddaz, M., Debat, P., Nikiéma, S., 2007. Geochemistry of Upper Birimian sediments (major and trace elements and Nd-Sr isotopes) and implications for weathering and tectonic setting of the Late Paleoproterozoic crust. Precambrian Res. 159, 197–211.

Roser, B.P., Korsch, R.J., 1988. Provenance signatures of sandstone-mudstone suites determined using discriminant function analysis of major-element data. Chem. Geol. 67, 119–139.

Roy, A., Dubey, C.P., Prasad, M., 2021a. Gravity inversion for heterogeneous sedimentary basin with b-spline polynomial approximation using differential evolution algorithm. Geophysics 86, 1–63.

Roy, A., Dubey, C.P., Prasad, M., 2021b. Gravity inversion of basement relief using particle swarm optimization by automated parameter selection of Fourier coefficients. Comput. Geosci. 156, 104875.

Sageman, B.B., Lyons, T.W., 2003. Geochemistry of fine-grained sediments and sedimentary rocks. Treatise Geochem. 7–9, 115–158.

Saito, M.A., Moffett, J.W., Chisholm, S.W., Waterbury, J.B., 2002. Cobalt limitation and uptake in Prochlorococcus. Limnol. Oceanogr. 47, 1629–1636.

Salonen, J.S., Seppä, H., Luoto, M., Bjune, A.E., Birks, H.J.B., 2012. A North European pollen–climate calibration set: analysing the climatic responses of a biological proxy using novel regression tree methods. Quat. Sci. Rev. 45, 95–110.

Salonen, J.S., Luoto, M., Alenius, T., Heikkilä, M., Seppä, H., Telford, R.J., Birks, H.J.B., 2014. Reconstructing palaeoclimatic variables from fossil pollen using boosted regression trees: comparison and synthesis with other quantitative reconstruction methods. Quat. Sci. Rev. 88, 69–81.

Salonen, J.S., Verster, A.J., Engels, S., Soininen, J., Trachsel, M., Luoto, M., 2016. Calibrating aquatic microfossil proxies with regression-tree ensembles: cross-validation with modern chironomid and diatom data. The Holocene 26, 1040–1048.
Salonen, J.S., Korpela, M., Williams, J.W., Luoto, M., 2019. Machine-learning based reconstructions of primary and secondary climate variables from North American and European fossil pollen data. Sci. Rep. 9, 1–13.
Sanchez, M.R., 2019. Global Assessment of Marine Paleoproductivity Proxies. PhD thesis University: Universitat Autònoma de Barcelona, p. 184.
Sandeep, K., Shankar, R., Warrier, A.K., Aravind, G.H., 2022. The geochemical and pedogenic signatures of Shantisagara lake sediments, southern India: implications for weathering, terrigenous influx, and provenance during the Holocene. Geol. J., 1–13.
Sannigrahi, P., Ingall, E., 2005. Polyphosphates as a source of enhanced P fluxes in marine sediments overlain by anoxic waters: evidence from 31 P NMR. Geochem. Trans. 6, 1–8.
Scheiderich, K., Zerkle, A.L., Helz, G.R., Farquhar, J., Walker, R.J., 2010. Molybdenum isotope, multiple sulfur isotope, and redox-sensitive element behavior in early Pleistocene Mediterranean sapropels. Chem. Geol. 279, 134–144.
Scheuvens, D., Schütz, L., Kandler, K., Ebert, M., Weinbruch, S., 2013. Bulk composition of northern African dust and its source sediments—a compilation. Earth-Science Rev. 116, 170–194.
Schneider, S., Hornung, J., Hinderer, M., Garzanti, E., 2016. Petrography and geochemistry of modern river sediments in an equatorial environment (Rwenzori Mountains and Albertine rift, Uganda)—implications for weathering and provenance. Sediment. Geol. 336, 106–119.
Shaji, J., Banerji, U.S., Maya, K., Joshi, K.B., Dabhi, A.J., Bharti, N., Bhushan, R., Padmalal, D., 2022. Holocene monsoon and sea-level variability from coastal lowlands of Kerala, SW India. Quat. Int. https://doi.org/10.1016/j.quaint.2022.03.005. In press.
Singh, S.K., Rai, S.K., Krishnaswami, S., 2008. Sr and Nd isotopes in River sediments from the Ganga basin: sediment provenance and spatial variability in physical erosion. Case Rep. Med. 113, 1–18.
Slabunov, A., Singh, V.K., Joshi, K.B., Li, X., 2017. Paleoarchean zircons from quartzite of South Bundelkhand Supracrustal Complex: origin and implications for crustal evolution in Bundelkhand Craton, Central India. Curr. Sci. 112 (4), 794–801.
Slack, J.F., Stevens, B.P.J., 1994. Clastic metasediments of the Early Proterozoic Broken Hill Group, New South Wales, Australia: geochemistry, provenance, and metallogenic significance. Geochim. Cosmochim. Acta 58, 3633–3652.
Song, L., Jie, D., Gao, G., Liu, L., Liu, H., Li, D., Liu, Y., 2022. Application of a topsoil phytolith dataset to quantitative paleoclimate reconstruction in Northeast China. Palaeogeogr. Palaeoclimatol. Palaeoecol. 601, 111108.
Spears, D.A., Kanaris-Sotiriou, R., 1976. Titanium in some carboniferous sediments from Great Britain. Geochim. Cosmochim. Acta 40, 345–351.
Sudom, M.D., Arnaud, R.J., 1971. Use of quartz, zirconium and titanium as indices in pedological studies. Can. J. Soil Sci. 51, 385–396.
Sun, Q., Wang, S., Zhou, J., Chen, Z., Shen, J., Xie, X., Wu, F., Chen, P., 2010. Sediment geochemistry of Lake Daihai, north-central China: implications for catchment weathering and climate change during the Holocene. J. Paleolimnol. 43, 75–87. https://doi.org/10.1007/s10933-009-9315-x.
Sun, L., Gui, H., Chen, S., 2012. Geochemistry of sandstones from the Neoproterozoic Shijia Formation, northern Anhui Province, China: implications for provenance, weathering and tectonic setting. Geochemistry 72, 253–260.
Taylor, S.R., McLennan, S.M., 1985. The Continental Crust: Its Composition and Evolution: An Examination of the Geochemical Record Preserved in Sedimentary Rocks. Oxford, Oxford, United States.

Ter Braak, C.J.F., Juggins, S., 1993. Weighted averaging partial least squares regression (WA-PLS): an improved method for reconstructing environmental variables from species assemblages. Springer, pp. 485–502.
Ter Braak, C.J.F., Prentice, I.C., 2004. A theory of gradient analysis. Adv. Ecol. Res. 34, 235–282.
Tribovillard, N., Algeo, T.J., Lyons, T., Riboulleau, A., 2006. Trace metals as paleoredox and paleoproductivity proxies: an update. Chem. Geol. 232, 12–32.
Tyson, R.V., Pearson, T.H., 1991. Modern and ancient continental shelf anoxia: an overview. Geol. Soc. Lond. Spec. Publ. 58, 1–24.
Warrier, A.K., Shankar, R., 2009. Geochemical evidence for the use of magnetic susceptibility as a paleorainfall proxy in the tropics. Chem. Geol. 265, 553–562.
Wedepohl, K.H., Correns, C.W., Shaw, D.M., Turekian, K.K., Zemann, J., 1969. Handbook of Geochemistry. Springer, Berlin.
Wehrli, B., Stumm, W., 1989. Vanadyl in natural waters: adsorption and hydrolysis promote oxygenation. Geochim. Cosmochim. Acta 53, 69–77.
Wei, G., Liu, Y., Li, X.H., Shao, L., Fang, D., 2004. Major and trace element variations of the sediments at ODP Site 1144, South China Sea, during the last 230 ka and their paleo climate implications. Palaeogeogr. Palaeoclimatol. Palaeoecol. 212, 331–342.
Wei, G., Peng, C., Zhu, Q., Zhou, X., Yang, B., 2021. Application of machine learning methods for paleoclimatic reconstructions from leaf traits. Int. J. Climatol. 41, E3249–E3262.
Wersin, P., Höhener, P., Giovanoli, R., Stumm, W., 1991. Early diagenetic influences on iron transformations in a freshwater lake sediment. Chem. Geol. 90, 233–252.
Whitfield, M., 2001. Interactions between phytoplankton and trace metals in the ocean. Adv. Mar. Biol. 41, 1–128.
Williams, M., Dunkerley, D., Deckker, P.D., Kershaw, P., Chappell, J., 1998. Quaternary sea level changes. In: Quaternary Environments. Arbold, London, pp. 107–125.
Wilson, S.A., Weber, J.H., 1979. An EPR study of the reduction of vanadium (V) to vanadium (IV) by fulvic acid. Chem. Geol. 26, 345–354.
Wu, L., Zhang, X., Hao, F., Wu, Y., Li, C., Xu, Y., 2020. Evaluating the contributions of climate change and human activities to runoff in typical semi-arid area, China. J. Hydrol. 590, 125555.
Xi, J., He, L.Y., Huang, Z., Sheng, X.F., 2014. Bacillus qingshengii sp. nov., a rock-weathering bacterium isolated from weathered rock surface. Int. J. Syst. Evol. Microbiol. 64, 2473–2479.
Yang, S.Y., Li, C.X., Yang, D.Y., Li, X.S., 2004. Chemical weathering of the loess deposits in the lower Changjiang Valley, China, and paleoclimatic implications. Quat. Int. 117, 27–34.
Zhang, X., Yan, Y., Huang, C.-Y., Chen, D., Shan, Y., Lan, Q., Chen, W., Yu, M., 2014. Provenance analysis of the Miocene accretionary prism of the Hengchun Peninsula, southern Taiwan, and regional geological significance. J. Asian Earth Sci. 85, 26–39.
Zhang, Z., Wagner, S., Klockmann, M., Zorita, E., 2022. Evaluation of statistical climate reconstruction methods based on pseudoproxy experiments using linear and machine learning methods. Clim. Past Discuss., 1–31.
Zhao, M.-Y., Zheng, Y.-F., 2015. The intensity of chemical weathering: geochemical constraints from marine detrital sediments of Triassic age in South China. Chem. Geol. 391, 111–122.
Zhu, L., Wu, Y., Wang, J., Lin, X., Ju, J., Xie, M., Li, M., Mäusbacher, R., Schwalb, A., Daut, G., 2008. Environmental changes since 8.4 ka reflected in the lacustrine core sediments from Nam Co, central Tibetan Plateau, China. Holocene 18, 831–839.

CHAPTER TWO

Stratigraphy in the Greenland/Iceland/Norwegian (GIN) seas: A multiproxy approach on Pleistocene sediments

Marjolaine E.J. Sabine-Lamoureux[a,†], Frédérique Eynaud[a,*,†], Sébastien Zaragosi[a,†], Jacques Giraudeau[a], Maxime Debret[b], Linda Rossignol-Malaize[a], Karine Charlier[a], Isabelle Billy[a], Bruno Malaize[a], Jimmy Daynac[a], Elodie Marches[c], and Thierry Garlan[c]

[a]Laboratoire Environnements et Paléoenvironnements Océaniques et Continentaux (EPOC) UMR CNRS 5805, Université de Bordeaux, Pessac, France
[b]Département de Géosciences et Environnement Morphodynamique Continentale et Côtière, Sciences et Techniques, Université de Rouen, Mont-Saint-Aignan, France
[c]Service hydrographique et océanographique de la Marine (SHOM), Brest, France
*Corresponding author e-mail address: frederique.eynaud@u-bordeaux.fr

Contents

Abstract

A multiproxy sedimentological study was conducted on five sediment cores retrieved between 67 and 79°N in the Greenland/Iceland/Norwegian seas in order to infer a

† These authors contributed equally.

Stratigraphy & Timescales, Volume 7
ISSN 2468-5178
https://doi.org/10.1016/bs.sats.2022.09.004

common chronostratigraphic model for the selected cores. This model is based on the use of a series of geochemical, physical and micropaleontological proxies which are routinely measured in sedimentological investigations of marine sediment cores: the major and minor element content derived from X-Ray Fluorescence (XRF) core scanner analyses; the lightness and color of the sediment derived from spectrophotometry analyses; the magnetic susceptibility of the sediment; the distribution of planktonic foraminiferal assemblages combined to coccolith stratigraphy (acme zones) providing preliminary stratigraphic tie points. All those proxies are correlated independently between the five sediment cores. Our results demonstrate that high resolution studies using those standard paleoceanographical tools are powerful for establishing core-to-core correlation. A chronostratigraphical framework is proposed based on the correlation of the planktonic $\delta^{18}O$ isotopic record obtained in core M17KC03 with the LR04 benthic $\delta^{18}O$ reference stack. A comparison between the M17KC03 stratigraphy and other dated cores from the sub-polar North Atlantic confirms the robustness of the approach. Our study suggests that XRF core scanner-derived elemental ratios (especially those related to Ca), the lightness L* and the coccolith-based biostratigraphy provide robust stratigraphic tie-points at the scale of the whole Greenland/Iceland/Norwegian seas. The elemental ratio Sr/sum, the magnetic susceptibility, and planktonic foraminiferal records have a more local stratigraphic significance in the GIN seas. Over the last 1100 ka BP, our datasets are indicative of: (1) an excursion of Ca content and planktonic foraminiferal abundances during MIS 11, (2) an overall decrease toward present of both the detrital content (with the exception of values characterizing MIS 12) and of the magnetic susceptibility of sediments deposited during Glacials.

Highlights
Multiproxy approach for a robust stratigraphy in the Nordic seas.
New Pleistocene subboreal records.
Sedimentological expression of marine isotopic stages.

1. Introduction

Stratigraphy is the backbone of every marine sedimentological investigation during the Pleistocene where glacial and interglacial climate shifts paced paleoceanographic changes. In the northernmost basins of the North Atlantic (NAtl), i.e., the Greenland/Iceland/Norwegian (GIN) seas, the alternation of glacial and interglacial periods at the orbital scale drives drastic environmental shifts and readjustments, those seas moving from cold-polar to milder-subpolar climates at both extremes. The GIN seas also constitute a crucial component of the Atlantic Meridional Overturning Circulation (AMOC), and thus conversely can drive global climatic changes (e.g., Becker et al., 2018; Chauhan et al., 2014; Müller and Stein, 2014; Thibodeau et al., 2017). Despite contrasted contexts registered between glacials and interglacials, the establishment of a robust stratigraphy is still a prime challenge in those seas, with the main limitations being attributed to: (1) meager sedimentation rates (1–3 cm/ka out of the channels and troughs, e.g., Laberg et al., 2000), (2) the microfossils scarcity and (3) strongly variable oceanographic conditions in these strongly reactive areas. Indeed, climatic changes impact on diverse components of this complex oceanographic system: e.g., on the thickness and lateral extension of the surrounding ice-sheets, the sea-level, the sea-ice cover extension and duration, the water column structure and dynamics (including the pathway of currents, deep, intermediate and surface as well), and also the photic zone thickness (e.g., Andrews, 2000; Blindheim et al., 2000; Broecker et al., 1990; Rahmstorf, 1999, 2002, 2013; Samtleben et al., 1995; Sejrup et al., 2011). Those subpolar seas are additionally strongly affected by the dissolution of pelagic carbonates, with the lysocline (or Carbonate Compensation Depth—CCD) which can be very close to the surface according to regional environmental and climatic conditions (e.g., Huber et al., 2000 and references therein; Henrich et al., 2002). The sediment cores from the GIN seas thus generally present several intervals of calcareous microfossils-free sediments (Bylinskaya et al., 2016; Chauhan et al., 2016; this study). This total absence might be explained by specific environmental conditions (such as a perennial sea-ice cover or cold and fresh surface currents over the studied area, e.g., Huber et al., 2000), or by a substantial dissolution (at the bottom by post-depositional dissolution or in the whole water column). The influence of variable sea-surface conditions, with seasonally driven freshwater advection, melting ice and/or sea-ice and brine rejection, can also modify the water $\delta^{18}O$ ratio and, in consequences, the foraminifera test chemical composition, making difficult the use of the $\delta^{18}O$ solely to construct a reliable stratigraphic scale (e.g., Hillaire-Marcel and de Vernal, 2008; Rohling, 2013). In consequence, paleoceanographic works conducted in the GIN seas with stratigraphies exceeding the ^{14}C measuring limits are hampered by all these limitations, and published ages can be questionable. However, thanks to the recovery of long marine archives over the last five decades, some coherent protocols have emerged crossing multiproxy data to refine age scales at their most reliable possible degree (e.g., Caricchi et al., 2020; Channell and Hodell, 2013; Chapman and Shackleton, 1998; Helmke et al., 2002; Hodell et al., 2008, 2015; Lucchi et al., 2013; Martrat et al., 2014). These studies provide relevant and widely used tools for stratigraphic correlation and dating, giving much more comprehensive palaeoclimatic studies, especially regarding the coupling between the ocean and other reservoirs (e.g., Clark et al., 2006). Actually, those dating protocols give access to reservoir phasing or dephasing and to the processes and feedbacks involved in such environmental dynamics at orbital scales. In the NAtl, a consensus exists regarding the sedimentary typicity of interglacials (and conversely glacials) with high carbonate contents (associated to high contents of calcareous microfossils and whiter sediments), low terrigenous supplies (ice-rafted debris -IRD included) and low stable isotopic benthic oxygen ratio occurring at that times and defining plateau during warmth optima (e.g., Bauch et al., 2000; Bylinskaya et al., 2016; Knies et al., 2009).

This study was conducted on a series of new long marine cores with the aim to further document the paleoceanographical history of the GIN seas in a consolidated stratigraphical approach, taking into account a panel of selected proxies (in complement to the classical $\delta^{18}O$ tool) providing the possibility of robust core correlations over the studied region. With this article, we finally conceptualize a practical tutorial to use as a paleoclimatic and stratigraphic guideline in these sub-boreal domains. This work is based on five sediment cores retrieved in 2017 during the MOCOSED cruise on the R/V *Pourquoi Pas*?, and on reference archives from the subpolar NAtl. The MOCOSED 2017 cores were selected because of their geographical position along contrasted contexts of the GIN seas, in order to illustrate and construct a coherent regional scenario of climate induced changes in those basins. Also, their distant position from the surrounding trough mouth fans (i.e., TMF, which can have an average sedimentation rate much higher, until 400 cm/ka according to, e.g., Baeten et al., 2014) gives more chance to register the most possible prolonged time scale. The analyses were performed following the same methodological protocol in order to obtain comparable datasets to be crossed core by core. The following paragraphs: (1) detail our approach and discuss its suitability to address stratigraphical problems classically met by paleoceanographers in the subboreal and boreal realms; and (2) discuss time intervals that our study recovers in the GIN seas at the light of other long records retrieved in this basin but also in the NAtl.

2. Materials, methods and proxy significance

Five sediment cores (Fig. 1, Table 1) were retrieved with a CALYPSO piston corer on the R/V *Pourquoi Pas?* during the MOCOSED cruise in September 2017. Cores M17KC01, M17KC02, and M17KC03 were retrieved from the southern GIN seas (south of Jan Mayen Fracture Zone) in the Norwegian/Icelandic seas. They were selected based on subsurface seismic CHIRP data which indicated a sediment stacking assuming the possibility

Fig. 1 Physiographic map of the North Atlantic and location of marine cores and drillings studied and/or discussed in the text. (For interpretation of the references to color in the figure legends, the reader is referred to the Web version of this article.)

Table 1 List of the marine cores and of reference archives (deep-sea drillings; ice core) used in this study to build a coherent stratigraphy.

	Core	Type	Altitude (m)	Recovery/ length (m)	Latitude N (decimal degrees)	Longitude E (decimal degrees)	Retrieved in:	Location
Reference cores	ODP site 644	sediment	−1227	238.4	66.667	4.567	1985	Voring basin
	ODP site 985	sediment	−2788	129.44	66.942	−6.450	1996	Northeastern Iceland Sea
	M23414	sediment	−2196	9.08	53.530	−17.000	1991	Rockall Plateau
	PS1243	sediment	−2721	7.67	69.380	−6.510	1984	Norwegian Sea
	MD99–2277	sediment	−2800	36.65	69.250	−6.321	1999	Jan Mayen Fracture Zone, Norwegian Sea
	MD01–2448	sediment	−3460	28.8	44.780	−11.275	2001	Bay of Biscay
	MD03–2705	sediment	−3000	37	18.097	−21.153	2003	Eastern Tropical Atlantic Ocean
	NGRIP	ice core	2917		71.500	−42.320	1999–2003	Center Greenland Ice Sheet
Studied cores	MOCO17-KC01	sediment	−2625	9.16	67.612	−6.659	2017	Norwegian Sea
	MOCO17-KC02	sediment	−3144	11.035	69.701	−5.079	2017	Jan Mayen Fracture Zone, Norwegian Sea
	MOCO17-KC03	sediment	−2750	20.66	70.176	−5.240	2017	Jan Mayen Fracture Zone, Norwegian Sea
	MOCO17-KC14	sediment	−2674	9.17	79.215	−0.727	2017	Fram Strait
	MOCO17-KC15	sediment	−3095	8.46	79.334	0.591	2017	Fram Strait

of collecting a long time series. Cores M17KC15 and M17KC14 were retrieved from the central Fram Strait, in the northern GIN seas, this area being at the transition between the arctic ocean and the GIN seas, which made it a key area to study the paleoclimatic changes.

2.1 Sedimentary analyses

The cores were cut in 1-m section on board, themselves divided into two half-sections, split lengthways. Each split section was analyzed with the spectrophotometer Minolta Cm-2600d UV-visible at a 1-cm step interval, with a 10 nm increment over the 360–740 nm bandwidth. The lightness (L*) scale from 0 to 100, can be compared to the gray-scale reflectance and can be used as a paleoclimatic indicator, as does the color a*, indicative of red-green color of the sediment (i.e., Chapman and Shackleton, 1998; Debret et al., 2011, see Table 2). Then, 1-cm-step magnetic susceptibility measurements were performed with a Barington MS2E.

Each split section was sampled by 7.5-cm-wide, 1-cm-thick aluminum plate in the central part of the section, giving slice of sediment. Then, sediment samples were collected from those aluminum plates (which minimize the effects of the core walls). They were analyzed for sedimentological characteristics and to determine their microfossil contents, focusing on coccolithophorids and planktonic and benthic foraminifera at *EPOC (Environnement et Paléoenvironnement Océaniques et Continentaux)* laboratory.

X-ray fluorescence (XRF) core-scan analyses were performed at a 1-cm-step resolution on each core, using an Avaatech instrument at 10 and 30 kV setting (*EPOC Laboratory PACS instruments: https://www.epoc.u-bordeaux.fr/index.php?lang=fr&page=pf_carottes*). For the purpose of this study, the Ca, Ti, Mn, Sr, and Fe contents were especially identified as significant. The following thus shortly summarizes key information identified for each studied area (in accordance with what is known for these elements, see Table 2 and Supplementary information in the online version at https://doi.org/10.1016/bs.sats.2022.09.004 SI-1). The two southernmost cores, M17KC01 and M17KC02, present a prevalent calcium content in the sediment which is usually a marker of the calcareous primary production (e.g., Hodell et al., 2008; Richter et al., 2006). The M17KC03, near Jan Mayen fracture zone, shows iron as prevalent element, followed by calcium. Then, in the two northernmost cores, M17KC14 and M17KC15, from the central Fram Strait, iron and silicon are prevalent in the sediment, as it is commonly observed in the Arctic Ocean (e.g., Stokes et al., 2005). In a recent study,

Table 2 List of the sedimentary parameters/proxies used in this study with respect to their significance for stratigraphical purposes.

Parameters	State of art/Stratigraphical interpretation and usage	Key references	Threshold of interest in our study
Lightness L*	Lightness L* (Black/white variation) is the capacity of sediment to reflect the light. Light (/white) sediments traduce a high carbonate content and few dilution by the terrigenous components, they are attributed to warm climatic periods	Bauch et al. (2000), Chapman and Shackleton (1998), Debret et al. (2011)	Values >57 are usual during interglacials (*this study*)
Color a*	a* (red/green variation) is primarily controlled by the deposition of Fe and Mn-rich components, like red-colored hematite minerals	Helmke et al. (2002), Nagao and Nakashima (1992)	Low values can highly be correlated to iron in the sediment
^{XRF} Ca/sum ratio	Ca is part of marine carbonates but also derives from reworked carbonaceous material (depending upon the age of the weathered rocks)	Hodell et al. (2008), Kaboth-Bahr et al. (2019), Kolling and Bauch (2017), Rothwell and Croudace (2015), Richter et al. (2006)	Values >0.3 are usual during interglacials (*this study*)
^{XRF} Ti/Ca ratio	Ti is derived from continental rocks. In sediment, it is interpreted as a continental detrital material	Kaboth-Bahr et al. (2019), Richter et al. (2006), Rothwell and Croudace (2015)	Values <0.5 are linked to high planktonic content (*this study*)
^{XRF} Mn/sum ratio	Mn follows glacial/interglacial cyclicity in the central Artic ocean	Helmke et al. (2002), Jakobsson et al. (2000)	Values >0.01 are linked to deep-sea diamict (*this study*)

Table 2 List of the sedimentary parameters/proxies used in this study with respect to their significance for stratigraphical purposes.—cont'd

Parameters	State of art/Stratigraphical interpretation and usage	Key references	Threshold of interest in our study
Absolute abundance of planktonic foraminifera (#/g)	High abundance of planktonic foraminifera implies favorable conditions for primary production and relatively warm and salty waters in the studied area	Bylinskaya et al. (2016), Chauhan et al. (2014)	Values >5 individual/g of dry sediment is usual during interglacials (*this study*)
Large Lithic Grains/LLG (#/g)	LLG correspond to sediment grains coarser than 1000 μm (independently of their morphology and nature)	Rorvik et al. (2010); Bylinskaya et al. (2016)	Values >2 LLG/g of dry sediment is usual during glacials (*this study*)
% > 125 μm	Coarse sediments are usually interpreted as the result of strong current transport or drift-ice transport	Rorvik et al. (2010), Bylinskaya et al. (2016)	Values >15% are usual during interglacials in the SOUTHERN GIN seas // Values >10% are usual during glacials in the NORTHERN GIN seas (*this study*)
MS	The MS signal traduces the magnetic properties of sedimented particles: it depends of the magnetic source distance (advection from Iceland or other magnetic provinces for instance) and of the pathway of the currents which have played on sites. Its use in the Pleistocene sediments of the GIN seas has lead to the construction of north hemisphere reference stack (NAPIS)	Channell and Raymo (2003), Channell et al. (2009) Kissel et al. (1997), Kissel et al. (1999), Stoner et al. (2000), (2002)	–

Carbonara et al. (2016) used Ca/Ti ratio from a mid-depth sediment core from the Kveithola trough, southern Svalbard, to describe biogenic carbonate sedimentation. Consequently, they interpreted XRF calcium content as related to a biogenic content. This interpretation seems coherent with our results (see following Section 3). In that way, the Ca/sum ratio, with "sum" corresponding to the sum of all the elements registered during the measurements for each level, are interpreted as mainly signing calcareous biogenic materials. Strontium (Sr) is usually fixed by marine microorganisms at the same time as Ca, making this element an additional biogenic marker for the marine primary production specifically (e.g., Richter et al., 2006). Titanium (Ti) is prevalent in detrital material (Richter et al., 2006; Rothwell and Croudace, 2015) and is frequently used to identify terrigenous input (Carbonara et al., 2016). Manganese (Mn) is generally used to identify the climatic cyclicity in the central Arctic Ocean (e.g., Helmke et al., 2002; Jakobsson et al., 2000).

The cores are composed of alternated brown, light-gray and dark-gray muds bearing IRD, and silty to sandy laminated IRD-free muds. Grain-size analyses (not shown here) were performed through 4657 samples collected within the five sediment cores, with facies dependants selected intervals. The grain sizes were analyzed using a laser diffraction grain-size analyzer Malvern Mastersizer 2000.

2.2 Micropalaeontological analyses

The five sediment cores were sampled for micropalaeontological analyses taking into account the sedimentological facies changes with a 10 cm mean step, giving a total of 588 samples. Those sediment samples were dried in a heat chamber at 40 °C, weighed (average weight: 7.7 g), wet sieved using two sieves, 125 μm and 1000 μm. In polar basins, the 125 μm sieve is considered as keeping enough planktonic foraminifera being representative of the adult faunal assemblage. Within the GIN seas, each large lithic grain >1000 μm found in a clayey matrix is considered as brought to the deepest environments by glacial processes (sea-ice and iceberg loadings) (e.g., Amundsen et al., 2015; Haflidason et al., 2007). Those large lithic grains are thus interpreted as ice-rafted debris (IRD) in the following. The sieved material 125–1000 μm and >1000 μm was dried and weighed and the <125 μm sediment was not kept. The planktonic foraminifera were counted through the 125–1000 μm fraction and the IRD were counted through the >1000 μm fraction.

Coccolithophorids were used to establish a preliminary biostratigraphy based on their acme zones (coccolithophorid acme zones—CAZ) (Figs. 2 and 3A) (e.g., Pujos and Giraudeau, 1993; Wefer et al., 1998a). Sixty-three smears were collected and analyzed on the studied cores, without a regular sampling step but actually determined by lithological changes, according to sediment color and texture variation: ten smears were collected from the M17KC14, nine from the M17KC15, twenty-two from the M17KC03, thirteen from the M17KC02 and nine smears were collected from the M17KC01. The smears were observed under a standard light microscope with a magnification x1,000. The quantitative variations of different calcareous nannofossils species permit the identification of specific periods of time (Thierstein et al., 1977). Thus, Thierstein et al. (1977) describe *Pseudoemiliania lacunosa* disappearance after 450 ka BP, *Emiliania huxleyi* appearance 275 ka BP, and they determine dominance changes among *E. huxleyi* and *Gephyrocapsa caribbeanica* in between 85 and 73 ka BP. The reference framework used in this work is the Neogene period's biostratigraphy defined by Wefer et al. (1998b) (Fig. 2).

Stable isotope ratios were measured on monospecific samples of foraminifera shells from the M17KC03 (benthic and planktonic) fossil assemblages. The measurements were done following standard laboratory techniques at the *EPOC Laboratory Isotopic Facility Platform* with a Thermo Scientific Kiel IV Carbonate Device connected to a MAT253 isotope ratio mass spectrometer. All the measurements were calibrated on the Pee Dee Belemnite isotope standard (PDB). The $\delta^{18}O$ analytical precision is better than 0.08‰. The planktonic $\delta^{18}O$ signal has been obtained on monospecific samples of *Neogloboquadrina pachyderma* (*Nps*) using 8–16 specimens (selected from the 125–250 μm size). Because of the microfossils paucity, only 161 (upon 251) samples contained planktonic foraminifera and have produced isotopic measurements (Fig. 2B and E). The last two meters of sediment were totally free of calcareous microfossils.

$\delta^{18}O$ measurements were also performed on monospecific samples of the benthic species *Planulina wuellerstorfi* (selected from the 250–350 μm size). A +0.64‰ correction factor was applied to *P. wuellerstorfi* $\delta^{18}O$ values, according to previous calibration studies (Duplessy et al., 1984; Shackleton and Opdyke, 1973). The microfossils scarcity limited the measurements to only 100 (upon 251) samples (Fig. 2C and D). Because of unexpected values, six points were double measured, the points 6, 56, 95, 157, 262 and 281 cm depth in the core. Each measurement was reproducible. However, the discontinuous specimen

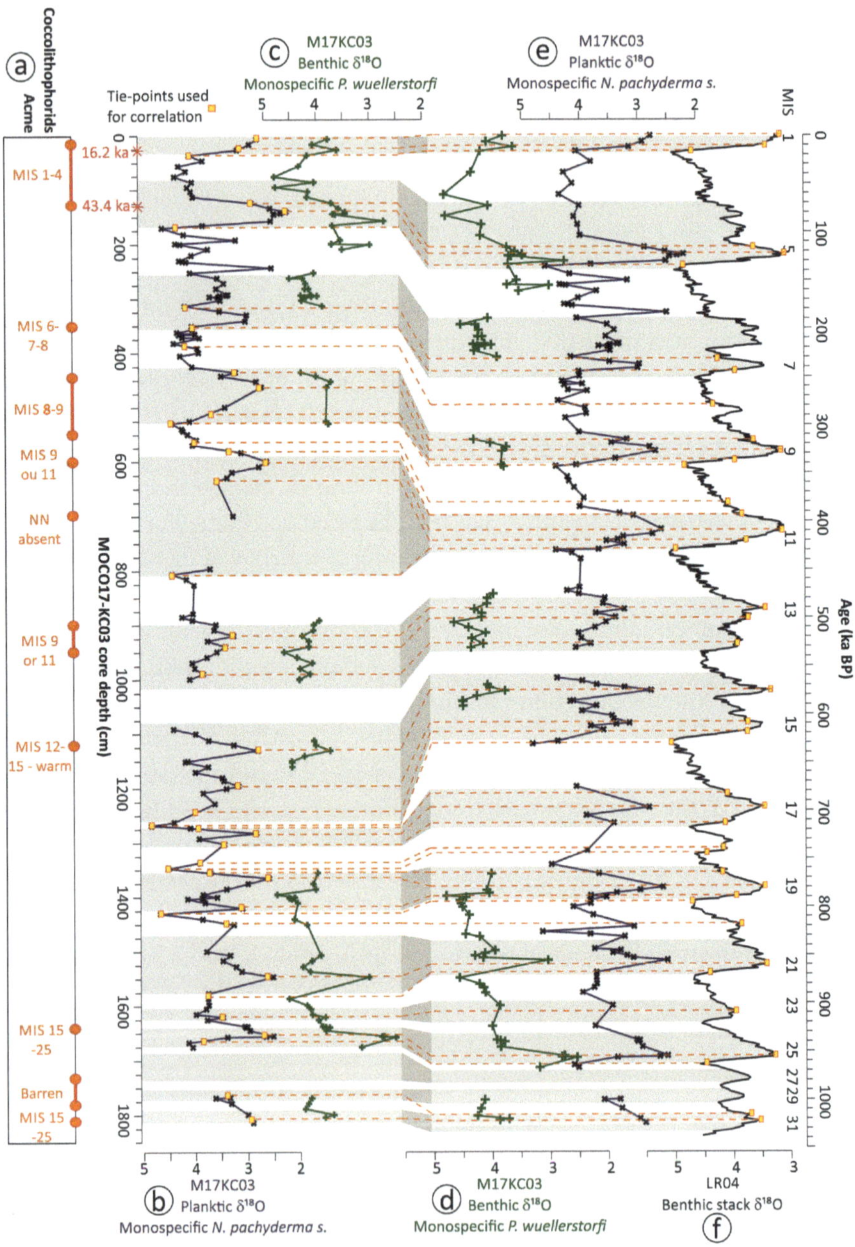

Fig. 2 M17KC03 age model construction steps (deriving from $\delta^{18}O$ data correlated with the LR04 $\delta^{18}O$ benthic stack), with: (A) identified Coccolithophorids acme zones, (B) M17KC03 planktonic $\delta^{18}O$ (*Nps*) vs. core depth, (C) M17KC03 benthic $\delta^{18}O$ (*P. wuellerstorfi*) vs. core depth, (D) M17KC03 benthic $\delta^{18}O$ vs. age, (E) M17KC03 planktonic $\delta^{18}O$ vs. age, (F) $\delta^{18}O$ LR04 benthic stack (Lisiecki and Raymo, 2005).

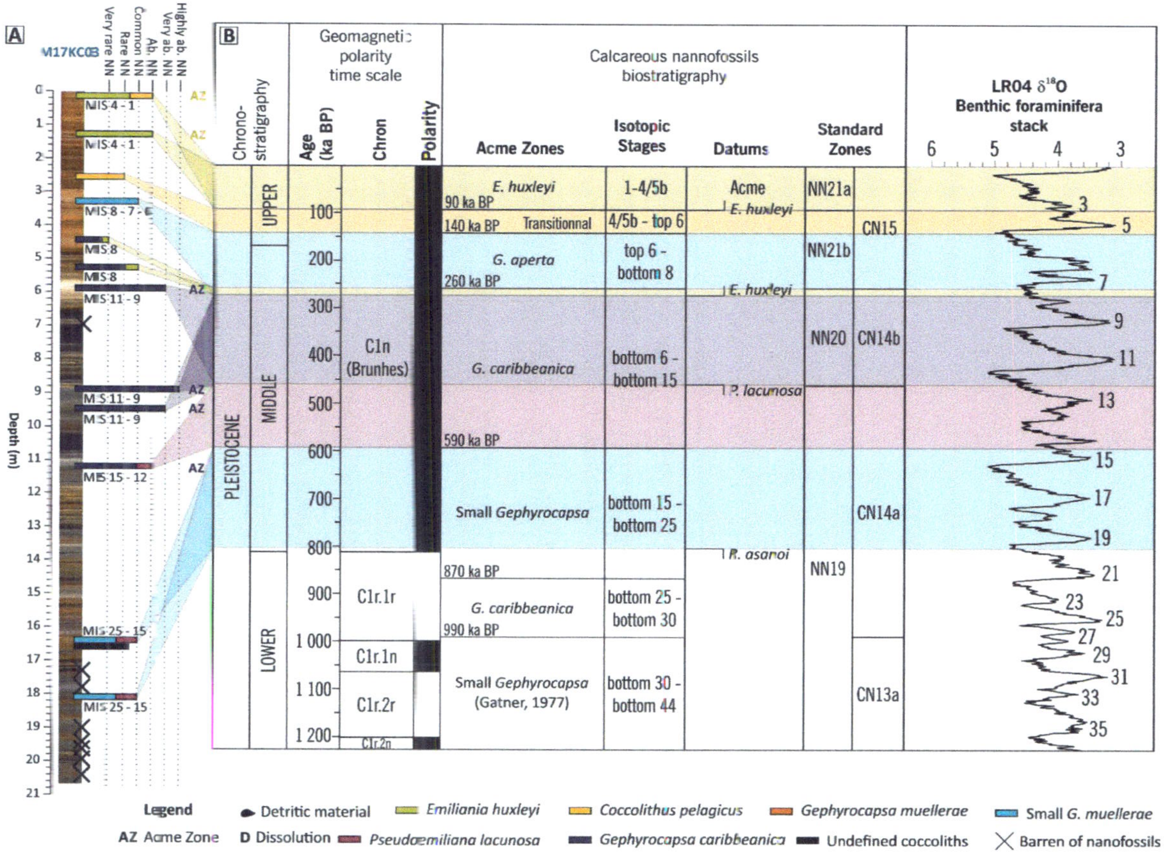

Fig. 3 M17KC03 nannofossil stratigraphy based on coccolithophorids acme zones. (A): identified coccolithophorids acme zones. (B): corresponding stratigraphy as defined by from Pujos and Giraudeau (1993) in the frame of the known chronostratigraphy, the geomagnetic polarity time scale, the calcareous nannofossil biostratigraphy and compared to the LR04 d18O benthic stack (Lisiecki and Raymo, 2005).

recovery over the core did not allowed us to produce a complete robust signal. Considering published works in the same area, we decided to follow approaches introduced by previous authors, i.e., to mainly rely on planktonic measurements for the construction of isotopic stratigraphic frameworks (see Bauch et al., 2000; Helmke and Bauch, 2002).

2.3 Data analytical procedures

Because the M17KC03 core (20.66 m long, southern GIN seas) recovers the temporally longest record, it was used as the reference core for our work. Then, with the AnalySeries software (Paillard et al., 1996), a core-to-core correlation was undertaken using a multiproxy method, fixing recurrent tie-points traceable in a maximum of cores. It facilitates the correlation of different graphical data (curves) peak-to-peak, and/or transition-to-transition, with the generation of tie-points. We considered scrupulously, all along the process, the stratigraphical framework defined by the CAZ (as illustrated in Fig. 2A and 3) as a basic timeline.

In complement, we conducted two radiocarbon datings on the reference core M17KC03 based on planktonic foraminifera shells. They were measured on monospecific samples of *Neogloboquadrina pachyderma* (sinistral)/*Nps* selected in the 125–1000 μm fraction (Table 3). Those datings were calibrated using the Calib 8.1.0 software, with the Marine 20 calibration set (Heaton et al., 2020; Reimer et al., 2013; Stuiver and Reimer, 1993) and indicated into our correlations (Fig. 2) even if their coherency is disputable.

3. Toward a coherent stratigraphy

3.1 Building the M17KC03 core stratigraphy

Our stratigraphy is based on a peak-to-peak correlation in between the M17KC03 *Nps* $\delta^{18}O$ record with the LR04 $\delta^{18}O$ benthic stack (Lisiecki et al., 2008; Lisiecki and Raymo, 2005). This correlation is shown on Fig. 2, both in depth (Fig. 2B and C) and age (Fig. 2D and F) to provide an objective view of the obtained tuning. Most of the $\delta^{18}O$ spikes correlate peak-by-peak to the glacial-interglacial interplay, reflecting the Marine Isotopic Stages 1 to 31.

Forty-one tie-points were needed for this correlation (Table 4, Figs. 2 and 4), taking into account the coherency with the CAZ obtained information (Fig. 3). This correlation, thus using independent microfossil tools, gives an age of around 1000 ka BP to the first 19 m of the M17KC03 sediment core.

Table 3 Radiocarbon dates obtained from the M17KC03 sedimentary core.

Sample reference	Material	Depth (cm) in core	Laboratory number	Conventional radiocarbon age (a BP)	Error (yr)	Calibrated dates (cal a BP)/CALIB 8.1.0 Cal. Set: Marine 20		
						1 sigma ranges [sart:end] relative area	2 sigma ranges [sart:end] relative area	Mid age (a cal BP) 1 sigma
MOCO_KC03_26 cm (Ech 1)	Monospecific samples of *Neogloboquadrina pachyderma* (sinistral) form the 125–1000 μm fraction	26 cm	Beta-529,357	14,050	±40	[cal BP 16031: cal BP 16262] 1.	[cal BP 15912: cal BP 16386] 1	16,146.5
MOCO_Cc03_124 cm (Ech 2)		124 cm	Beta-529,358	41,210	±490	[cal BP 42981: cal BP 43821] 1.	[cal BP 42708: cal BP 44201] 1	43,401

Ages were calibrated with the CALIB 8.1.0 software (marine reservoir correction database Marine 20 after Reimer et al., 2013; Stuiver and Reimer, 1993; Heaton et al., 2020).

Table 4 Table listing the forty-one tie-points resulting from the correlation of the M17KC03 planktonic $\delta^{18}O$ record with the benthic $\delta^{18}O$ stacked record LR04 (Lisiecki and Raymo, 2005) using the Analyseries software.

	M17KC03	**LR04**
Tie-points	**Core-depth (cm from the top)**	**Age (ka BP)**
1	6	0
2	24	11
3	37	17
4	124	116
5	140	123
6	170	135
7	317	232
8	353	245
9	388	280
10	436	316
11	463	328
12	512	337
13	529	343
14	564	381
15	583	394
16	601	411
17	635	420
18	810	430
19	919	491
20	942	501
21	990	528
22	1130	576
23	1196	610
24	1243	620
25	1268	631
26	1273	680

Table 4 Table listing the forty-one tie-points resulting from the correlation of the M17KC03 planktonic $\delta^{18}O$ record with the benthic $\delta^{18}O$ stacked record LR04 (Lisiecki and Raymo, 2005) using the Analyseries software.—cont'd

	M17KC03	**LR04**
Tie-points	**Core-depth (cm from the top)**	**Age (ka BP)**
27	1282	697
28	1337	741
29	1365	780
30	1400	790
31	1403	800
32	1423	824
33	1547	856
34	1576	868
35	1617	890
36	1626	928
37	1654	954
38	1667	964
39	1769	1000
40	1785	1016
41	1807	1022

According to this chronostratigraphy, the sedimentation rate values are low, around 3 cm/ka, except for the Marine Isotopic Stage (MIS) 11 record which is the thickest of the core, presenting a sedimentation rate of 19 cm/ka and for the MIS 16 which is almost absent (Fig. 4). The MIS 11 thickness explains the plateau observed on the age model (Fig. 4B). This anomaly strongly questioned our correlation at a first glance, but was constrained by the CAZ (Fig. 3). We consider the possibility of a *Pseudoemiliania lacunosa* disappearance occurring during the MIS 13 instead of the MIS 11 in the GIN seas, explaining the MIS 9–11 CAZ defined until 9.5 m depth in the sediment core (Pujos and Giraudeau, 1993, Thierstein et al., 1977).

To validate the obtained stratigraphy, the dated M17KC03 planktonic $\delta^{18}O$ was compared with the planktonic $\delta^{18}O$ record of the dated core PS1243 (the closest core) (Bauch et al., 2000), the planktonic $\delta^{18}O$ record

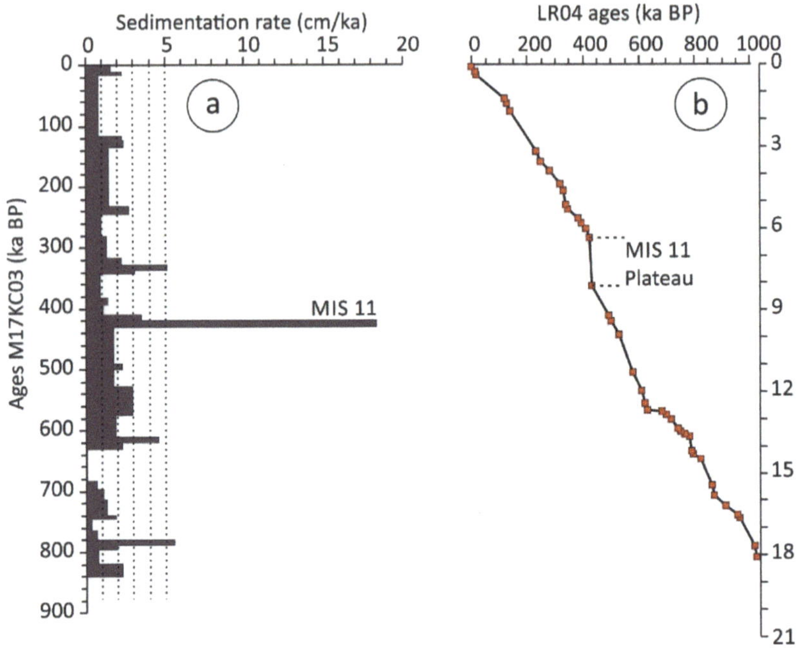

Fig. 4 M17KC03 constructed age model with the: (A) derived sedimentation rates in the core and (B) the tie graph between the M17KC03 signal and the LR04 scale.

of the ODP site 644 (Eldholm et al., 1987; Myhre and Eldholm, 1988), and the North GRIP $\delta^{18}O$ ice core data (Gkinis et al., 2014). Despite resolution differences, this comparison (Fig. 5) shows that the isotopic data measured on core M17KC03 are coherent with those records over the considered intervals. To go further, multiproxy data from the M17KC03 core, now put in age (Fig. 5A), have been compared with other datasets from those reference records when available (Fig. 6). Unsurprisingly, the Ca/sum ratio correlates well with the benthic $\delta^{18}O$ stack LR04 (Fig. 6A) (Lisiecki and Raymo, 2005). A slight delay is however observed between interglacial peaks and Ca/sum ratio peaks during some intervals. The Ca/sum ratio correlates also quite well with the ODP site 985 $CaCO_3$ (Fig. 6B) (Jansen et al., 1996) but a dephasing is observed before 400 ka BP. It can be explained by the fact that the chronostratigraphy of this latter record was established in 1999 by Raymo et al. on the basis of magnetostratigraphy and on dinocysts. Worth is noting that it thus shows also a delay with the more recent benthic $\delta^{18}O$ stack LR04 curve. The L* is impressively reproducible over more than 450 ka BP between the M17KC03 core and the PS1243 and M23414

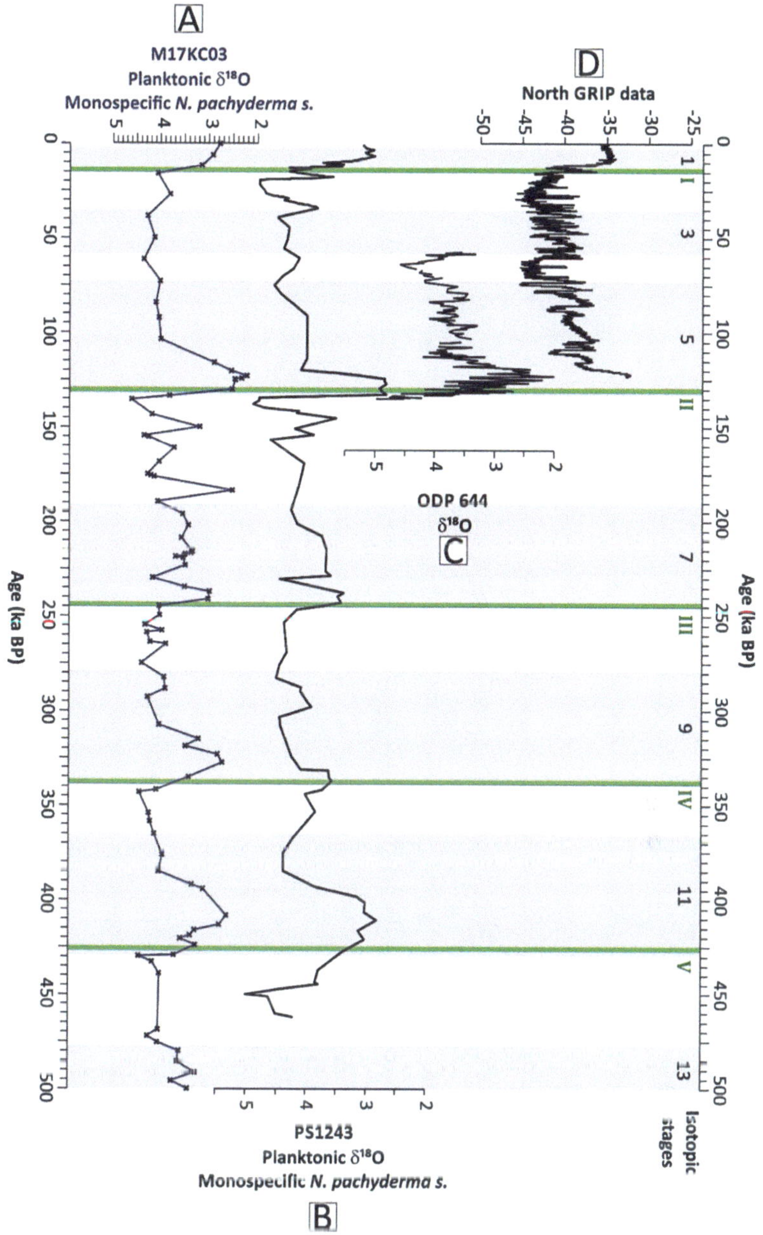

Fig. 5 (A) M17KC03 planktonic *Nps* $\delta^{18}O$ record in age, compared with: (B) the planktonic *Nps* $\delta^{18}O$ of the proximal PS1243 sediment core, and, for the last climatic cycle, with: (C) the planktonic $\delta^{18}O$ measurements of ODP site 644, and (D) the North GRIP glaciological $\delta^{18}O$ data (Bauch et al., 2000; Eldholm et al., 1987; Gkinis et al., 2014; Myhre and Eldholm, 1988).

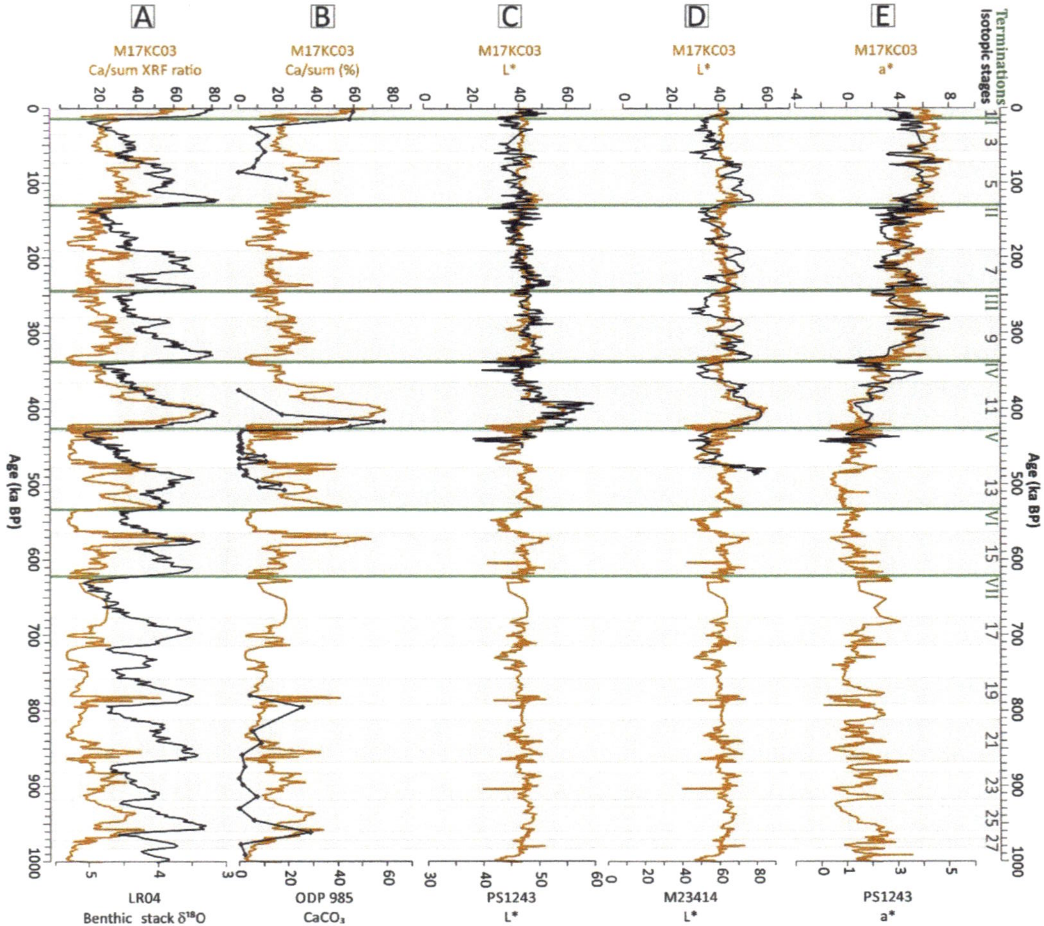

Fig. 6 See figure legend on opposite page.

sediment cores (Fig. 6C and D) (Bauch et al., 2000; Helmke et al., 2002). In the same way, the color a* is very well correlated between the M17KC03 and the PS1243 (Fig. 6E) (Bauch et al., 2000). All those correlations taken together reinforce and give confidence to the approach we have chosen to construct the M17KC03 core stratigraphy.

3.2 Core-to-core correlations (in depth) from the MOCOSED 2017 cruise

The six multiproxy parameters listed in Table 2 were used for the depth-to-depth correlations, generating 7 to 14 common tie-points (Table 5 and Fig. 7) between the studied cores. M17KC01 and M17KC02 (southern GIN seas) data are very similar to the M17KC03 core taken as a reference. M17KC14 and M17KC15 (northern GIN seas) show less similarity and three different parameters presented below (Figs. 8–10, other parameters are shown as Supplementary information in the online version at https://doi.org/10.1016/bs.sats.2022.09.004—Fig. SI-1 and SI-2) allowed us to produce a good correlation between the northern and southern GIN basins.

The first used parameter, the Ca/sum ratio, shows a very good match between the five sediment cores (Fig. 8). A very high peak of Ca/sum is identified around 5–7 m depth in the cores M17KC03, M17KC02, M17KC01, M17KC15 and around 9 m depth in the core M17KC14. The Coccolithophorids Acme Zones locate this positive excursion between the MIS 11 and 9 for the three southern cores (Fig. 8). The two northern cores are almost barren of coccolithophorids (Fig. 8).

The Ti/Ca ratio is one of the easiest parameters to correlate through the entire GIN seas (Fig. 9). Around 4–3 m depth in the southern cores, a succession of short positive peaks defines the MIS 6–8 according to the CAZ.

Chapman and Shackleton (1998), Bauch et al. (2000), and Debret et al. (2011), among others, demonstrated the importance of the lightness (L*) as a paleoclimatic indicator with whitest sediments marking warmest periods and darkest sediments marking coldest ones. Here, this parameter depicts good match for the three southern cores (Fig. 10). The correlation is less obvious

Fig. 6 Comparison of key data from reference cores identified in the vicinity of the GIN seas (black curves) with those of the studied core M17KC03 (orange curves). (A) ^{XRF}Ca/sum ratio in M17KC03 compared with the $\delta^{18}O$ LR04 benthic stack (Lisiecki and Raymo, 2005), (B) ^{XRF}Ca/sum ratio in M17KC03 compared with the equivalent carbonate concentration in the ODP site 985 (Raymo et al., 1999), (C) lightness parameter L* in M17KC03 and PS1243 (Bauch et al., 2000). (D) Lightness parameter L* in M17KC03 and M23414 (Bauch et al., 2000). (E) Color parameter a* in M17KC03 and PS1243 (Bauch et al., 2000). Note that we used comparable X scales to strengthen the correlations.

Table 5 Tie-points used for correlating M17KC03 with the four other MOCOSED 2017 cores.

Reference Core		Correlated cores			
	M17KC03	M17KC01	M17KC02	M17KC14	M17KC15
Pointers	Core depth (m from top)	Core depth (m from top)	Core depth (m from top)	Core depth (m from top)	Core depth (m from top)
1	0.04	0.04	0.04	0.00	0.04
2	0.23	0.21	0.47	0.04	0.31
3	0.79	0.67	1.56	0.10	1.00
4	2.56	2.19	3.80	0.60	3.06
5	4.75	4.26	6.00	2.60	5.83
6	5.55	5.26	6.92	4.19	6.82
7	6.47	6.04	8.01	8.10	7.51
8	7.08	6.72	8.51	9.05	7.72
9	7.70	7.01	9.01		7.94
10	8.03	7.38	9.34		8.14
11	8.23	7.73	9.56		8.35
12	8.82	8.39	9.73		
13	9.22	8.67	10.61		
14	9.75	9.14			

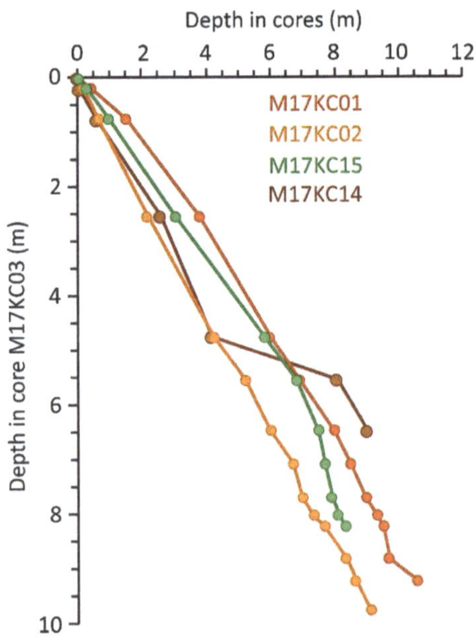

Fig. 7 Original depths in cores M17KC01, M17KC02, M17KC14 and M17KC15 with reference to their equivalent depth after their tuning with core M17KC03 using the 14 tie-points cited in Table 5.

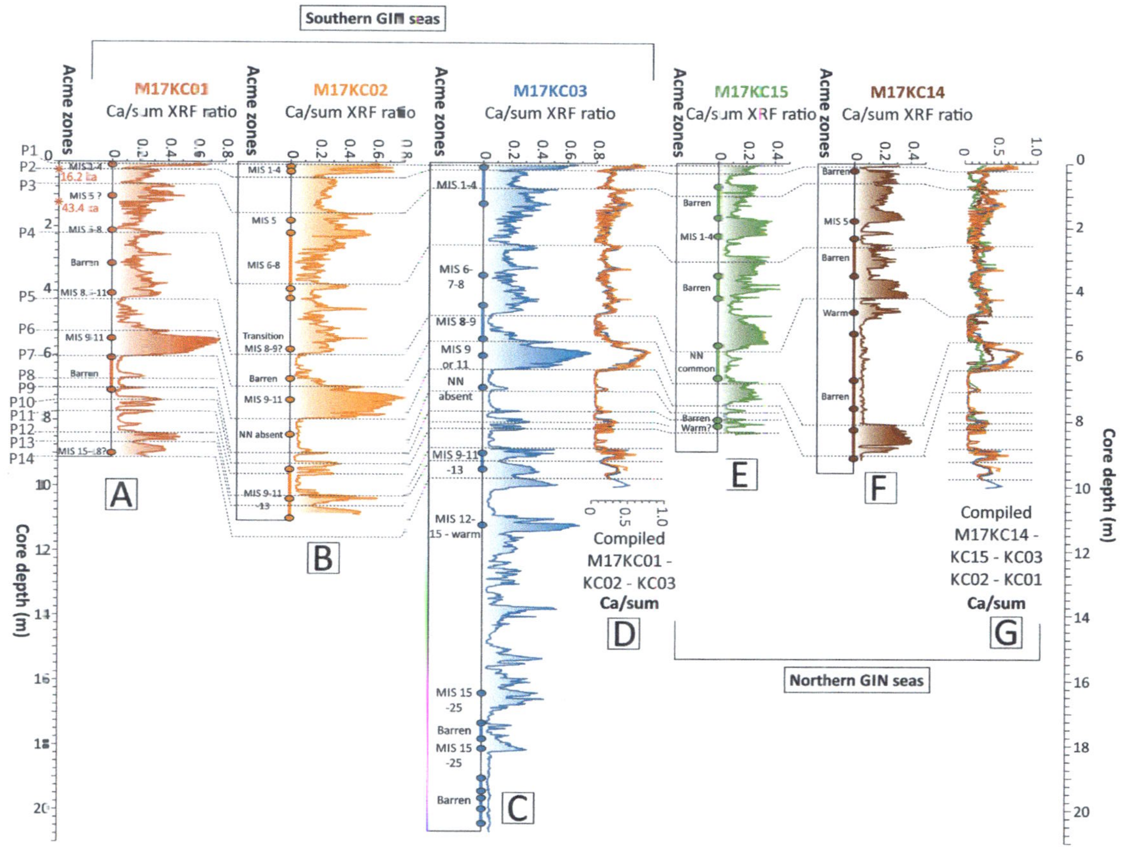

Fig. 8 ^{XRF}Ca/sum records from the five MOCO17 sedimentary cores (presented in depth) along a south-north gradient. Correlation (dashed lines) are based on the 14 tie-points (P1–P14) listed on the left of the figure and in Table 5. The coccolithophorids acme zones (CAZ) are presented on the left side of each curve. (A) ^{XRF}Ca/sum in M17KC01 (red curve). (B) ^{XRF}Ca/sum ratio M17KC02 (orange curve). (C) ^{XRF}Ca/sum ratio in M17KC03 (blue curve, reference curve). (D) Stacked ^{XRF}Ca/sum ratio in cores M17KC01, M17KC02 and M17KC03 vs. core depth in reference core M17KC03. (E) ^{XRF}Ca/sum ratio in M17KC15 (green curve). (F) ^{XRF}Ca/sum ratio in M17KC14 (brown curve). (G) Stacked ^{XRF}Ca/sum ratio in the five MOCO17 sedimentary cores vs. core depth in reference core M17KC03.

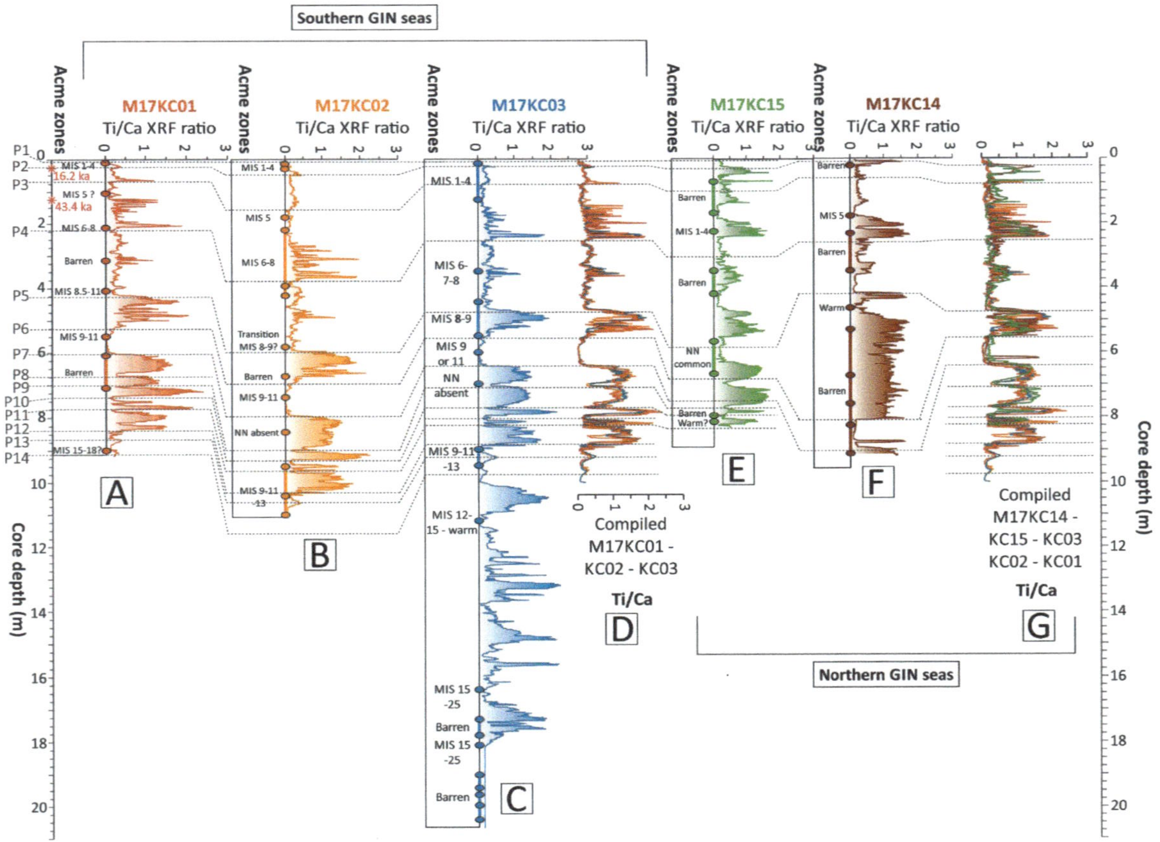

Fig. 9 $^{XRF}Ti/^{XRF}Ca$ ratio records from the five MOCO17 sedimentary cores (presented in depth) along a south-north gradient. Correlation (dashed lines) are based on the 14 tiepoints (P1–P14) listed on the left of the figure and in Table 5. The coccolithophorids acme zones (CAZ) are presented on the left side of each curve. (A) $^{XRF}Ti/^{XRF}Ca$ ratio in M17KC01 (red curve). (B) $^{XRF}Ti/^{XRF}Ca$ ratio in M17KC02 (orange curve). (C) $^{XRF}Ti/^{XRF}Ca$ ratio in M17KC03 (blue curve, reference curve). (D) Stack $^{XRF}Ti/^{XRF}Ca$ ratio in cores M17KC01, M17KC02 and M17KC03 vs. core depth in reference core M17KC03. (E) $^{XRF}Ti/^{XRF}Ca$ ratio in M17KC15 (green curve). (F) $^{XRF}Ti/^{XRF}Ca$ ratio in M17KC14 (brown curve). (G) Stack $^{XRF}Ti/^{XRF}Ca$ ratio in the five MOCO17 sedimentary cores vs. core depth in reference core M17KC03.

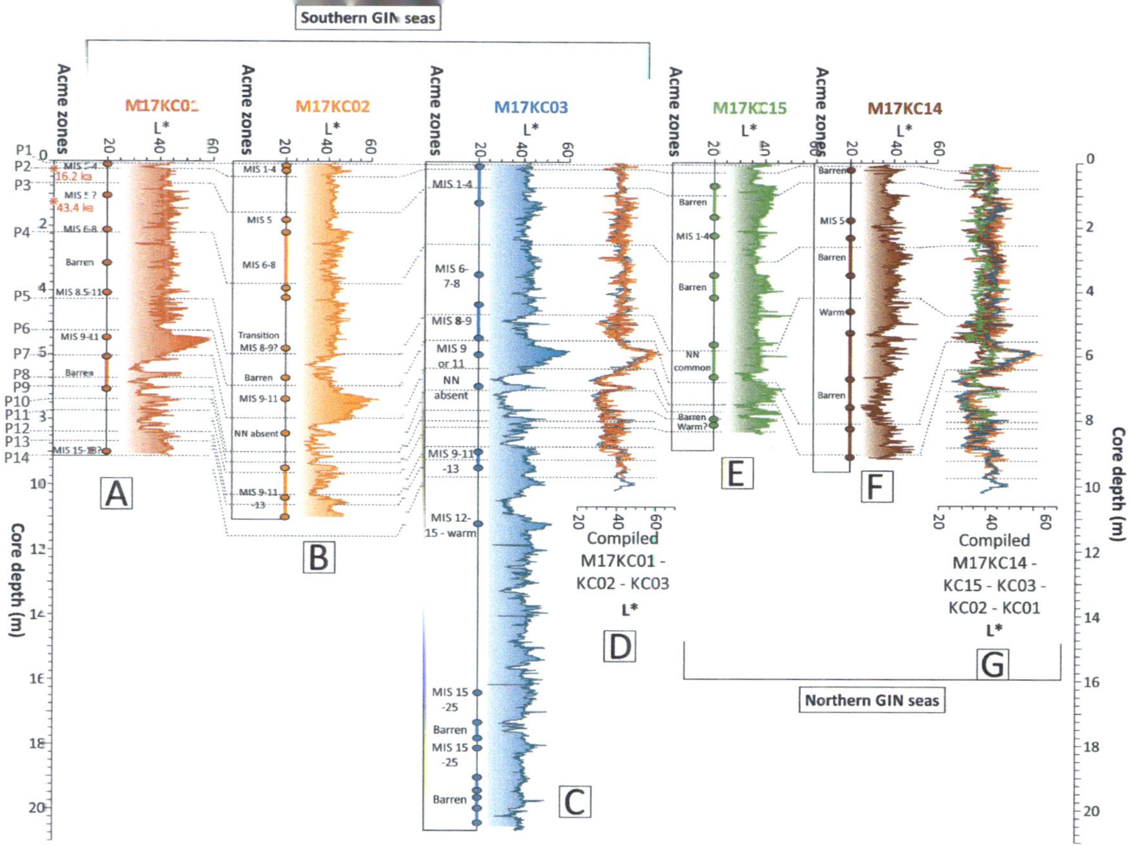

Fig. 10 Lightness L* records from the five MOCO17 sedimentary cores (presented in depth) along a south-north gradient. Correlation (dashed lines) are based on the 14 tiepoints (P1–P14) listed on the left of the figure and in Table 5. The coccolithophorids acme zones (CAZ) are presented on the left side of each curve. (A) Lightness L* in M17KC01 (red curve). (B) Lightness L* in M17KC02 (orange curve). (C) Lightness L* in M17KC03 (blue curve, reference curve). (D) Compilation of the three firsts sedimentary s (M17KC01, M17KC02 and M17KC03) presented here in correlated depth. (E) Lightness L* in M17KC15 (green curve). (F) Lightness L* in M17KC14 (brown curve). (G) Stack L* in the five MOCO17 sedimentary cores vs. core depth in reference core M17KC03.

between the two northern cores, but the L* variations are similar. The CAZ MIS 6–8, around 4–3 m depth, presents a low L*, indicative of dark sediments, whereas the CAZ MIS 9–11, around 6 m depth in sediment, presents very high L* values, characteristic of whiter sediments.

The Sr/sum ratio parameters depict good parallelism through the southern GIN seas (Supplementary information in the online version at https://doi.org/10.1016/bs.sats.2022.09.004—Fig. SI-2). Similarly to the Ca/sum ratio, there is a high peak of Sr/sum identified around 5–7 m depth in the cores M17KC03, M17KC02, M17KC01, showing the CAZ MIS 9–11. Reversely, the CAZ MIS 8 shows low values of Sr/sum ratio. Finally, planktonic foraminifera count, in this article depicted by their absolute abundances in the sediment (number of specimens per gram of dried sed.), reveal a very good regional consistency (Supplementary information in the online version at https://doi.org/10.1016/bs.sats.2022.09.004—Fig. SI-3). Some coherent peaks are identified through the southern cores. Around 2–1 m depth in the sediment, a positive peak is dated from MIS 5 according to the CAZ, and the strongest abundance peak, around 6 m depth, marks the CAZ MIS 9–11.

4. Discussion

4.1 Coherent stratigraphy over GIN seas

The studied cores have been correlated on the basis of the age model first established on the core M17KC03 (Figs. 12–14), providing the depositional history over the last 500 ka BP, thus since the end of the Mid-Pleistocene Transition (MPT).

The comparison of the multiproxy record during this time interval reveals strong regional patterns with typical signatures over glacials (G) as well as terminations (T) and interglacials (IG), worth to be noted here as potential stratigraphical landmarks in the GIN seas. The pulses of sedimentation rates during terminations are synchronous with maximum of obliquity and summer and winter insolation at 65°N (Fig. 11). A latitudinal gradient is observed in the Ti/Ca and Sr/Ca ratio with northern records being much more imprinted by such high values during the nearly complete duration of glacials (Fig. 12B and D). The possible causality of such a sedimentary specificity in the GIN seas will be discussed in details (Section 4.2), taking into account other pan-Atlantic similar long records and potential forcings.

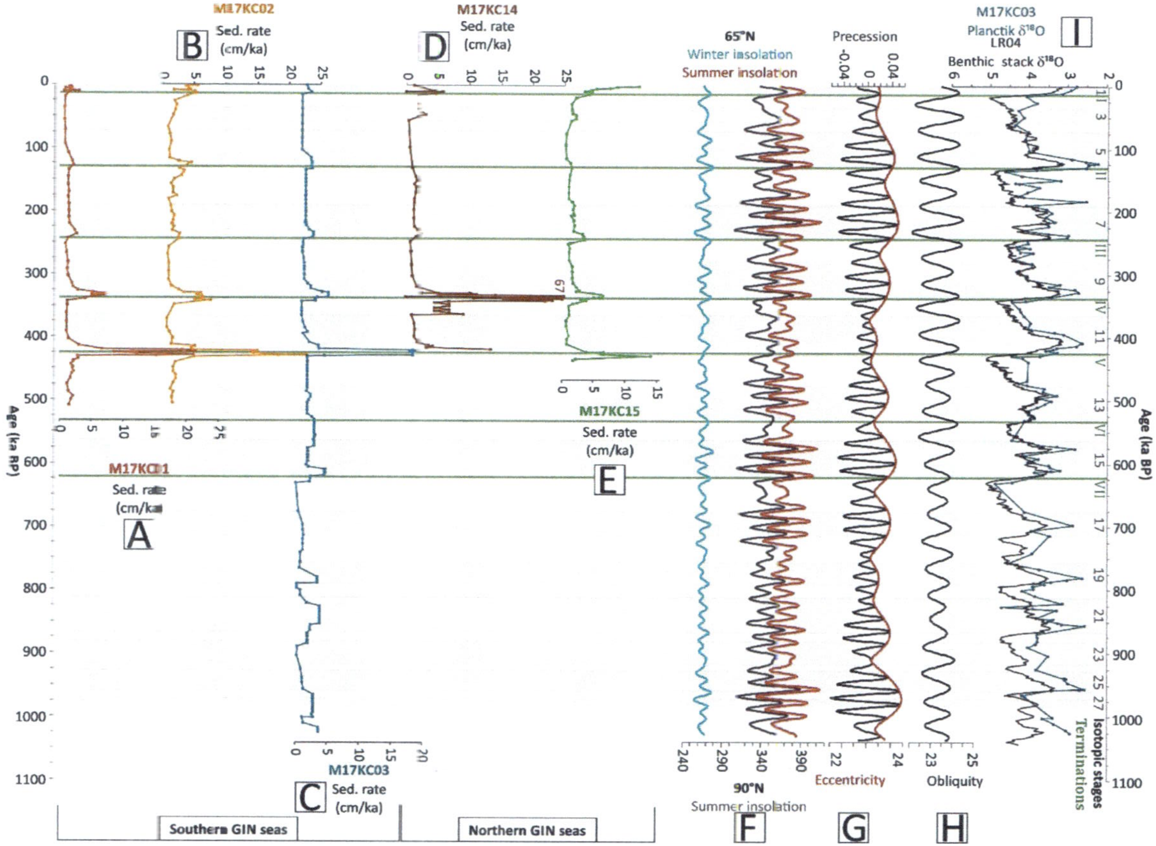

Fig. 11 Derived sedimentation rates in the five sediment cores from MOCOSED 2017 along a south-north gradient. (A) in M17KC01. (B) in M17KC02. (C) in M17KC03. (D) in M17KC14. (E) in M17KC15. They are put in the frame of the: (F) Winter insolation at 65°N (light blue curve, from Laskar et al., 2004) and summer insolation at 65°N and 90°N (respectively red curve and dark gray curve, from Laskar et al., 2004), (G) Precession parameter (black curve) and eccentricity parameter (red curve) from Laskar et al., 2004, (H) Obliquity parameter (black curve) from Laskar et al., 2004, (I) Benthic stack LR04 (black curve) and the tuned M17KC03 planktonic $\delta^{18}O$ (light blue curve).

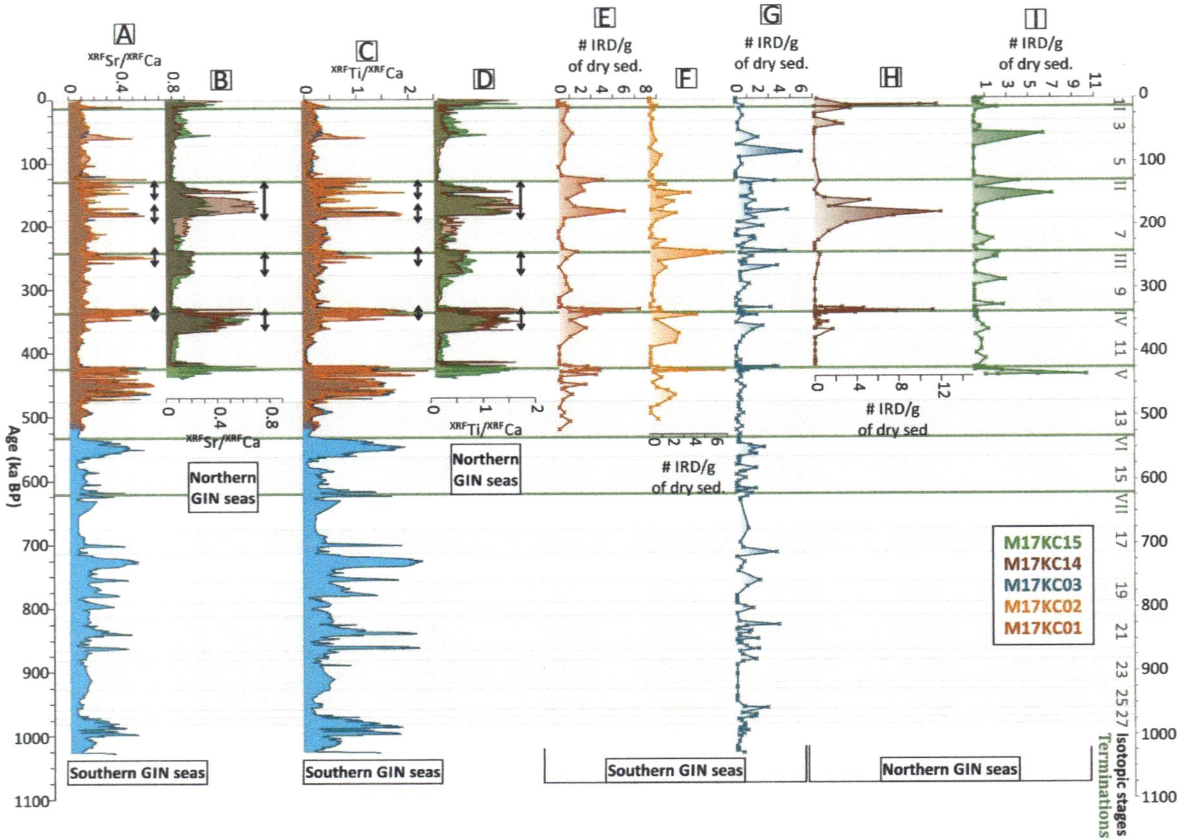

Fig. 12 Multiproxy compilation for the five MOCOSED 2017 sedimentary cores in age. (A) Southern cores $^{XRF}Sr/^{XRF}Ca$ compilation. (B) Northern cores $^{XRF}Sr/^{XRF}Ca$ compilation. (C) Southern cores $^{XRF}Ti/^{XRF}Ca$ compilation. (D) Northern cores $^{XRF}Ti/^{XRF}Ca$ compilation. Those data are compared to the IRD counts (number per gram of dry sediment) for each respective core: (E) M17KC01, (F) M17KC02, (G) M17KC03, (H) M17KC14, (I) M17KC15.

Over the last 1000 ka BP, each IG records evidences of highest positive excursions in the set of selected parameters we have used for correlation (Figs. 11–13). The main outstanding and coherent feature for the investigated IG is their record of signatures typifying a dominant calcareous content (obvious with all related indices, i.e.: Ca/sum, Ca/Sr, L* for the entire GIN seas and Sr/sum, Ti/Ca, Sr/Ti, and absolute concentrations of foraminifera for the southern GIN seas, Figs. 12 and 13). A gradient is clearly expressed in the set of cores we have used, with more pronounced positive excursions recorded in the southern GIN seas. This feature (i.e., high calcareous amounts during IG) is of course not a surprise for subpolar basins and was at the basis of stratigraphical approaches for many cores from the northern NAtl (e.g., Carbonara et al., 2016; Helmke and Bauch, 2002; Lucchi et al., 2013). However, tie points generated for core to core correlation, if chosen on the basis of Ca proxies, were dated thanks to the crossing of the M17KC03 planktonic $\delta^{18}O$ signal compared with the LR04 and constrained by CAZ. Our findings thus confirm a better preservation of carbonates during IG, either related to enhanced carbonaceous pelagic production, and/or minored dissolution processes as well.

Each IG is however strictly different from the other and some high values occur also at IG/G transitions (and G/IG conversely). At orbital scale, the Ca/sum, Ca/Sr, L*, planktonic foraminifera counting and Sr/Ti signatures recorded for MIS 9 and MIS 7 show moderate excess in amplitude in comparison to MIS 11 and 5 (Figs. 12 and 13). This variation in the amplitude probably reflects some kind of threshold in the warming intensities, necessary to generate a better preservation of carbonates (e.g., Helmke and Bauch, 2002; Huber et al., 2000). The modulation of the amplitude from a MIS to another echoes observations done in comparable long records from the proximal Atlantic (Channell et al., 2012). Interestingly, in the GIN sediments, except for the sedimentation rate rises (Fig. 11), the modulation does not seem to be a direct imprint of insolation forcings (documented here by winter and summer values at 65°N, and summer insolation at 90°N). No coherent pattern occurs even if for the last 500 ka BP, Ca/sum highest values (i.e., peaks, interpreted as warm events after Bauch and collaborator's works, light green bands on Fig. 13) are related to high obliquity values before 500 ka BP and rather to low ones after, and also, to low or decreasing summer insolation at 90°N and winter insolation at 65°N. This observation tends to suggest that Ca preservation/warmth peaks in the GIN seas may occur during polar conditions with limited melting during summer (with impacts on the bottom and surface circulation and on the existence of polynyas) (Kaboth-Bahr et al., 2019).

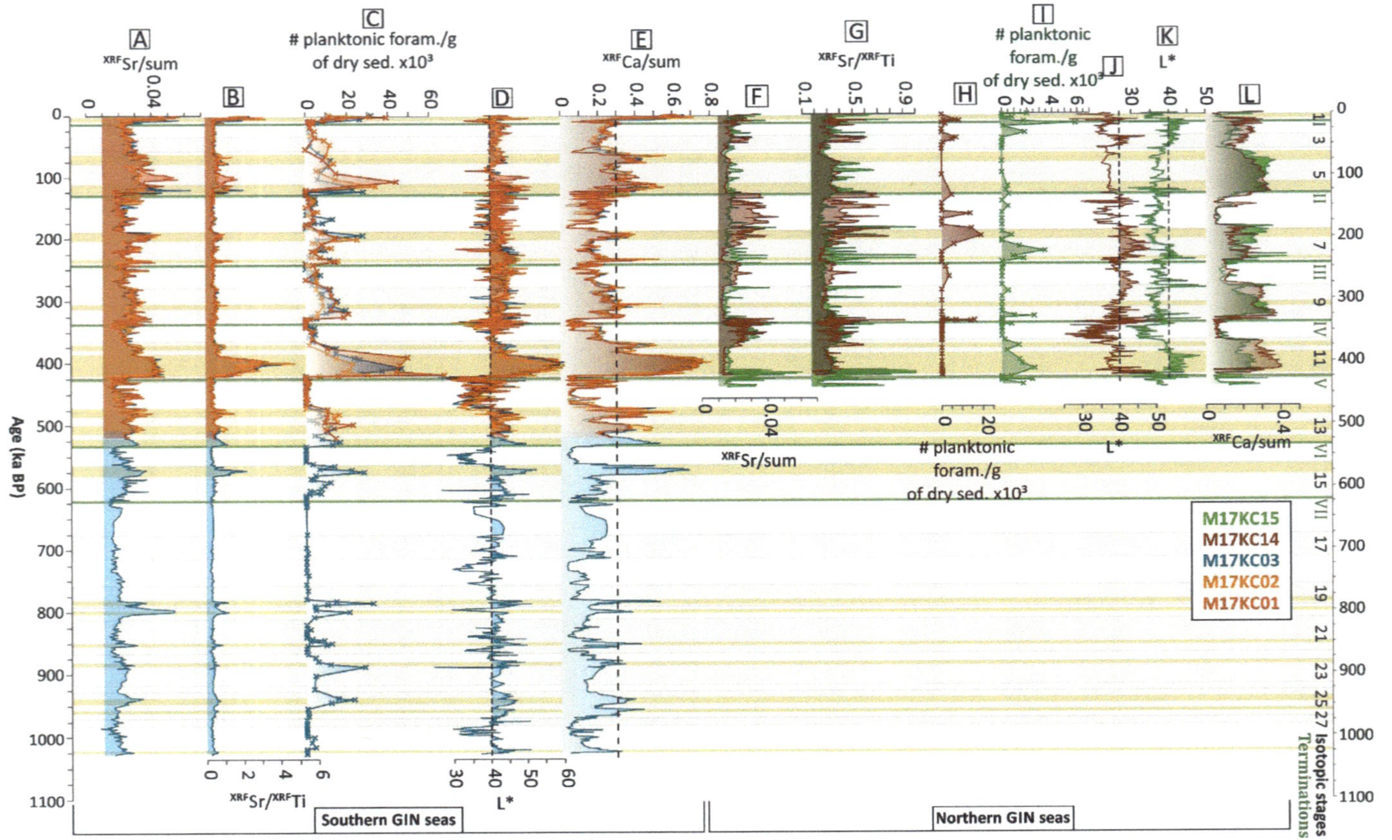

Fig. 13 Multiproxy compilation for the five MOCOSED 2017 sedimentary cores in age. (A) Southern cores $^{XRF}Sr/sum$ compilation. (B) Southern cores $^{XRF}Sr/^{XRF}Ti$ compilation. (C) Southern cores planktonic foraminifera counts compilation. (D) Southern cores lightness compilation. (E) Southern cores $^{XRF}Ca/sum$ compilation. (F) Northern cores $^{XRF}Sr/sum$ compilation. (G) Northern cores $^{XRF}Sr/^{XRF}Ti$ compilation. (H) M17KC14 planktonic foraminifera counts per gram of dry sediment. (I) M17KC15 planktonic foraminifera counts per gram of dry sediment. (J) M17KC14 lightness. (K) M17KC15 lightness. (L) Northern cores $^{XRF}Ca/sum$ compilation.

As previously introduced, Glacials are featured by high Ti/Ca, Sr/Ca values (Fig. 12A–D), and by high IRD amount (Fig. 12E–I) but again (as also seen during IG but with conversely low values), each of them depicts different ranges in the recorded amplitudes. Over the last 1000 ka BP, the most prominent signals are related to MIS 6, MIS 12 and MIS 18. The MIS 6 is known to be marked by a large geographical extension of European ice-sheets (e.g., Ehlers and Gibbard, 2007; Svendsen, 2004), MIS 12 by a very intense decay of ice-rafted material up to the Iberian margin latitudes (Ehlers and Gibbard, 2007; Kandiano et al., 2012; Rodríguez-Tovar et al., 2015), and for MIS 18, the oldest, by the lying of the Brunhes-Matuyama reversal, previous the firsts Heinrich Like Events really identified in the boreal seas (Channell et al., 2010; Naafs et al., 2013). Interestingly, the Saalian glaciation attributed to MIS 6 has been suggested by modeling exercises (Colleoni et al., 2016), following geological evidences on the sea-floor (e.g., Hughes et al., 1977; Jakobsson et al., 2010, 2016), to be characterized by one of the largest extensions of boreal ice-shelves over the Arctic domain. Few constrains exist on the dimension of MIS 12 ice-sheets (related to the Elsterian/Anglian glaciation) but evidences of large terrigenous advection to the boreal Atlantic have already been documented, and this period has been described as one of the most severe Glacials of the Quaternary (Koutsodendris et al., 2019; Toucanne et al., 2009; Wagner et al., 2017). The data presented in this paper suggest major changes and reorganizations in the drainage configuration of boreal ice-sheets at that time, potentially calling for mechanistic effects related to a post-MPT period, initiated with MIS 16, after which Heinrich events began to appear under a context of largest thickness of ice-sheets and glacial durations (Hodell et al., 2008; Naafs et al., 2013). We could thus make an extrapolation in the interpretations and suggest that Ti/Ca and Sr/Ca values -associated to high amount of IRD in the GIN seas are proportional to the boreal ice-sheet sizes and/or decay modes. This latter extrapolation is coherent with the regolith layer hypothesis (Willeit et al., 2019) which supports a change in the intensity and modality of glacial erosion after MIS 18.

If we detect some coherent timings of Ti/Ca high values from a core to another, peaks are however not perfectly synchronously recorded and some appear to be isolated on specific intervals (Fig. 12C and D). This is especially the case when comparing the northern and southern basins, as for instance during glacial terminations. Termination IV, before MIS 9, is restricted to a narrow time period in the southern basin in comparison to what is recorded for northern records. Does this observation sign continuous terrigenous advections (from ice-sheets) during MIS 10 in the north whereas it does occur more briefly in the south? Or following the argumentation of Hodell et al. (2008) for the initiation of Heinrich events, does it mean that glacial purges are related to critical thresholds in the size of ice-sheets (or even ice-shelves?), and thus that northern glaciers were thicker, as proposed by Hughes et al. (1977) and Jakobsson et al. (2016)?

4.2 The GIN seas at orbital scales: Forcings, teleconnections and typicity?

The temporal cover of our study, which thanks to core M17KC03, extends over 1000 ka BP with no detected major gap, offers the possibility to test teleconnections existing in between the GIN seas and the changes at hemispheric and global scales. Over the last decade, long and well-resolved records, covering the MPT at least, have been produced in several basin of the NAtl and worldwide. Even if the comparison is sometime hampered by stratigraphical issues, changes marking the last 22 marine isotopic stages can be discussed robustly when limiting our observations to plurimillennial trends. We have thus selected some records along a latitudinal gradient from the tropics to the temperate belt to tentatively identify forcings driving the modulation of the recorded signals in the GIN seas, with a focus on the calcareous sedimentation especially. Fig. 14 thus compiles those records, together with the EPICA dome C methane record, the latter being considered as representing a global signal, thanks to the short residence time of the greenhouse gas in the atmosphere, and furthermore coeval with changes in the hydrological budget in spite of diversified sources (e.g., Landais et al., 2005; Parrenin and Paillard, 2012; Parrenin et al., 2012, 2013, 2015, 2017; Tzedakis et al., 2006, 2009; and reference therein).

This compilation highlights striking phasing when considered the Ca peaks (values of the ratio Ca/sum overpassing 0.3 in the GIN seas, see pink bands on Fig. 14 and light green bands on Fig. 13), suggesting that not only our constructed age scales are robust, but also that recorded changes really traduce pan-Atlantic modulations with warmings detected in the GIN seas corresponding to global warmings and pulses of high methane atmospheric contents. This latter observation could be in favor of a strong link in between boreal ice-sheet retreats and methane source changes, already observed at the scale of Heinrich events (e.g., Tzedakis et al., 2006, 2009). One very interesting result is the strong correlation of the calcium excursions detected in our records with the percentage of biogenic silica registered in the Lake Baikal (Fig. 14I and J, Prokopenko et al., 2006). In between MIS 14 and

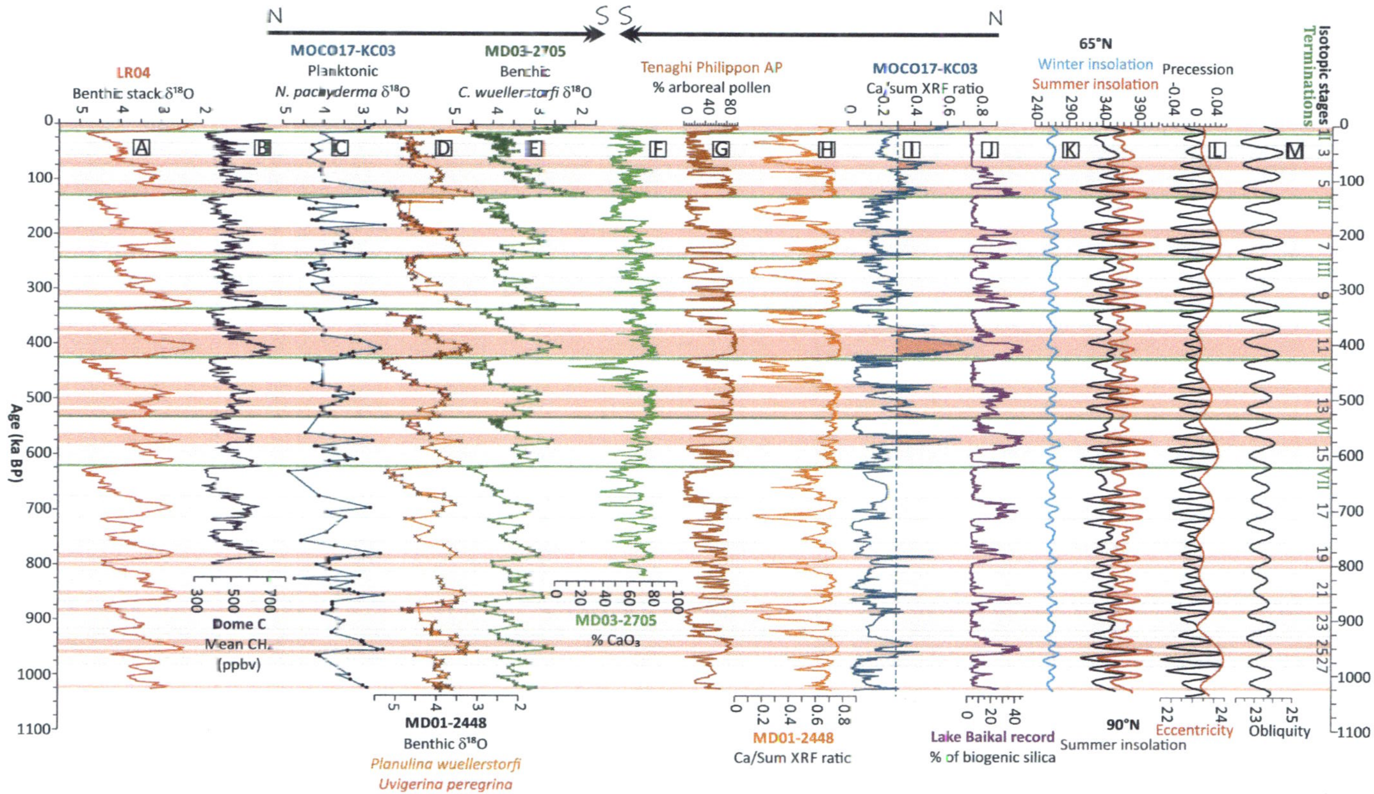

Fig. 14 See figure legend on next page.

MIS 1, nearly all the peaks of Ca detected echo maxima of Ca content in other southern records and occur during IG. The discrimination is less obvious in the tropical records where a continuous pelagic and planktonic production of carbonate does occur. The combination of detrital carbonates to the Ca signal is especially expected in the Ruddiman belt (Bond et al., 1993; Hemming et al., 2002; Hodell et al., 2008; McCave and Andrews, 2019; Ruddiman and McIntyre, 1981), thus potentially affecting the MD01-2448 record, but regarding the GIN seas, our observations tend to infer an overprint from the biogenic carbonate production. Currently, detrital Ca sources around the GIN basins are restricted to North-East Greenland and Siberia areas (Andrews et al., 2017; Spencer et al., 2011). However, Sr/Ca ratios correlates well with the Ti/Ca ones (Fig. 12). The correlation between Ti/Ca and Sr/Ca can be explained by the enhanced detrital/continental supplies which dilute partially the bio-productivity and thus the link between calcium and strontium. In the northern GIN seas, the reduced primary production coupled to the vicinity of Greenland and Svalbard are responsible of the same mechanism to a larger scale: the Sr in calcareous foraminifera is invisible in the sediment, making only the detrital Sr as a marker in the sediment. This explains the two different patterns showed by Sr/sum ratio values during interglacials for the last 500 ka BP: a positive correlation with the Ca/sum peaks, thus to biogenic production, in the southern GIN seas (Fig. 3A–E); and a negative correlation between those parameters in the northern GIN seas

Fig. 14 Pan Atlantic comparison in between high and low latitudes records for the last 1 million years. Light gray bands correspond to interglacial periods after Lisiecki and Raymo, 2005. Pink bands correspond to M17KC03 Ca/sum ratio > 0.3. (A) $\delta^{18}O$ LR04 benthic stack (black curve, Lisiecki and Raymo, 2005). (B) Dôme C mean CH_4 values from Antarctica (ice core, EDC99, Loulergue et al., 2008). (C) M17KC03 monospecific planktonic $\delta^{18}O$, GIN seas (blue curve, this study). (D) Compilation of two benthic $\delta^{18}O$ from the sedimentary core MD01-2448, Bay of Biscay, mid-latitude (orange curve for *P. wuellerstorfi*, red curves for *U. peregrina*; Toucanne et al., 2009). (E) Monospecific *P. wuellerstorfi* benthic $\delta^{18}O$ from the sedimentary core MD03-2705, west-Africa, low latitude (green curve, Malaizé et al., 2012). (F) Carbonate percentage in the core MD03-2705 west-Africa, low latitude (green curve, Malaizé et al., 2012). (G) Percentage of arboreal pollens from Greece, Tenaghi Philippon AP (Tzedakis et al., 2006). (H) Ca/sum ratio from the MD01-2448 core, Bay of Biscay, mid-latitude (yellow curve, Toucanne et al., 2009). (I) Ca/sum ratio from the M17KC03, GIN seas (blue curve, this study). (J) Percentage of biologic silica in the sediment from the Lake Baikal (Prokopenko et al., 2006). (K) Winter insolation at 65°N (light blue curve, from Laskar et al., 2004) and summer insolation at 65°N and 90°N (respectively red curve and dark gray curve, from Laskar et al., 2004). (L) Precession parameter (black curve) and eccentricity parameter (red curve) from Laskar et al., 2004. (M) Obliquity parameter (black curve) from Laskar et al., 2004.

(Fig. 13F–I). Moreover, high Sr amounts most probably call for a detrital source in or around the Fram Strait when ice-sheets are at their maximum extension, maybe linked with coastal related deposits thus submitted to glacier erosion. This hypothesis is corroborated in the Fig. 13 which presents a positive correlation between planktonic foraminifera amount in the southern GIN seas and Sr/Ti ratio. This correlation exists in the northern GIN seas, but with a very little amplitude, implying rather a detrital source for the Sr.

Interestingly, carbonate excursions in the GIN seas were previously related to IG rather to G (Helmke and Bauch, 2002). The latter authors noted enhanced carbonate dissolution during IG in relation to more productive sea-surface waters which in return conducted to more aggressive corrosion processes at the bottom. Their observations were especially well illustrated with MIS 11, with a high index of weight loss for *N. pachyderma* shells noted for cores retrieved in the vicinity of the Jan Mayen Island. Surprisingly using the same and other proximal marine archives, they also observed a high Ca content during MIS 11 (e.g., Helmke et al., 2005; Helmke and Bauch, 2002; Kandiano et al., 2012, 2016, 2017), similarly to what we have observed in the M17KC03 core (Figs. 8 and 13). This "carbonate shoot" was also detected in several ODP drillings from the south-eastern GIN seas by Henrich et al. (2002), the latter author attributing this feature to "*an extraordinary high carbonate production in surface waters,*" in a post-MPT context of growing carbonate contents preserved in the youngest sediments. Further extensive works conducted by the same research group (e.g., Kandiano et al., 2012, 2016, 2017 and references therein) on the MIS 11 oceanographic conditions lead them to conclude that this stage was cooler in the GIN seas than today. They explained those lower Sea Surface Temperature (SST) values by intense and steady melting of the Greenland Ice-Sheet (GIS), even after the end of the Termination V, at 420 ka BP. This melting would imply the formation of a thick and south-western extended fresh surface water in the GIN seas, until 68°N and, at least, underlying the MD99 2277 core site at 6°W (Kandiano et al., 2012). The meltwater surface layer would favor a large-extent of winter sea-ice cover even during the MIS 11 sensu stricto (ss) (410–398 ka BP, Kandiano et al., 2012), enhancing the summer primary productivity and increasing the carbonate export to the bottom (Kandiano et al., 2012, 2016). MIS 11 is a period of high sea-level (6–9 m above the modern level, Raymo and Mitrovica, 2012) implying a larger collapse of polar ice sheets during this interglacial. Hillaire-Marcel and de Vernal (2008) and Reyes et al. (2014) also demonstrated a peculiar artifact in southern Greenland at that time: the southern GIS totally melted, allowing the development of a boreal forest according to their findings. It is interesting to note that our "carbonate shoot" maximum occurs at 406 ka BP, during the MIS 11 ss, at the time of the highest SST recorded in the GIN seas (Kandiano et al., 2012). Our study thus validates their model with a large delay before the onset of the MIS 11 ss and the development of a high primary productivity in the GIN seas. All those observations converge to make MIS 11 a key period regarding climate modulations within the boreal basins: a change does happen in the water column which favored a better preservation of carbonates at that time and during recent IG. Further works focusing on that period are obviously needed in the GINS seas to better document what can be the consequences of a long interglacial in those reactive areas.

5. Conclusions

Our study allowed the determination of a coherent stratigraphic framework for recent sediments (last 500 ka BP at least) over the whole GIN seas. We demonstrated that taking into account the same systematic series of proxies, age scales can be constructed robustly. In those polar basins which suffer from low sedimentation rates and carbonate dissolution, the planktonic $\delta^{18}O$ can, but not solely, be used to generate age control points: it should be used in combination with additional proxies, especially elemental ratios obtained after XRF core-scanning and the stratigraphical control of coccolithophore acme zones are recommended tools. Correlation to the benthic stack LR04, followed by a core-to-core extrapolation, revealed coherent patterns in the sedimentation of the GIN seas at orbital scales: interglacials vs. glacials are clearly modulated at least along the last 1000 ka BP by cryospheric-oceanic-climatic tight interactions and their signals can even be linked to the middle and low latitude similarly long archives, showing a good match in the global climatic teleconnections between high and lower latitudes.

The sedimentation rates obtained from the stratigraphical framework highlight the Terminations by recording sharp rises, and especially for the Termination V which presents the strongest sedimentation rates (around 19 cm/ka) of the whole record from north to south.

Thanks to the set of studied cores distributed in the GIN seas from the south-eastern Jan Mayen plateau to the Fram strait, we have identified a N-S gradient over the last common history of our archives, i.e., over 500 ka BP, with more intense glacial conditions in the northern GIN seas than in the southern GIN seas (especially expressed throughout XRF ratios and those implying the Titanium). In the same way, we have noted a deconvolution of the Sr and Ca signals in the sediment with, for the southern GIN seas, the Sr which seems to be linked to the Ca and the biogenic production, whereas in the northern GIN seas, the Sr is dissociated from the Ca and planktonic foraminifera, but presents a ratio around 1 with the Ti, giving a more detrital origin of the northern Sr.

The interglacial MIS 11 is quite peculiar in the southern GIN seas. Indeed, it is strongly highlighted by the biogenic signals, making this warm period the most prominent of the last 1000 ka BP. Interestingly, at 406 ka BP, a huge carbonate spike is recorded in the cores, during the MIS 11 sensu stricto, recognized with full-interglacial conditions in the northern hemisphere, and with an extended winter sea-ice cover in the GIN seas (Kandiano et al., 2012, 2016). Conversely, in the northern GIN seas, this very warm interglacial is under-represented, implying different N-S oceanographic conditions and processes, explained by the different vicinity of the ice-front. Those features make this warm period a very interesting one to investigate further in the GIN seas for future studies.

The two glacials MIS 12 and MIS 6 are particularly prominent in our cores. Those stages are already known as intense glacials, with very extended ice-sheets over Europe and North America, and with some proof of ice-stream extensions over the Arctic Ocean (Jakobsson et al., 2016; Koutsodendris et al., 2019; Wagner et al., 2017). Our data strongly support those arguments because of the intense terrigenous discharges associated to those glacials, interpreted as the result of the high amount of detrital material produced by the extremely extended ice-sheets.

Acknowledgments

The authors are grateful to the Service Hydrographique et Océanographique de la Marine (SHOM) for funding this study. We thank GENAVIR and the crew of the R/V Pourquoi Pas? for their assistance during the cruises MOCOSED 2014 and MOCOSED 2017. We are grateful to P. Lebleu, S. Bujan, B. Martin, M-C. Perello, O. Ther, L. Devaux, M-H. Castera, A. Cogné, and M. Sabourdy for their technical assistance.

References

Amundsen, H.B., Laberg, J.S., Vorren, T.O., Haflidason, H., Forwick, M., Buhl-Mortensen, P., 2015. Late Weichselian–Holocene evolution of the high-latitude Andøya submarine Canyon, North-Norwegian continental margin. Mar. Geol. 363, 1–14. https://doi.org/10.1016/j.margeo.2015.02.002.

Andrews, J., 2000. Icebergs and iceberg rafted detritus (IRD) in the North Atlantic: facts and assumptions. Oceanography 13, 100–108. https://doi.org/10.5670/oceanog.2000.19.

Andrews, J.T., Dunhill, G., Vogt, C., Voelker, A.H.L., 2017. Denmark Strait during the late glacial maximum and marine isotope stage 3: sediment sources and transport processes. Mar. Geol. 390, 181–198. https://doi.org/10.1016/j.margeo.2017.07.003.

Baeten, N.J., Laberg, J.S., Vanneste, M., Forsberg, C.F., Kvalstad, T.J., Forwick, M., Vorren, T.O., Haflidason, H., 2014. Origin of shallow submarine mass movements and their glide planes-sedimentological and geotechnical analyses from the continental slope off northern Norway: glide plane of a shallow mass movement. Case Rep. Med. 119, 2335–2360. https://doi.org/10.1002/2013JF003068.

Bauch, H.A., Erlenkeuser, H., Helmke, J.P., Struck, U., 2000. A Paleoclimatic Evaluation of Marine Oxygen Isotope Stage 11 in the High-Northern Atlantic žNordic Seas/13.

Becker, L.W.M., Sejrup, H.P., Hjelstuen, B.O., Haflidason, H., Dokken, T.M., 2018. Ocean-ice sheet interaction along the SE Nordic seas margin from 35 to 15 ka BP. Mar. Geol. 402, 99–117. https://doi.org/10.1016/j.margeo.2017.09.003.

Blindheim, J., Borovkov, V., Hansen, B., Malmberg, S.-A., Turrell, W.R., Østerhus, S., 2000. Upper layer cooling and freshening in the Norwegian Sea in relation to atmospheric forcing. Deep Sea Res Part I Oceanogr. Res. Pap. 47, 655–680. https://doi.org/10.1016/S0967-0637(99)00070-9.

Bond, G.C., Devlin, W.J., Kominz, M.A., Beavan, J., McManus, J., 1993. Evidence of astronomical forcing of the earth's climate in Cretaceous and Cambrian times. Tectonophysics 222, 295–315. https://doi.org/10.1016/0040-1951(93)90356-O.

Broecker, W.S., Bond, G., Klas, M., Bonani, G., Wolfli, W., 1990. A salt oscillator in the glacial Atlantic? 1. The concept. Paleoceanography 5, 469–477. https://doi.org/10.1029/PA005i004p00469.

Bylinskaya, M.E., Golovina, L.A., Radionova, E.P., Pokrovskii, B.G., Lavrushin, V.Y., 2016. Paleoenvironments in the Fram Strait during marine isotope stages 2–6 based on planktonic paleobiological and stable-isotope proxies and ice-rafted debris. Quat. Int. 420, 272–279. https://doi.org/10.1016/j.quaint.2015.09.086.

Carbonara, K., Mezgec, K., Varagona, G., Musco, M.E., Lucchi, R.G., Villa, G., Morigi, C., Melis, R., Caffau, M., 2016. Palaeoclimatic changes in Kveithola, Svalbard, during the late Pleistocene deglaciation and Holocene: evidences from microfossil and sedimentary records. Palaeogeogr. Palaeoclimatol. Palaeoecol. 463, 136–149. https://doi.org/10.1016/j.palaeo.2016.10.003.

Caricchi, C., Sagnotti, L., Campuzano, S.A., Lucchi, R.G., Macrì, P., Rebesco, M., Camerlenghi, A., 2020. A refined age calibrated paleosecular variation and relative paleointensity stack for the NW Barents Sea: implication for geomagnetic field behavior during the Holocene. Quat. Sci. Rev. 229, 106133. https://doi.org/10.1016/j.quascirev.2019.106133.

Channell, J.E.T., Hodell, D.A., 2013. Magnetic signatures of Heinrich-like detrital layers in the quaternary of the North Atlantic. Earth Planet. Sci. Lett. 369–370, 260–270. https://doi.org/10.1016/j.epsl.2013.03.034.

Channell, J.E.T., Hodell, D.A., Singer, B.S., Xuan, C., 2010. Reconciling astrochronological and $^{40}Ar/^{39}Ar$ ages for the Matuyama-Brunhes boundary and late Matuyama Chron. Geochem. Geophys. Geosyst. 11, Q0AA12. https://doi.org/10.1029/2010GC003203.

Channell, J.E.T., Hodell, D.A., Romero, O., Hillaire-Marcel, C., de Vernal, A., Stoner, J.S., Mazaud, A., Röhl, U., 2012. A 750-kyr detrital-layer stratigraphy for the North Atlantic (IODP sites U1302–U 1303, orphan knoll, Labrador Sea). Earth Planet. Sci. Lett. 317–318, 218–230. https://doi.org/10.1016/j.epsl.2011.11.029.

Channell, J.E.T., Raymo, M.E., 2003. Paleomagnetic record at ODP Site 980 (Feni Drift, Rockall) for the past 1.2 Myrs. Geochem. Geophys. Geosyst. 4. https://doi.org/10.1029/2002GC000440.

Channell, J.E.T., Xuan, C., Hodell, D.A., 2009. Stacking paleointensity and oxygen isotope data for the last 1.5 Myr (PISO-1500). Earth Planet. Sci. Lett. 283, 14–23. https://doi.org/10.1016/j.epsl.2009.03.012.

Chapman, M.R., Shackleton, N.J., 1998. What level of resolution is attainable in a deep-sea core? Results of a spectrophotometer study. Paleoceanography 13, 311–315. https://doi.org/10.1029/98PA01067.

Chauhan, T., Rasmussen, T.L., Noormets, R., Jakobsson, M., Hogan, K.A., 2014. Glacial history and paleoceanography of the southern Yermak plateau since 132 ka BP. Quat. Sci. Rev. 92, 155–169. https://doi.org/10.1016/j.quascirev.2013.10.023.

Chauhan, T., Rasmussen, T.L., Noormets, R., 2016. Palaeoceanography of the Barents Sea continental margin, north of Nordaustlandet, Svalbard, during the last 74 ka. Boreas 45, 76–99. https://doi.org/10.1111/bor.12135.

Clark, P.U., Archer, D., Pollard, D., Blum, J.D., Rial, J.A., Brovkin, V., Mix, A.C., Pisias, N.G., Roy, M., 2006. The middle Pleistocene transition: characteristics, mechanisms, and implications for long-term changes in atmospheric pCO2. Quat. Sci. Rev. 25, 3150–3184. https://doi.org/10.1016/j.quascirev.2006.07.008.

Colleoni, F., Kirchner, N., Niessen, F., Quiquet, A., Liakka, J., 2016. An east Siberian ice shelf during the late Pleistocene glaciations: numerical reconstructions. Quat. Sci. Rev. 147, 148–163. https://doi.org/10.1016/j.quascirev.2015.12.023.

Debret, M., Sebag, D., Desmet, M., Balsam, W., Copard, Y., Mourier, B., Susperrigui, A.-S., Arnaud, F., Bentaleb, I., Chapron, E., Lallier-Vergès, E., Winiarski, T., 2011. Spectrocolorimetric interpretation of sedimentary dynamics: the new "Q7/4 diagram". Earth Sci. Rev. 109, 1–19. https://doi.org/10.1016/j.earscirev.2011.07.002.

Duplessy, J.-C., Shackleton, N.J., Matthews, R.K., Prell, W., Ruddiman, W.F., Caralp, M., Hendy, C.H., 1984. d13C record of benthic foraminifera in the last Interglacial Ocean: implications for the carbon cycle and the global deep water circulation. Quatern. Res. 21, 225–243. https://doi.org/10.1016/0033-5894(84)90099-1.

Ehlers, J., Gibbard, P.L., 2007. The extent and chronology of Cenozoic global glaciation. Quat. Int. 164–165, 6–20. https://doi.org/10.1016/j.quaint.2006.10.008.

Eldholm, O., Thiede, J., Taylor, E., 1987. Evolution of the Norwegian Continental Margin: Background and Objectives. Ocean Drilling Program, pp. 5–25. https://doi.org/10.2973/ODP.PROC.IR.104.101.1987.

Gkinis, V., Simonsen, S.B., Buchardt, S.L., White, J.W.C., Vinther, B.M., 2014. Water isotope diffusion rates from the NorthGRIP ice core for the last 16, 000 years—glaciological and paleoclimatic implications. Earth Planet. Sci. Lett. 405, 132–141. https://doi.org/10.1016/j.epsl.2014.08.022.

Haflidason, H., de Alvaro, M.M., Nygard, A., Sejrup, H.P., Laberg, J.S., 2007. Holocene sedimentary processes in the Andøya canyon system, North Norway. Mar. Geol. 246, 86–104. https://doi.org/10.1016/j.margeo.2007.06.005.

Heaton, T.J., Köhler, P., Butzin, M., Bard, E., Reimer, R.W., Austin, W.E.N., Bronk Ramsey, C., Grootes, P.M., Hughen, K.A., Kromer, B., Reimer, P.J., Adkins, J., Burke, A., Cook, M.S., Olsen, J., Skinner, L.C., 2020. Marine20—the marine radiocarbon age calibration curve (0–55,000 cal BP). Radiocarbon 62, 779–820. https://doi.org/10.1017/RDC.2020.68.

Helmke, J.P., Bauch, H.A., 2002. Glacial–interglacial carbonate preservation records in the Nordic seas. Global Planet. Change 33, 15–28. https://doi.org/10.1016/S0921-8181(02)00058-9.

Helmke, J.P., Schulz, M., Bauch, H.A., 2002. Sediment-color record from the Northeast Atlantic reveals patterns of millennial-scale climate variability during the past 500,000 years. Quatern. Res. 57, 49–57. https://doi.org/10.1006/qres.2001.2289.

Helmke, J.P., Bauch, H.A., Röhl, U., Mazaud, A., 2005. Changes in sedimentation patterns of the Nordic seas region across the mid-Pleistocene. Mar. Geol. 215, 107–122. https://doi.org/10.1016/j.margeo.2004.12.006.

Hemming, S.R., Vorren, T.O., Kleman, J., 2002. Provinciality of ice rafting in the North Atlantic: application of 40Ar/39Ar dating of individual ice rafted hornblende grains. Quat. Int. 95–96, 75–85. https://doi.org/10.1016/S1040-6182(02)00029-0.

Henrich, R., Baumann, K.-H., Huber, R., Meggers, H., 2002. Carbonate preservation records of the past 3 Myr in the Norwegian–Greenland Sea and the northern North Atlantic: implications for the history of NADW production. Mar. Geol. 184, 17–39. https://doi.org/10.1016/S0025-3227(01)00279-1.

Hillaire-Marcel, C., de Vernal, A., 2008. Stable isotope clue to episodic sea ice formation in the glacial North Atlantic. Earth Planet. Sci. Lett. 268, 143–150. https://doi.org/10.1016/j.epsl.2008.01.012.

Hodell, D.A., Channell, J.E.T., Curtis, J.H., Romero, O.E., Röhl, U., 2008. Onset of "Hudson Strait" Heinrich events in the eastern North Atlantic at the end of the middle Pleistocene transition (~640 ka)? Paleoceanography 23, PA4218. https://doi.org/10.1029/2008PA001591.

Hodell, D., Lourens, L., Crowhurst, S., Konijnendijk, T., Tjallingii, R., Jiménez-Espejo, F., Skinner, L., Tzedakis, P.C., Abrantes, F., Acton, G.D., Alvarez Zarikian, C.A., Bahr, A., Balestra, B., Barranco, E.L., Carrara, G., Ducassou, E., Flood, R.D., Flores, J.-A., Furota, S., Grimalt, J., Grunert, P., Hernández-Molina, J., Kim, J.K., Krissek, L.A., Kuroda, J., Li, B., Lofi, J., Margari, V., Martrat, B., Miller, M.D., Nanayama, F., Nishida, N., Richter, C., Rodrigues, T., Rodríguez-Tovar, F.J., Roque, A.C.F., Sanchez Goñi, M.F., Sierro Sánchez, F.J., Singh, A.D., Sloss, C.R., Stow, D.A.V., Takashimizu, Y., Tzanova, A., Voelker, A., Xuan, C., Williams, T., 2015. A reference time scale for site U1385 (Shackleton site) on the SW Iberian margin. Global Planet. Change 133, 49–64. https://doi.org/10.1016/j.gloplacha.2015.07.002.

Huber, R., Meggers, H., Baumann, K.-H., Henrich, R., 2000. Recent and Pleistocene carbonate dissolution in sediments of the Norwegian–Greenland Sea. Mar. Geol. 165, 123–136. https://doi.org/10.1016/S0025-3227(99)00138-3.

Hughes, T., Denton, G.H., Grosswald, M.G., 1977. Was there a late-Würm Arctic ice sheet? Nature 266, 596–602.

Jakobsson, M., Løvlie, R., Al-Hanbali, H., Arnold, E., Backman, J., Mörth, M., 2000. Manganese and color cycles in Arctic Ocean sediments constrain Pleistocene chronology. Geology 28, 23–26.

Jakobsson, M., Nilsson, J., O'Regan, M., Backman, J., Löwemark, L., Dowdeswell, J.A., Mayer, L., Polyak, L., Colleoni, F., Anderson, L.G., Björk, G., Darby, D., Eriksson, B., Hanslik, D., Hell, B., Marcussen, C., Sellén, E., Wallin, Å., 2010. An Arctic Ocean ice shelf during MIS 6 constrained by new geophysical and geological data. Quat. Sci. Rev. 29, 3505–3517. https://doi.org/10.1016/j.quascirev.2010.03.015.

Jakobsson, M., Nilsson, J., Anderson, L., Backman, J., Björk, G., Cronin, T.M., Kirchner, N., Koshurnikov, A., Mayer, L., Noormets, R., O'Regan, M., Stranne, C., Ananiev, R., Barrientos Macho, N., Cherniykh, D., Coxall, H., Eriksson, B., Flodén, T., Gemery, L., Gustafsson, Ö., Jerram, K., Johansson, C., Khortov, A., Mohammad, R., Semiletov, I., 2016. Evidence for an ice shelf covering the Central Arctic Ocean during the penultimate glaciation. Nat. Commun. 7, 10365. https://doi.org/10.1038/ncomms10365.

Jansen, E., Raymo, M.E., Blum, P., 1996. Site 985. In: Jansen, E., Raymo, M.E., Blum, P., et al. (Eds.), Proc. ODP, Init. Repts. vol. 162. Ocean Drilling Program, College Station, TX, p. 33, https://doi.org/10.2973/odp.proc.ir.162.108.1996.

Kaboth-Bahr, S., Denis, V., Su, C.-C., O'Regan, M., Gyllencreutz, R., Jakobsson, M., Löwemark, L., 2019. Deciphering ~45.000 years of Arctic Ocean lithostratigraphic variability through multivariate statistical analysis. Quat. Int. 514, 141–151. https://doi.org/10.1016/j.quaint.2018.11.043.

Kandiano, E.S., Bauch, H.A., Fahl, K., Helmke, J.P., Röhl, U., Pérez-Folgado, M., Cacho, I., 2012. The meridional temperature gradient in the eastern North Atlantic during MIS 11 and its link to the ocean–atmosphere system. Palaeogeogr. Palaeoclimatol. Palaeoecol. 333–334, 24–39. https://doi.org/10.1016/j.palaeo.2012.03.005.

Kandiano, E.S., van der Meer, M.T.J., Bauch, H.A., Helmke, J., Damsté, J.S.S., Schouten, S., 2016. A cold and fresh ocean surface in the Nordic seas during MIS 11: significance for the future ocean. Geophys. Res. Lett. 43, 10,929–10,937. https://doi.org/10.1002/2016GL070294.

Kandiano, E.S., van der Meer, M.T.J., Schouten, S., Fahl, K., Sinninghe Damsté, J.S., Bauch, H.A., 2017. Response of the North Atlantic surface and intermediate ocean structure to climate warming of MIS 11. Sci. Rep. 7, 46192. https://doi.org/10.1038/srep46192.

Kissel, C., Laj, C., Labeyrie, L., Dokken, T., Voelker, A., Blamart, D., 1999. Rapid climatic variations during marine isotopic stage 3: magnetic analysis of sediments from Nordic Seas and North Atlantic. Earth Planet. Sci. Lett. 171, 489–502.

Kissel, C., Laj, C., Lehman, B., Labyrie, L., Bout-Roumazeilles, V., 1997. Changes in the strength of the Iceland–Scotland Overflow Water in the last 200, 000 years: Evidence from magnetic anisotropy analysis of core SU90–33. Earth Planet. Sci. Lett. 152, 25–36. https://doi.org/10.1016/S0012-821X(97)00146-5.

Knies, J., Matthiessen, J., Vogt, C., Laberg, J.S., Hjelstuen, B.O., Smelror, M., Larsen, E., Andreassen, K., Eidvin, T., Vorren, T.O., 2009. The Plio-Pleistocene glaciation of the Barents Sea–Svalbard region: a new model based on revised chronostratigraphy. Quat. Sci. Rev. 28, 812–829. https://doi.org/10.1016/j.quascirev.2008.12.002.

Kolling, H., Bauch, H.A., 2017. A stratigraphical-sedimentological study of the last interglacial period in the central Nordic seas on the basis of XRF core scanning. Open Access Polarforschung 87 (1), 15–22. https://doi.org/10.2312/polarforschung.87.1.15.

Koutsodendris, A., Kousis, I., Peyron, O., Wagner, B., Pross, J., 2019. The marine isotope stage 12 pollen record from Lake Ohrid (SE Europe): investigating short-term climate change under extreme glacial conditions. Quat. Sci. Rev. 221, 105873. https://doi.org/10.1016/j.quascirev.2019.105873.

Laberg, J.S., Vorren, T.O., Dowdeswell, J.A., Kenyon, N.H., Taylor, J., 2000. The Andøya Slide and the Andøya Canyon, north-eastern Norwegian–Greenland Sea. Mar. Geol. 162, 259–275. https://doi.org/10.1016/S0025-3227(99)00087-0.

Landais, A., Jouzel, J., Masson-Delmotte, V., Caillon, N., 2005. Large temperature variations over rapid climatic events in Greenland: a method based on air isotopic measurements. C. R. Geosci. 337, 947–956. https://doi.org/10.1016/j.crte.2005.04.003.

Laskar, J., Correia, A.C.M., Gastineau, M., Joutel, F., Levrard, B., Robutel, P., 2004. Long term evolution and chaotic diffusion of the insolation quantities of Mars. Icarus 170, 343–364. https://doi.org/10.1016/j.icarus.2004.04.005.

Lisiecki, L.E., Raymo, M.E., 2005. A Pliocene-Pleistocene stack of 57 globally distributed benthicd18O records. Paleoceanography 20. https://doi.org/10.1029/2004PA001071.

Lisiecki, L.E., Raymo, M.E., Curry, W.B., 2008. Atlantic overturning responses to late Pleistocene climate forcings. Nature 456, 85–88. https://doi.org/10.1038/nature07425.

Loulergue, L., Schilt, A., Spahni, R., Masson-Delmotte, V., Blunier, T., Lemieux, B., Barnola, J.-M., Raynaud, D., Stocker, T.F., Chappellaz, K., 2008. Orbital and millennial-scale features of atmospheric CH4 over the past 800,000 years. Nature 453, 383–386. https://doi.org/10.1038/nature06950.

Lucchi, R.G., Camerlenghi, A., Rebesco, M., Colmenero-Hidalgo, E., Sierro, F.J., Sagnotti, L., Urgeles, R., Melis, R., Morigi, C., Bárcena, M.-A., Giorgetti, G., Villa, G., Persico, D., Flores, J.-A., Rigual-Hernández, A.S., Pedrosa, M.T., Macri, P., Caburlotto, A., 2013. Postglacial sedimentary processes on the Storfjorden and Kveithola trough mouth fans: significance of extreme glacimarine sedimentation. Global Planet. Change 111, 309–326. https://doi.org/10.1016/j.gloplacha.2013.10.008.

Malaizé, B., Jullien, E., Tisserand, A., Skonieczny, C., Grousset, E.F., Eynaud, F., Kissel, C., Bonnin, J., Karstens, S., Martinez, P., Bory, A., Bout-Roumazeilles, V., Caley, T., Crosta, X., Charlier, K., Rossignol, L., Flores, J.A., Schneider, R., 2012. The impact

of African aridity on the isotopic signature of Atlantic deep waters across the Middle Pleistocene Transition. Quat. Res. 77, 182–191.
Martrat, B., Jimenez-Amat, P., Zahn, R., Grimalt, J.O., 2014. Similarities and dissimilarities between the last two deglaciations and interglaciations in the North Atlantic region. Quat. Sci. Rev. 99, 122–134. https://doi.org/10.1016/j.quascirev.2014.06.016.
McCave, I.N., Andrews, J.T., 2019. Distinguishing current effects in sediments delivered to the ocean by ice. I. Principles, methods and examples. Quat. Sci. Rev. 212, 92–107. https://doi.org/10.1016/j.quascirev.2019.03.031.
Müller, J., Stein, R., 2014. High-resolution record of late glacial and deglacial sea ice changes in Fram Strait corroborates ice–ocean interactions during abrupt climate shifts. Earth Planet. Sci. Lett. 403, 446–455. https://doi.org/10.1016/j.epsl.2014.07.016.
Myhre, A.M., Eldholm, O., 1988. The western Svalbard margin (74°–80°N). Mar. Pet. Geol. 5, 134–156. https://doi.org/10.1016/0264-8172(88)90019-0.
Naafs, B.D.A., Hefter, J., Stein, R., 2013. Millennial-scale ice rafting events and Hudson Strait Heinrich(−like) events during the late Pliocene and Pleistocene: a review. Quat. Sci. Rev. 80, 1–28. https://doi.org/10.1016/j.quascirev.2013.08.014.
Nagao, S., Nakashima, S., 1992. The factors controlling vertical color variations of North Atlantic Madeira abyssal plain sediments. Mar. Geol. 109, 83–94. https://doi.org/10.1016/0025-3227(92)90222-4.
Paillard, D., Labeyrie, L., Yiou, P., 1996. Macintosh program performs time-series analysis. Eos Trans. Am. Geophys. Union 77, 379. https://doi.org/10.1029/96EO00259.
Parrenin, F., Barker, S., Blunier, T., Chappellaz, J., Jouzel, J., Landais, A., Masson-Delmotte, V., Schwander, J., Veres, D., 2012. On the gas-ice depth difference (Δdepth) along the EPICA Dome C ice core. Climate Past Discuss. 8, 1089–1131. https://doi.org/10.5194/cpd-8-1089-2012.
Parrenin, F., Masson-Delmotte, V., Köhler, P., Raynaud, D., Paillard, D., Schwander, J., Barbante, C., Landais, A., Wegner, A., Jouzel, J., 2013. Synchronous change of atmospheric CO2 and Antarctic temperature during the last Deglacial warming. Science 339, 1060–1063. https://doi.org/10.1126/science.1226368.
Parrenin, F., Cavitte, M.G.P., Blankenship, D.D., Chappellaz, J., Fischer, H., Gagliardini, O., Masson-Delmotte, V., Passalacqua, O., Ritz, C., Roberts, J., Siegert, M.J., Young, D.A., 2017. Is there 1.5-million-year-old ice near Dome C, Antarctica? The Cryosphere 11, 2427–2437. https://doi.org/10.5194/tc-11-2427-2017.
Parrenin, F., Paillard, D., 2012. Terminations VI and VIII (~530 and ~720 kyr BP) tell us the importance of obliquity and precession in the triggering of deglaciations. Clim. Past 8, 2031–2037. https://doi.org/10.5194/cp-8-2031-2012.
Prokopenko, A.A., Hinnov, L.A., Williams, D.F., Kuzmin, M.I., 2006. Orbital forcing of continental climate during the Pleistocene: a complete astronomically tuned climatic record from Lake Baikal, SE Siberia. Quat. Sci. Rev. 25, 3431–3457. https://doi.org/10.1016/j.quascirev.2006.10.002.
Pujos, A., Giraudeau, J., 1993. Répartition des Noelaerhabdaceae (nannofossiles calcaires) dans le Quaternaire moyen et supérieur des océans Atlantique et Pacifique. Oceanol. Acta 16, 349–362.
Rahmstorf, S., 1999. Shifting seas in the greenhouse? Nature 399, 523–524.
Rahmstorf, S., 2002. Ocean circulation and climate during the past 120,000 years. Nature 419, 207–214. https://doi.org/10.1038/nature01090.
Rahmstorf, S., 2013. Thermohaline Circulation. Elsevier, pp. 737–747.
Raymo, M.E., Mitrovica, J.X., 2012. Collapse of polar ice sheets during the stage 11 interglacial. Nature 483, 453–456. https://doi.org/10.1038/nature10891.
Raymo, M.E., Jansen, E., Blum, P., Herbert, T.D. (Eds.), 1999. Proceedings of the Ocean Drilling Program, 162 Scientific Results, Proceedings of the Ocean Drilling Program. Ocean Drilling Program, https://doi.org/10.2973/odp.proc.sr.162.1999.

Reimer, P.J., Bard, E., Bayliss, A., Beck, J.W., Blackwell, P.G., Ramsey, C.B., Buck, C.E., Cheng, H., Edwards, R.L., Friedrich, M., Grootes, P.M., Guilderson, T.P., Haflidason, H., Hajdas, I., Hatté, C., Heaton, T.J., Hoffmann, D.L., Hogg, A.G., Hughen, K.A., Kaiser, K.F., Kromer, B., Manning, S.W., Niu, M., Reimer, R.W., Richards, D.A., Scott, E.M., Southon, J.R., Staff, R.A., Turney, C.S.M., van der Plicht, J., 2013. IntCal13 and Marine13 radiocarbon age calibration curves 0–50,000 years cal BP. Radiocarbon 55, 1869–1887. https://doi.org/10.2458/azu_js_rc.55.16947.

Reyes, A.V., Carlson, A.E., Beard, B.L., Hatfield, R.G., Stoner, J.S., Winsor, K., Welke, B., Ullman, D.J., 2014. South Greenland ice-sheet collapse during marine isotope stage 11. Nature 510, 525–528. https://doi.org/10.1038/nature13456.

Richter, T.O., van der Gaast, S., Koster, B., Vaars, A., Gieles, R., de Stigter, H.C., Haas, H.D., van Weering, T.C.E., 2006. The Avaatech XRF Core scanner: technical description and applications to NE Atlantic sediments. Geol. Soc. Lond. Spec. Publ. 267, 39–50. https://doi.org/10.1144/GSL.SP.2006.267.01.03.

Rodríguez-Tovar, F.J., Dorador, J., Martin-Garcia, G.M., Sierro, F.J., Flores, J.A., Hodell, D.A., 2015. Response of macrobenthic and foraminifer communities to changes in deep-sea environmental conditions from Marine Isotope Stage (MIS) 12 to 11 at the "Shackleton Site.". Global Planet. Change 133, 176–187. https://doi.org/10.1016/j.gloplacha.2015.08.012.

Rohling, E.J., 2013. PALEOCEANOGRAPHY, PHYSICAL AND CHEMICAL PROXIES | oxygen isotope composition of seawater. In: Encyclopedia of Quaternary Science. Elsevier, pp. 915–922, https://doi.org/10.1016/B978-0-444-53643-3.00293-4.

Rorvik, K.-L., Laberg, J.S., Hald, M., Ravna, E.K., Vorren, T.O., 2010. Behavior of the northwestern part of the Fennoscandian ice sheet during the last glacial maximum – a response to external forcing. Quat. Sci. Rev. 29, 2224–2237. https://doi.org/10.1016/j.quascirev.2010.05.022.

Rothwell, R.G., Croudace, I.W., 2015. Twenty years of XRF Core scanning marine sediments: what do geochemical proxies tell us? In: Croudace, I.W., Rothwell, R.G. (Eds.), Micro-XRF Studies of Sediment Cores. Springer, Netherlands, Dordrecht, pp. 25–102, https://doi.org/10.1007/978-94-017-9849-5_2.

Ruddiman, W.F., McIntyre, A., 1981. The mode and mechanism of the last deglaciation: oceanic evidence. Quatern. Res. 16, 125–134. https://doi.org/10.1016/0033-5894(81)90040-5.

Samtleben, C., Schafer, P., Andruleit, H., Baumann, A., Baumann, K.-H., Kohly, A., Matthiessen, J., Schrader-Ritzrau, A., Synpal Working Group, 1995. Plankton in the Norwegian-Greenland Sea: from living communities to sediment assemblages? An actualistic approach. Geol. Rundsch. 84, 108–136. https://doi.org/10.1007/BF00192245.

Sejrup, H.P., Haflidason, H., Andrews, J.T., 2011. A Holocene North Atlantic SST record and regional climate variability. Quat. Sci. Rev. 30, 3181–3195. https://doi.org/10.1016/j.quascirev.2011.07.025.

Shackleton, N.J., Opdyke, N.D., 1973. Oxygen isotope and Palaeomagnetic stratigraphy of equatorial Pacific Core V28-238: oxygen isotope temperatures and ice volumes on a 10" year and 10" year scale. Quatern. Res. 3, 39–55. https://doi.org/10.1016/0033-5894(73)90052-5.

Spencer, A.M., Embry, A.F., Gautier, D.L., Stoupakova, A.V., Sørensen, K., 2011. Chapter 1 an overview of the petroleum geology of the Arctic. Geol. Soc. Lond. Mem. 35, 1–15. https://doi.org/10.1144/M35.1.

Stokes, C., Clark, C., Darby, D., Hodgson, D., 2005. Late Pleistocene ice export events into the Arctic Ocean from the M'Clure Strait ice stream, Canadian Arctic Archipelago. Global Planet. Change 49, 139–162. https://doi.org/10.1016/j.gloplacha.2005.06.001.

Stoner, J.S., Channell, J.E.T., Hillaire-Marcel, C., Kissel, C., 2000. Geomagnetic paleointensity and environmental record from Labrador Sea core MD95-2024: global marine sediment and ice core chronostratigraphy for the last 110 kyr. Earth Planet. Sci. Lett. 183, 161–177. https://doi.org/10.1016/S0012-821X(00)00272-7.

Stoner, J.S., Laj, C., Channell, J.E.T., Kissel, C., 2002. South Atlantic and North Atlantic geomagnetic paleointensity stacks (0–80ka): implications for inter-hemispheric correlation. Quat. Sci. Rev. 21, 1141–1151. https://doi.org/10.1016/S0277-3791(01)00136-6.

Stuiver, M., Reimer, P.J., 1993. Extended 14C Data Base and revised CALIB 3.0 14C age calibration Program | Radiocarbon | Cambridge Core. Radiocarbon 35, 215–230. https://doi.org/10.1017/S0033822200013904.

Svendsen, J., 2004. Late quaternary ice sheet history of northern Eurasia. Quat. Sci. Rev. 23, 1229–1271. https://doi.org/10.1016/j.quascirev.2003.12.008.

Thibodeau, B., Bauch, H.A., Pedersen, T.F., 2017. Stratification-induced variations in nutrient utilization in the polar North Atlantic during past interglacials. Earth Planet. Sci. Lett. 457, 127–135. https://doi.org/10.1016/j.epsl.2016.09.060.

Thierstein, H.R., Geitzenauer, K.R., Molfino, B., Shackleton, N.J., 1977. Global synchroneity of late Quaternary coccolith datum levels Validation by oxygen isotopes. Geology 5 (7), 400–404. https://doi.org/10.1130/0091-7613(1977)5%3C400:GSOLQC%3E2.0.CO;2.

Toucanne, S., Zaragosi, S., Bourillet, J.F., Cremer, M., Eynaud, F., Van Vliet-Lanoë, B., Penaud, A., Fontanier, C., Turon, J.L., Cortijo, E., 2009. Timing of massive 'Fleuve Manche' discharges over the last 350kyr: insights into the European ice-sheet oscillations and the European drainage network from MIS 10 to 2. Quat. Sci. Rev. 28, 1238–1256. https://doi.org/10.1016/j.quascirev.2009.01.006.

Tzedakis, P.C., Hooghiemstra, H., Pälike, H., 2006. The last 1.35 million years at Tenaghi Philippon: revised chronostratigraphy and long-term vegetation trends. Quat. Sci. Rev. 25, 3416–3430. https://doi.org/10.1016/j.quascirev.2006.09.002.

Tzedakis, P.C., Raynaud, D., McManus, J.F., Berger, A., Brovkin, V., Kiefer, T., 2009. Interglacial diversity. Nat. Geosci. 2, 751–755. https://doi.org/10.1038/ngeo660.

Wagner, B., Wilke, T., Francke, A., Albrecht, C., Baumgarten, H., Bertini, A., Combourieu-Nebout, N., Cvetkoska, A., Dapos Addabbo, M., Donders, T.H., Föller, K., Giaccio, B., Grazhdani, A., Hauffe, T., Holtvoeth, J., Joannin, S., Jovanovska, E., Just, J., Kouli, K., Koutsodendris, A., Krastel, S., Lacey, J.H., Leicher, N., Leng, M.J., Levkov, Z., Lindhorst, K., Masi, A., Mercuri, A.M., Nomade, S., Nowaczyk, N., Panagiotopoulos, K., Peyron, O., Reed, J.M., Regattieri, E., Sadori, L., Sagnotti, L., Stelbrink, B., Sulpizio, R., Tofilovska, S., Torri, P., Vogel, H., Wagner, T., Wagner-Cremer, F., Wolff, G.A., Wonik, T., Zanchetta, G., Zhang, X.S., 2017. The environmental and evolutionary history of Lake Ohrid (FYROM/Albania): interim results from the SCOPSCO deep drilling project. Biogeosciences 14, 2033–2054. https://doi.org/10.5194/bg-14-2033-2017.

Wefer, G., Berger, W.H., Richter, C., et al. (Eds.), 1998a. Proceedings of the ocean drilling program 175 initial reports. In: Proceedings of the Ocean Drilling Program. Ocean Drilling Program., https://doi.org/10.2973/odp.proc.ir.175.1998.

Wefer, G., Berger, W.H., Richter, C., Grout, S., Anderson, B., Jansen, L., Maslin, P., Pufahl, V., Christensen Giraudeau, J., Hermelin, L., Motoyama, F., Yamazaki, A., Laser, M., Wigley, M., Gorgas, C., 1998b. Explanatory notes, shipboard scientific party. In: Proc. Ocean Drill. Program Initial Rep. vol. 175, p. 20.

Willeit, M., Ganopolski, A., Calov, R., Brovkin, V., 2019. Mid-Pleistocene transition in glacial cycles explained by declining CO2 and regolith removal | science advances. Sci. Adv. 5. https://doi.org/10.1126/sciadv.aav7337.

CHAPTER THREE

Quaternary Palynostratigraphy of Germany with special emphasis on the Late Pleistocene

Philipp Stojakowits[a,*] and Christoph Mayr[b,c,d]

[a]State Authority for Mining, Energy and Geology, Hannover, Germany
[b]Institute of Geography, Friedrich-Alexander-University Erlangen-Nürnberg, Erlangen, Germany
[c]Department of Earth and Environmental Sciences, Paleontology and Geobiology, Ludwig-Maximilians-University München, München, Germany
[d]GeoBio-Center, Ludwig-Maximilians-University München, München, Germany
*Corresponding author: e-mail address: philipp.stojakowits@lbeg.niedersachsen.de

Contents

Abstract

Here, we summarize and review the current state of knowledge on Quaternary palynostratigraphy in Germany. Our focus is especially on the Pleistocene and a comparison between northern and southern Germany, while we consider the Holocene only marginally, since many review articles already exist about this epoch. In the region under consideration several warm and cold stages were reported for the Early Pleistocene. The thermomeres are characterized by the partially frequent occurrence of relict Neogene elements. Concerning the last interglacial of the Early Pleistocene (= Osterholz Interglacial), for instance, *Eucommia* and *Azolla filiculoides* are important for stratigraphical purposes. The Middle Pleistocene encompasses the Cromerian Complex with five interglacials detected so far in northern Germany. Subsequently,

Stratigraphy & Timescales, Volume 7
ISSN 2468-5178
https://doi.org/10.1016/bs.sats.2022.09.001

the Elsterian and Saalian glacial cycles occurred intercalated by the Holsteinian interglacial. This warm stage was detected at many sites all over Germany with a characteristic vegetational development including the presence of *Fagus* and *Pterocarya*. In contrast, just a few records exist from the warm stages within the Cromerian Complex and the Saalian and their vegetational succession is poorly known. The interglacial preceding the Holocene, the Eemian Interglacial, in turn is the palynostratigraphically best defined Pleistocene warm stage. The subsequent alternation between interstadial and stadial conditions during the last glacial (i.e., Würmian or Weichselian) led to the formation and decline of boreal forests and shrub tundra, respectively. However, these vegetation changes are well understood only up to early Marine Isotope Stage (MIS) 3. The further course of the Middle Würmian and Weichselian vegetation development is still insufficiently reconstructed, since most pollen records provide only short time intervals and do not include the full range of stadials and interstadials found, for example, in Greenland ice cores. There are also few continuous pollen records for the Last Glacial Maximum. In contrast, the vegetation history of the Late Glacial and subsequent Holocene is again well known and in some places very precisely dated based on varve chronologies. We conclude that despite a long palynological tradition in Germany, some periods are underrepresented or hardly studied for various reasons and a correlation between north and south is often not given. The focus of future studies should therefore be on these periods and on a better correlation of pollen stratigraphy using state-of-the-art dating technologies.

1. Introduction

Palynostratigraphy is a type of stratigraphy that deals with the description, interpretation, and timing of vegetational successions based on the identification of pollen and spore assemblages in their stratigraphic context including lithostratigraphy. Although some reviews on palynostratigraphy have been published for Germany, but, so far no comparisons on the vegetation gradient from north to south during the Pleistocene. In addition, new findings and modern dating techniques necessitate a reinterpretation of the chronology of some older pollen records and interglacials, such as the Holsteinian. Despite a long tradition of Quaternary palynostratigraphy in Germany, some works have been largely forgotten or not noticed from a broader audience (e.g., the long pollen sequence from Gorleben; Müller, 1992), despite of possibly great importance for correlation with the palynostratigraphy of adjacent countries, such as the Netherlands and France. More recently, further publications document Quaternary vegetation and climate changes with unprecedented temporal resolution and dating accuracy. Here, we summarize the Quaternary palynostratigraphy of the entire Federal Republic of Germany making comparisons between the North and the South, where possible, and, where necessary, reinterpretating of some pollen records with respect to their stratigraphic position. The focus here is on the Pleistocene, as the Holocene palynostratigraphy of this area has been the subject of several recent reviews.

2. Brief research history

The first palynological studies just dealt with the qualitative description of pollen types recovered from Quaternary deposits. Weber (1893, 1896) was one of the first palynologists who counted a particular pollen sum in order to determine the quantities of the identified pollen and spores. The next milestone was the presentation of the first pollen diagram by Lennart von Post at a conference 1916 in Oslo (Lang, 1994), which was the precursor of modern quantitative pollen analysis with stratigraphical purposes. In the 1920s Bertsch (1924) and Stark (1923, 1924) published some of the first pollen diagrams from Holocene deposits in southern Germany. Jessen and Milthers (1928), Firbas (1928), and Gistl (1928) did pioneer work regarding pollen-analytical studies of last and penultimate interglacial deposits in northern Germany and several studies followed (e.g., Jonas, 1937). Baas (1932) studied an old Pleistocene sequence in Central Germany, in which he identified several Neogene elements. Up to the 1950s a considerable number of Late Glacial and Holocene studies was published, which were summep up by Firbas (1949, 1952). Additionaly, first regional palynostratigraphical outlines were compiled (Rein, 1955; von der Brelie, 1955). In the 1960s another important new interglacial stage was found, the so called Osterholz warm stage (Grüger, 1967), and the Rhume warm stage, known since the middle 1950s (Lüttig and Rein, 1954), was re-investigated by Müller (1965) in detail. In 1973, three reviews summarized the state of knowledge on Quaternary palynostratigraphy at that time: Menke and Behre (1973) described the biostratigraphy of the area of the Scandinavian glaciation and Frenzel (1973a, 1973b) the ones of the Alpine foreland and the periglacial area with reference to the glaciated areas. The investigation of interglacial deposits in southern Germany, except of occasional studies (e.g., Filzer, 1967; Filzer and German, 1968; German et al., 1965, 1967; Reich, 1953), was neglected for a long time. In the 1970s several fundamental studies shed more light in the Pleistocene vegetation history of southern Germany (e.g., Beug, 1979; Frenzel, 1978; Göttlich and Werner, 1974; Grüger, 1979a, 1979b; Jung et al., 1972). In 1994, an issue of the Geologisches Jahrbuch Reihe A was dedicated to new investigations on interglacial deposits in Lower Saxony. There, the paleobotanical study of another warm stage, the so-called Hunteburg thermomere, is included (Hahne et al., 1994a). Lang (1994) published an overview about the Quaternary vegetation history of Europe with emphasis on Germany. Therein, he also provided a detailed overview of the successive extinction of Neogene elements during the Quaternary. This compilation was adapted and updated by Hahne et al. (2008) and is of central importance for the palynostratigraphically classification of pollen sequences in conjunction with the frequently characteristic vegetational succession (Table 1). Since the turn of the millennium, further research was conducted (e.g., Kühl and Gobet, 2010; Koutsodendris et al., 2010; Stojakowits et al., 2020) including the publication of stratigraphical terms and palynostratigraphy in northern Germany and the periglacial area, respectively (Litt et al., 2007; Urban, 2007). Concerning southern Germany, a comprehensive review of the current state of knowledge, however, is still missing. (See Fig. 1.)

Table 1 Absence and presence of important arboreal taxa with diagnostic value in Central European pollen sequences from the Reuverian to the Holocene.

	Pliocene	**Early Pleistocene**			**Middle Pleistocene**		**Late Pleistocene**	
	Reuver	**Tiglian**	**Waalian**	**Bavelian**	**Cromerian**	**Holsteinian**	**Eemian**	**Holocene**
Sequoia-Habitus	•	o						
Taxodium-Habitus	•	o						
Sciadopytis	•	•						
Liquidambar	•	o						
Nyssa	•	o						
Carya	•	o	o	o				
Ostrya-type	•	o	o	o				
Tsuga	•	•	•	•	o			
Eucommia	•	•	•	•	•			
Celtis	•	•	•	o	•	o[a]		
Pterocarya	•	•	•	•	•	•		
Fagus	•	•	o	o	•	•		•

[a]Single occurrences in northern Germany.

Modified after Lang, G., 1994. Quartäre Vegetationsgeschichte Europas, Gustav Fischer, Stuttgart and Hahne, J., Ellwanger, D., Stritzke, R., 2008. Evidence for a Waalian thermomer pollen record from the borehole Heidelberg UniNord, Upper Rhine Graben, Baden-Württemberg. Eiszeitalter u. Gegenwart (E&G Quaternary Science Journal) 57, 403–410.

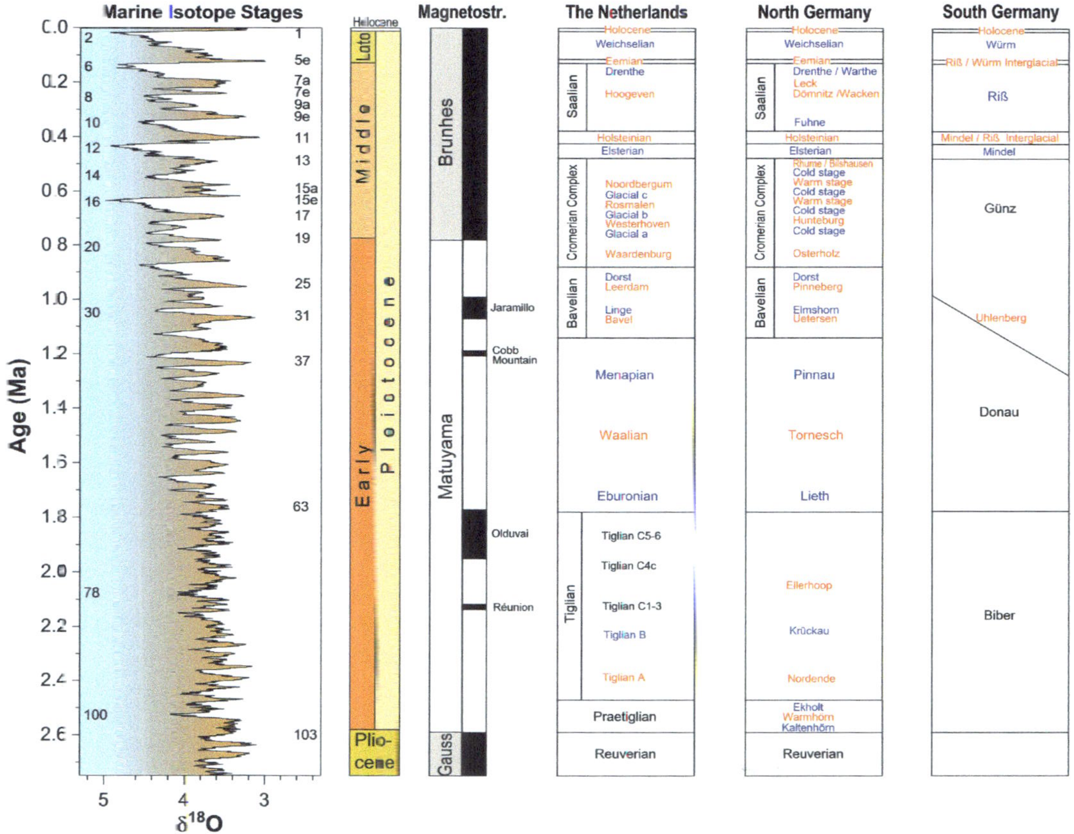

Fig. 1 See f gure legend on nex: page.

3. Plio-Pleistocene boundary

The Plio-Pleistocene boundary has been originally defined as the base of the Calabrian Stage at 1.8 Ma (Aguirre and Pasini, 1985). In 2009, the boundary has been re-defined to the base of the Gelasian Stage at 2.58 Ma (Gibbard and Head, 2010) in congruence with the onset of major glaciations in the northern hemisphere revealed by the marine isotope record (Lisiecki and Raymo, 2005; Zachos et al., 2001). In the course of the Late Pliocene (Reuverian B and C) the so-called 'Tertiary' elements (or better: Neogene elements), like *Carya*, *Liquidambar*, *Nyssa*, *Pterocarya*, *Sciadopytis*, *Sequoia*, *Taxodium*, and *Tsuga*, more or less disappeared gradually due to the climatic cooling towards the Pleistocene. This marked change in the forest composition was studied by Zagwijn (1957, 1960, 1974a, 1974b) in the Dutch-German border region. Later on, this vegetational change was also studied at sites in northwestern Germany (Menke, 1970, 1975; Proborukmi et al., 2017), the Lower Rhine area of Germany (e.g., Boenigk et al., 1974; Heumann and Litt, 2002; Urban, 1978a), and central Germany (e.g., Boenigk et al., 1977).

4. Early Pleistocene

The first section of the beginning Pleistocene is the Pretiglian (Zagwijn, 1957, 1960, 1963). It is a cryomere with a strongly reduced arboreal pollen flora. Just *Alnus*, *Betula*, and *Pinus* occurred, whereas taxa

Fig. 1 Stratigraphic subdivision of the Quaternary in the Netherlands and North and South Germany in comparison to the marine isotope record (Lisiecki and Raymo, 2005), the Quaternary Series and Subseries, and the simplified paleomagnetic record. The subdivision of the Quaternary follows Zagwijn (1989) for the Netherlands, Litt et al. (2005) and Stephan (2014) for North Germany, and Doppler et al. (2011) for South Germany. Note that the major ice advances of the Elsterian and Saalian Glaciations in northern Germany occurred during MIS 12 and 6 (Lauer and Weiss, 2018) and the major ice advances of the Günz, Mindel, and Riß cold stages in the Alpine foreland during MIS 16, 12, and 6 (Reitner, 2022). Correlations between northern and southern Germany are adapted from STG (2016) (German Stratigraphic Commission). *Modified after Litt, T., Behre, K.-E., Meyer, K.-D., Stephan, H.-J., Wansa, S., 2007. Stratigraphische Begriffe für das Quartär des norddeutschen Vereisungsgebietes. Eiszeitalter u. Gegenwart (E&G Quaternary Science Journal) 56, 7–65. After Cohen, K.M., Gibbard, P., 2011. Global chronostratigraphical correlation table for the last 2.7 million years. Subcommisions on Quaternary Stratigraphy (International Commission on Stratigraphy), Cambridge, England.*

demanding warm temperatures were completely absent. Herb pollen (especially Poaceae and Cyperaceae) prevailed together with Ericaceae. In the drill core of Oldenswort (northwestern Germany), Menke (1975) established the Pretiglian cryomere (syn. Kaltenhörn cryomere), which is a time-equivalent of the Pretiglian introduced by Zagwijn (1957, 1960). At Oldenswort and at the nearby outcrop of Lieth, Menke (1975) found a warm stage, which he termed Warmhörn. According to the pollen content, this thermomere could be correlative to a warm stage in the Meinweg drill core of Zagwijn (1960) and the pollen record of Rippersroda in eastern Germany is likely correlative, too (Litt et al., 2007). The Fortuna oscillation from the Lower Rhine area is also assumed as a correlative (Urban, 1978a, 1978b). Subsequently, another cold stage (= Ekholt cryomere) is found at the long sequence of Lieth (Menke, 1969, 1975). The Nordende Interglacial, which follows, is characterized by the presence of *Eucommia*, *Pterocarya*, and, of special interest, *Symplocos* as well (Menke, 1969, 1975). Besides that, high values of *Tsuga* and the presence of *Fagus* seem to be characteristic for this thermomere. Based on the pollen assemblages, the Tiglian A (Zagwijn, 1960, 1963) from the Netherlands and the Frechen Interglacial I (Urban, 1978a, 1978b) from the Lower Rhine area represent time-equivalents. The Krückau cold stage, located in superposition of the preceding Nordende thermomere, was only found at Lieth so far. According to Menke (1975), this cold stage dominated by non-arboreal pollen can be correlated with the Tiglian B sensu Zagwijn (1960, 1963). The following Ellerhoop thermomere (Menke, 1975) is assumed as a time-equivalent of the Tiglian C, which can be subdivided in 6 pollen zones (TC1 to TC6; Zagwijn, 1960, 1963). Menke (1975) divided the warm stage into 6 pollen zones, too. Pollen zone 4 reflects a climatic depression with the spread of boreal forests. The so called Frechen Interglacials II and III separated by a gravel layer (= Frechen cold stage III) are considered as time-equivalent of the Ellerhoop and Tiglian C. The occurrences of the water ferns *Azolla tegeliensis* and *Azolla filiculoides* have a strong diagnostic value (Florschütz, 1938; Zagwijn, 1963). *Azolla tegeliensis* was distributed in all the warm stages of the Tiglian, whereas *A. filiculoides* occurred from pollen zone TC4 onward (Zagwijn, 1963). Worth to mention is also the fact, that *Fagus* pollen is restricted to the Tiglian A and the Nordende Interglacial respectively.

The termination of the Tiglian is marked by the onset of the Lieth cold stage (Menke, 1975) which is correlative to the Eburonian cold stage introduced by Zagwijn (1957, 1960, 1963). During cold phases an open tundra vegetation occurred, whereas in the course of interstadial conditions boreal forests expanded. The subsequent warm stage (= Waalian sensu; Zagwijn, 1957, 1960, 1963 and Tornesch warm stage after Menke, 1969, 1975) can be divided in two interglacials (Waalian A and C) separated by a short cryomere. Within the warm stages some Neogene elements were distributed, such as *Carya*, *Eucommia*, *Pterocarya*, and *Tsuga*. *Tsuga*, however, is not represented at Lieth due to strong maritime conditions. From Heidelberg, Upper Rhine Graben, a Waalian pollen record was reported (Hahne et al., 2008) with frequent *Tsuga* and a continuous presence of *Carya*, *Celtis*, *Eucommia*, *Ostrya*-type, and *Pterocarya*.

The Waalian is followed by the Pinnau cold stage (Menke, 1970, 1975) and the time-equivalent Menapian (Zagwijn, 1960) and the subsequent Bavelian Complex. The Bavelian is subdivided in, at least, two warm stages and two cold stages. After the Menapian the Bavelian Interglacial started (Zagwijn and de Jong, 1984), which is dating in the Jaramillo event and characterized in the Netherlands by the given succession (here and in the following always listed from old to young):

1. *Pinus*-*Alnus*-*Ulmus*-Zone
2. *Alnus*-*Pinus*-*Ulmus*-*Taxus*-Zone
3. (a) *Carpinus*-*Eucommia*-Zone
   (b) *Carpinus*-*Eucommia*-*Tsuga*-Zone
4. *Tsuga*-*Abies*-*Carpinus*-Zone
5. *Pinus*-*Picea*-Zone.

According to Menke (1975), the Uetersen warm stage is correlative to the Bavelian Interglacial. At Gorleben, Müller (1992) (Fig. 2) detected this warm stage, too. In southern Germany, there are also some sites probably belonging to the Bavelian Stage. One of them is located in the Upper Rhine Graben at Iffezheim with frequent occurrences of *Tsuga* and *Pterocarya* during the climatic optimum of this thermomere (Bludau, 1995). Another site is the Uhlenberg in the Bavarian Alpine foreland first studied by Filzer and Scheuenpflug (1970). There, a three-partite vegetational succession was recorded (Schedler, 1979, 1981):

1. *Tsuga*-*Alnus*-*Pinus*-Zone with high values of *Tsuga*
2. *Alnus*-*Pinus*-*Picea*-Zone
3. *Picea*-*Pinus*-*Betula*-Zone.

According to the paleomagnetic evidence, the pollen record of the Uhlenberg dates into the Jaramillo event (Brunnacker et al., 1976; Koči in Scheuenpflug, 1979).

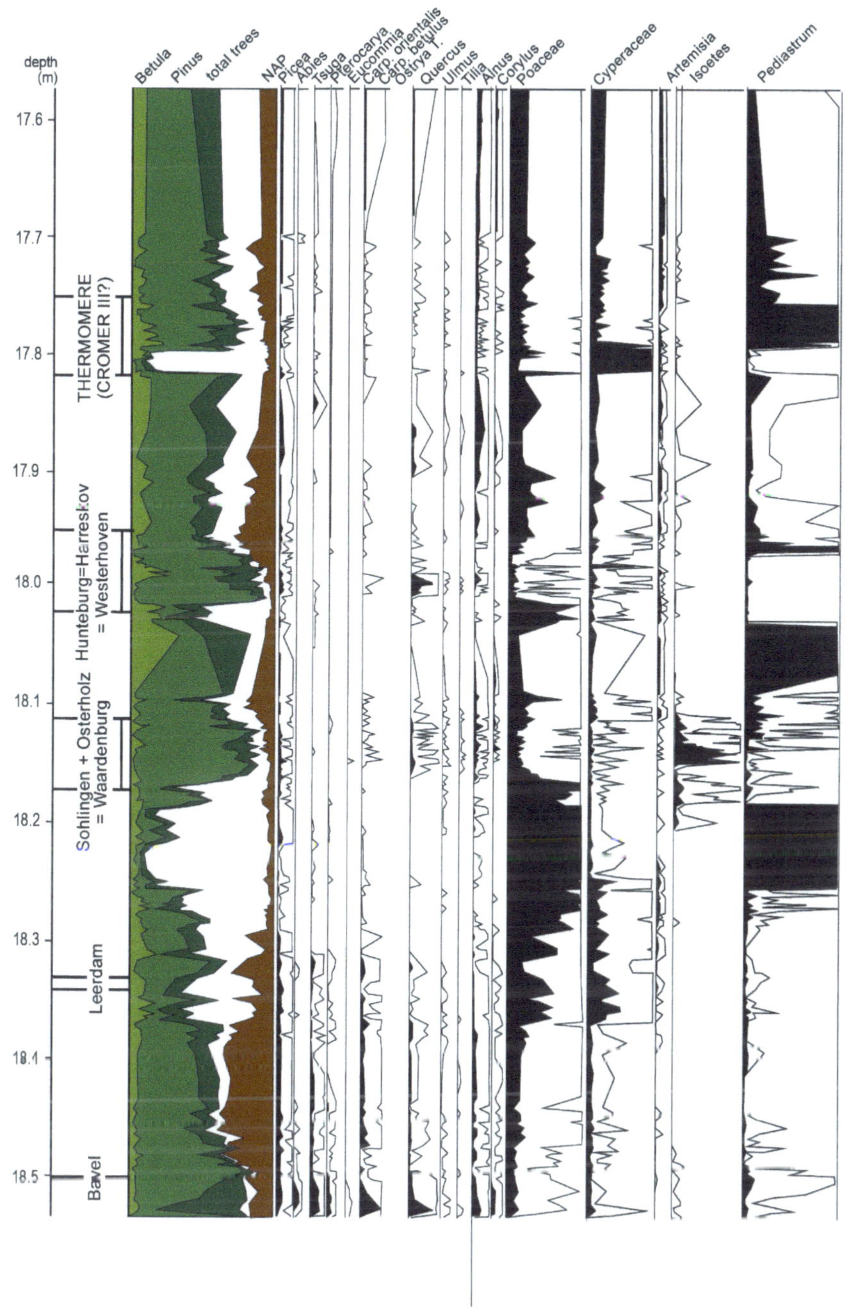

Fig. 2 Pollen diagram of an Early and Middle Pleistocene sequence from borehole GoHy 1270 of Gorleben, Part 1. *Modified after Müller, H., 1992. Climatic changes during and at the end of the interglacials of the Cromerian Complex. In: Kukla, G. J., Went, E. (Eds.), Start of a Glacial. NATO ASI Series I 3, Springer, Berlin, pp. 51–69.*

Within the next cryomere, the Linge cold stage (Zagwijn and de Jong, 1984) or Elmshorn cold stage (Menke, 1975), respectively, the pollen content is reduced to herbs and *Pinus*, *Betula*, and *Juniperus*. The second thermomere of the Bavelian Complex, also known as Leerdam Interglacial, shows also a clear forest succession (Zagwijn and de Jong, 1984):

1. *Betula*-Zone
2. (a) *Pinus-Ulmus*-Zone
   (b) *Pinus-Quercus-Carpinus-Ulmus*-Zone
3. *Pinus-Alnus-Betula*-Zone.

At Lieth the Pinneberg warm stage (Menke, 1975) is considered to be correlative to the Leerdam Interglacial (Litt et al., 2007). At Gorleben Müller (1992) detected this interglacial in superposition of the preceding Bavel Interglacial and Linge Glacial. The last stage belonging to the Bavelian is the Dorst cold stage with dominating herbs (Zagwijn and de Jong, 1984), which was also found at Gorleben (Müller, 1992; Fig. 2).

The youngest warm stage belonging to the Early Pleistocene in Northern Germany is the so called Osterholz Interglacial which was first described by Grüger (1967) at the 'Osterholz' site south of Hannover. It is characterized by a *Pinus-Betula* zone followed by a zone dominated by thermophilous trees consisting mainly of *Quercus* and *Ulmus* at the beginning. Later on, *Carpinus* became dominant. During the phase with dominating *Carpinus*, the relatively frequent occurrence of *Eucommia* and *A. filiculoides* is important for stratigraphical purposes. The Osterholz warm stage is the first interglacial within the Cromerian Complex (Litt et al., 2007). A time equivalent pollen record at Elm, southeast of Braunschweig, was also studied by Grüger (1967). Müller (1992) also detected this warm stage with the same vegetational succession and the characteristic occurrence of *Eucommia* in the long pollen record close to the salt dome at Gorleben (Fig. 2). The so-called Sohlingen warm stage, which was studied by Homann and Lepper (1994), is correlative to the Osterholz warm stage, too. In the Netherlands Zagwijn et al. (1971) found another warm stage (= Waardenburg) correlative to the Osterholz Interglacial (Zagwijn, 1996). The sediments of the Waardenburg warm stage showed a reversed direction within the Matuyama period. This also applies to the sediments of the Sohlingen warm stage (Fromm, 1994).

5. Middle Pleistocene

A cold stage marks the onset of the Middle Pleistocene. The first warm stage within the Middle Pleistocene is the subsequent Hunteburg Interglacial, which was found northeast of Osnabrück. This interglacial was divided into nine pollen zones (Hahne, 1996; Hahne et al., 1994a):

1. *Pinus*-Zone with relatively high NAP values indicating open pine forests.
2. *Ulmus*-Zone with low *Pinus* values and expanding *Quercus* and *Alnus* representing closed forests.
3. *Ulmus-Quercus*-Zone. Beside *Ulmus* and *Quercus*, *Tilia* and *Alnus* also occur comparatively frequent.
4. QM-*Corylus-Taxus*-Zone. Taxa of the Quercetum mixtum (= QM) dominate together with *Corylus* and *Taxus*. The pollen curve of *Celtis* becomes continuous.
5. QM-*Corylus*-Zone. Compared to the previous zone, the Quercetum mixtum and *Celtis* values are more or less the same, but *Taxus* has decreased and *Corylus* slightly increased.
6. *Corylus-Alnus*-QM-Zone. High *Alnus* und *Corylus* values together with decreasing *Ulmus* are characteristic for this zone. Furthermore, *Abies* has a continuous curve.
7. *Abies-Alnus*-QM-Zone. *Abies* has a maximum over 40%. The curves of *Alnus* und *Corylus* decrease, whereas the values of *Pinus*, *Picea* and *Betula* increase.
8. *Picea-Pinus-Carpinus*-Zone. This zone is dominated by *Pinus* and *Picea*. Additionally, *Carpinus* is present with low values. The taxa of the Quercetum mixtum are reduced during this time span and both *Pinus*, *Picea* and *Betula* spread.
9. *Pinus-Picea-Betula*-Zone. At the end of the interglacial boreal forests consisting mainly of *Pinus*, *Picea* and *Betula* prevail. In the course of this zone Poaceae, Cyperaceae and other non-arboreal pollen types become more important indicating the approaching cold phase.

The Hunteburg Interglacial is correlative to the second warm stage within the Cromerian Complex detected by Müller (1992) (Fig. 2). From a palynostratigraphical point of view, the Hunteburg Interglacial should be correlative to the interglacial sites Harreskov and Ølgod (Andersen, 1965) in Denmark and maybe to the Westerhoven Interglacial (Zagwijn and Zonneveld, 1956). According to Zagwijn (1996), the Westerhoven Interglacial is an equivalent of the Hunteburg Interglacial, whereas Hahne et al. (1994a) have doubts with this correlation. Based on paleomagnetic data, a correlation between the Hunteburg and Westerhoven warm stage would be possible (Fromm, 1994), but also an age of the Hunteburg Interglacial between the Waardenburg and Westerhoven Interglacials of the Dutch terminology seems to be possible. Knipping (2008) found an interglacial termed Schifferstadt in the northern Upper Rhine Graben, which can be likely assigned to the Hunteburg Interglacial.

The first stadial phase after the Hunteburg warm stage is characterized by an open vegetation without any forests (= 10. Lower unforested zone; Hahne et al., 1994a) followed by an interstadial, the so called Vennermoor Interstadial, with boreal forests composed of *Pinus*, *Picea*, *Betula* and *Larix* as climax vegetation (= 11. *Pinus-Larix*-Zone; Hahne et al., 1994a). The end of the interstadial is marked by a fall of the boreal forests and the spread of an open, unforested vegetation which dominated the last pollen zone of the Hunteburg record (= 12. Upper unforested zone).

In superposition of the cold stage after the Vennermoor Interstadial, Müller (1992) detected a probably third Cromerian warm stage (Fig. 2), which was cooler compared to the two previous interglacials of the Cromerian Complex. The climatic optimum of this warm stage, however, was likely eroded within the following cold stage. Additionally, in the long pollen record at Gorleben a fourth interglacial belonging to the Cromerian Complex was also found (Müller, 1992; Fig. 3). The forest succession is characterized by a spread of *Betula* first, followed by *Pinus* dominance and the immigration and increase of *Quercus* and *Alnus*. The occurrence of *Picea* is also witnessed, but with much less extent.

In a coring at Bilshausen three interstadials were registered preceeding the Rhume Interglacial as possibly fifth warm stage within the Cromerian Complex (Müller, 1992). These interstadials are characterized by open boreal forests dominated by *Pinus* and *Betula*. During the climatic optima, *Picea* and *Alnus* played a subordinate role in the forest composition, too.

The Rhume Interglacial (= syn. Bilshausen) was studied at the type locality Bilshausen (northeast of Göttingen, southern Lower Saxony) several times (Averdieck and von der Brelie, 1963; Chanda, 1962; Kühl and Gobet, 2010; Lüttig and Rein, 1954; Müller, 1965, 1992). The pollen analyses of Müller (1965, 1992), Diehl (2007), and Kühl and Gobet (2010) revealed reproducable results. The older studies, however, showed different results. The pollen record can be divided into twelve pollen zones (Müller, 1965, 1992; note that the two oldest zones are only represented in the core published in 1965; Fig. 4):

1. Unforested zone a between the last interstadial and the forthcoming interglacial.
2. Reforestation zone b with the spread of *Ulmus*, *Alnus*, *Picea*, and *Pinus*.
3. Early *Ulmus-Picea-Pinus*-Zone (= c) with intercalated steppe-like periods indicated by comparatively high NAP values.

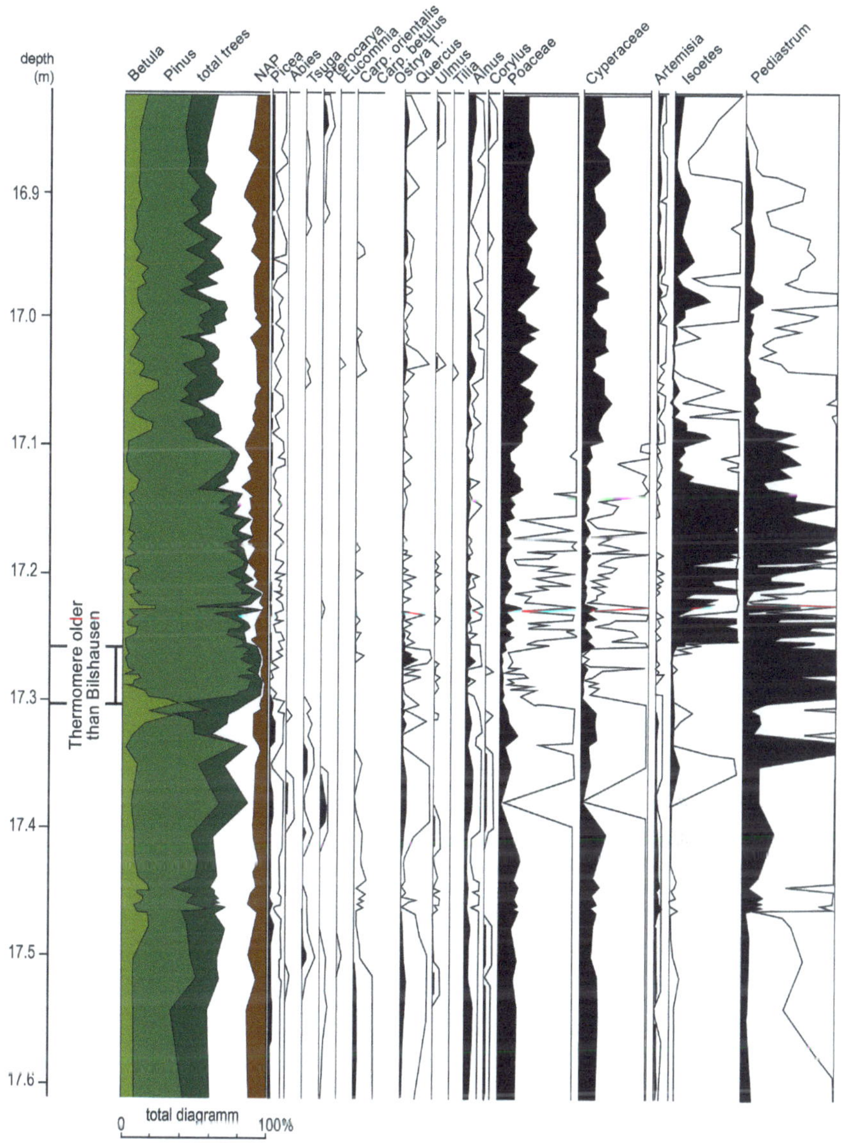

Fig. 3 Pollen diagram of an Early and Middle Pleistocene sequence from borehole GoHy 1270 of Gorleben, Part 2. *Modified after Müller, H., 1992. Climatic changes during and at the end of the interglacials of the Cromerian Complex. In: Kukla, G. J., Went, E. (Eds.), Start of a Glacial. NATO ASI Series I 3, Springer, Berlin, pp. 51–69.*

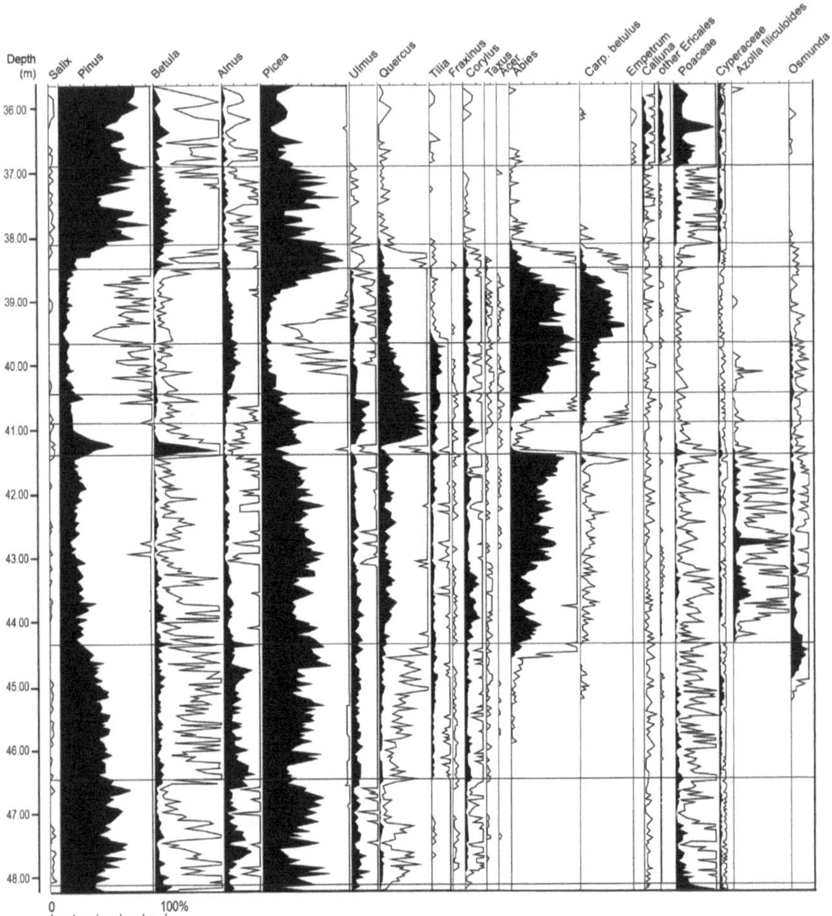

Fig. 4 Pollen diagram of the Rhume Interglacial in the boring Bilshausen 1/78. *Modified after Müller, H., 1992. Climatic changes during and at the end of the interglacials of the Cromerian Complex. In: Kukla, G. J., Went, E. (Eds.), Start of a Glacial. NATO ASI Series I 3, Springer, Berlin, pp. 51–69.*

4. Late *Ulmus-Picea-Pinus*-Zone (= d) with some *Tilia*.
5. *Picea-Abies*-QM-Zone (= e). Beside the dominating taxa *Picea* and *Abies*, *Quercus* and other thermophilious trees of the Quercetum mixtum (e.g., *Ulmus* and *Tilia*) occur relatively frequent as well as the water fern *A. filiculoides*.
6. *Betula-Pinus*-Zone (= f) as a result of a climatic deterioration. During this zone, especially *Picea*, *Quercus*, and *Ulmus* spread again. *Azolla* is still sporadically present at low values as well as in the subsequent zones g and h.

7. QM-Zone (= g) with dominating *Quercus* accompanied by *Ulmus*, *Tilia*, and *Picea*. In the course of this zone, *Abies* and *Carpinus* increase, whereas *Picea* shows decreasing proportions.
8. *Abies-Carpinus*-QM-Zone (= h) with dominating *Abies* and *Carpinus* and decreasing values of *Quercus* and *Ulmus*. At the boundary of the zones h and i a 4 mm thick tephra layer occurs, which, unfornunately, has not been dated yet.
9. *Quercus-Abies-Carpinus*-Zone (= i). Towards the end of the zone, *Abies* and *Carpinus* decrease and the Picea curve shows a strong rise.
10. *Picea*-Zone (= j) with further decreasing values of *Abies* and of deciduous thermophilous trees. In the course of this zone, *Pinus* and *Betula* gain more importance.
11. *Picea-Betula-Pinus*-Zone (= k). The pollen content is dominated by *Picea* and *Pinus* with subordinated *Betula*. Except of *Alnus*, all other deciduous trees have been disappeared.
12. Tundra vegetation representing the onset of the Elsterian glaciation (= Zone l).

The record of Bilshausen also contains a rich macroflora. Chanda (1962) identified several aquatic plants and tree remains, such as *Acer* spec., *Betula pendula*, *B. pubescens*, *Carpinus* cf. *betulus*, and *Populus tremula* as well as relatively frequent *Abies* cf. *alba* and *Picea* cf. *abies* belonging to the pollen zones i to k of Müller (1965, 1992). Based on the counting of the annually laminated lake deposits at Bilshausen (Müller, 1965, 1992), the Rhume warm stage lasted approximately 27,000 years. At Salzderhelden a correlative section of this interglacial was detected representing the last 5000 years of the Bilshausen record. Müller (1986) mentioned another time-equivalent record in a coring at Gorleben. According to Bittmann and Müller (1996), the Kärlich Interglacial (Neuwied Basin, Middle Rhine area) correlates well with the corresponding section of the Rhume record of Bilshausen. The Mannheim Interglacial, located in the northern Upper Rhine Graben, likely correlates with the pollen sequences of Bilshausen and Kärlich. All these three pollen records are characterized by the occurrences of *Fagus*, *Celtis*, and *Azolla* together with high values of *Abies*, *Carpinus*, *Quercus*, and *Corylus* (Knipping, 2008).

In the German Alpine foreland another interglacial was found at Unterpfauzenwald overlain by till of Mindelian age (Bibus et al., 1996; Göttlich and Werner, 1974). A correlation with other sequences is not possible so far. As this interglacial at Unterpfauzenwald is situated below Mindel till deposits, a pre-Holsteinian age can be derived.

The termination of the Cromerian Complex is defined by the onset of the Elsterian glaciation. The start of this cold period is represented in the tundra vegetation pollen zone 1 from Müller (1965, 1992), where a red clay up to 8 m thick was deposited overlying the interglacial sequence consisting of lacustrine lignitic clay (Lüttig, 1965). In top of the red clay another lacustrine grey silty clay was deposited. Two intercalated silty muds revealed interstadial conditions divided by a tundra vegetation period with lots of Poaceae and increasing values of *Artemisia* and *Thalictrum* towards the upper interstadial. The pollen content of these interstadials is dominated by *Betula* and subordinated *Pinus*. During a short period *Picea* is present. According to the current state of knowledge, these two interstadials should represent short warm periods within the Elsterian (Caspers et al., 1995). Unfortunately, both are not distinctive and have therefore no significant stratigraphic importance. The same holds true for three interstadials found by Urban et al. (1988) in the opencast mine of Schöningen (Lower Saxony) underlying Holsteinian sediments and therefore belonging to the Late Elsterian. The lowermost interstadial (= Offleben I) shows the following succession: *Pinus-Betula-Picea*-Zone, *Pinus-Picea-Betula*-Zone, *Pinus-Betula*-Poaceae-Zone, *Alnus-Pinus*-Zone, *Pinus*-Poaceae-Zone. The second one = Offleben II) encompasses just two samples and the third one (= Esbeck) lacks of the initial phase of interstadial, but two pollen zones could be distinguished. First a *Pinus-Betula*-Zone followed by a *Pinus-Betula*-Poaceae-Zone (Urban et al., 1988). Urban et al. (1991) studied another profile close to Schöningen at Esbeck. There, a nearly complete succession of an interstadial and the late phase of the preceding cold period is witnessed: Poaceae-Cyperaceae-*Artemisia*-Zone, *Betula-Artemisia*-NAP-Zone, Poaceae-NAP-*Betula*-Zone, *Pinus-Betula-Picea*-Zone, *Pinus*-Cyperaceae-NAP-Zone. Urban et al. (1991) correlated this pollen sequence with the Esbeck interstadial found in an outcrop of the opencast mine Schöningen. This correlation and the exact stratigraphic position of this interstadial vegetational succession are questionable (Litt et al., 2007).

At Surheide, in northwestern Germany, an interglacial was investigated by Behre (2004), which could not be clearly correlated with one of the known interglacials in Germany from the Middle Pleistocene. Based on the stratigraphic position, however, the pollen recrod corresponds to an Intra-Saalian or Elsterian warm stage.

After the ice retreat of the Elsterian glaciation, the Holstein Interglacial started. This interglacial is well-studied and known from many sites in northern (e.g., Hallik, 1960; Koutsodendris et al., 2010; Linke and Hallik, 1993; Meyer, 1974; Müller, 1974a; Müller and Höfle, 1994), eastern (e.g., Erd,

1970; Erd et al., 1987; Eißmann et al., 1995) and southern Germany (Grüger, 1983; Hahne et al., 2012). Last, but not least Diehl and Sirocko (2007) published a pollen record, which filled the gap between northern and southern Germany. Several regional zonations exist (e.g., Erd, 1970; Grüger, 1983; Müller, 1974a) reflecting the vegetation gradient from the south to the north and from the west to the east. The characteristic vegetational succession of the Holstein thermomere is simplified as follows: A *Pinus-Betula*-phase, in which already *Picea* and *Alnus* occurred, preceded a *Pinus-Alnus*-phase, in which thermophilous trees (*Quercus*, *Ulmus*, *Tilia* and *Fraxinus*) slightly spread and *Corylus* and *Taxus* were relatively frequent, too. Subsequently, *Abies* and *Carpinus* became more important. In southern Germany forests mainly consisting of *Abies* with an understory rich in *Buxus* became even dominant (Fig. 5). There, a distinctive phase with *Fagus* and *Pterocarya* followed. However, the Holsteinian record from Samerberg is incomplete, but has the same vegetational development as the almost complete pollen record from Thalgut in Switzerland investigated by Drescher-Schneider (2000a). In contrast, *Buxus*, *Fagus*, and *Pterocarya* played only a subordinate role in the vegetation composition of the northern German lowlands (Fig. 6). Additionally, *Azolla* and *Celtis* were present in northern Germany, too. The occurrence of *Pterocarya* in conjunction with the forest succession is assumed to be of special stratigraphic importance for the Holstein Interglacial. Another *Pinus-Betula*-phase with some *Larix* and increased NAP values indicates the climatic deterioration towards the end of this warm stage. According to varve counts from diatomites at the sites Hetendorf (Meyer, 1974) and Munster-Breloh (Müller, 1974a), the Holsteinian had probably a duration of 15.000 to 16.000 years, if both datasets are combined. Actually, there is an ongoing debate about the age position of the Holsteinian. U/Th datings from the section Bossel close to Hamburg within the type area of the Holsteinian (Jerz and Linke, 1987) yielded ages of 330–310 ka BP and was therefore correlated with MIS 9 (Geyh and Müller, 2005). In contrast, new luminescence datings place the Elsterian till around 400 ka and consequently the Holsteinian should be correlative to MIS 11 (Lauer and Weiss, 2018).

The transition from the Holsteinian to the next glaciation (= Saalian Complex) is marked by the demise of the last forests and the spread of a shrub tundra with *Betula* and *Juniperus* in the pollen record of Pritzwalk (Brandenburg, Erd, 1970). The shrub tundra is followed by a cold steppe consisting of Poaceae and many other heliophytes, like *Artemisia*, Chenopodiaceae, and *Thalictrum*, reflecting subarctic climatic conditions. Erd (1970) termed the beginning of the Saalian Complex Fuhne cold period. The Fuhne cold

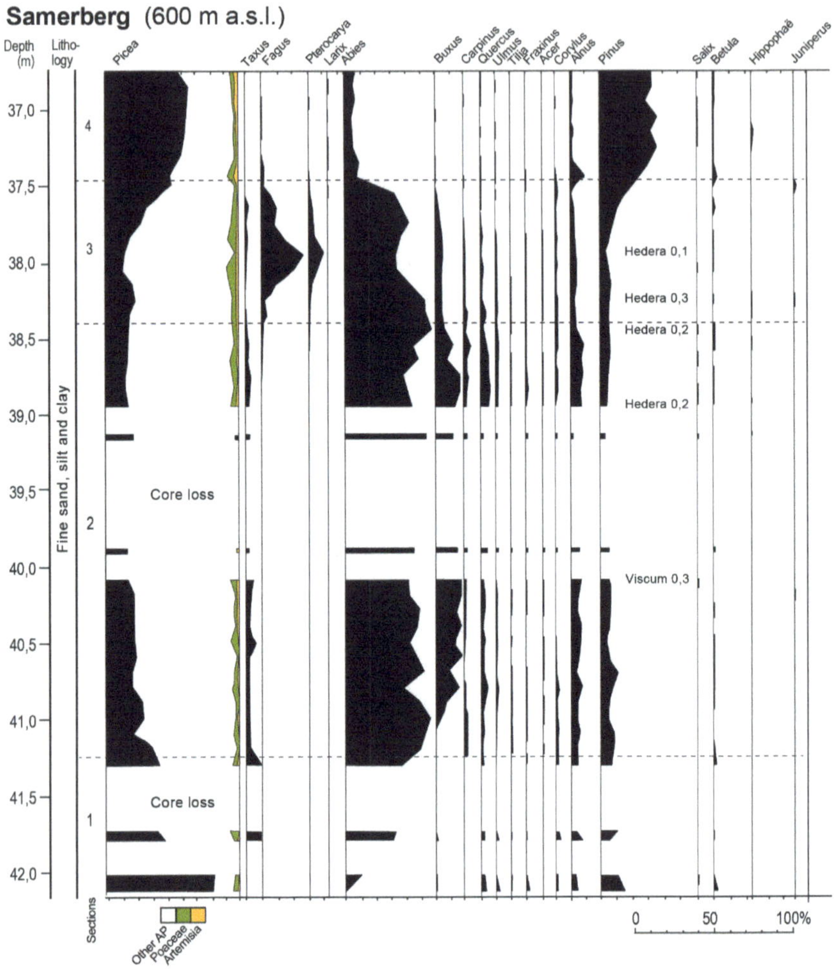

Fig. 5 Pollen diagram of the Holstein Interglacial in the boring Samerberg II. *Modified after Grüger, E., 1983. Untersuchungen zur Gliederung und Vegetationsgeschichte des Mittelpleistozäns am Samerberg in Oberbayern. Geologica Bavarica 84, 21–40.*

period was also detected by Strahl and Zwirner (2002) in northwestern Brandenburg and Urban et al. (1988) at the opencast mine Schöningen in superposition of Holsteinian beds. The Fuhne cold period can be subdivided into two zones separated by a short interstadial period called Pritzwalk Interstadial (e.g., Erd, 1973; Erd et al., 1987). This subdivision is probably also reensembled in a pollen record from northwestern Brandenburg (Strahl and Zwirner, 2002), where this interstadial is not completely represented. The characteristic succession of the Pritzwalk Interstadial reaches from a

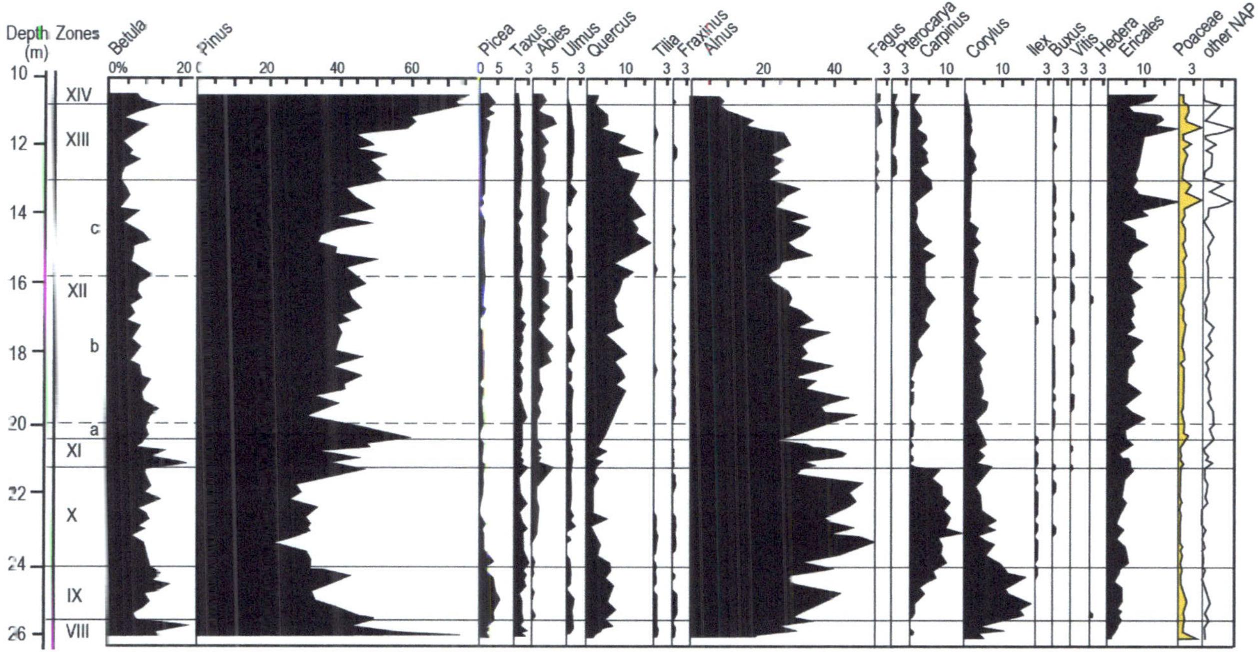

Fig. 6 Pollen diagram of the Holstein Interglacial in the boring KS 407/71 Munster-Breloh. *Modified after Müller, H., 1974a. Pollenanalytische Untersuchungen und Jahresschichtenzählungen an der holsteinzeitlichen Kieselgur von Munster-Breloh. Geol. Jb. A 21, 107–140.*

shrub phase with dominating *Juniperus* to an open forest with a *Betula* phase followed by a *Pinus* phase. Urban et al. (1991) described three interstadials in superposition of the Holstein Interglacial. The first interstadial (Mißaue Interstadial I) is characterized first by a phase rich in *Pinus* and subordinated *Picea* and *Betula* followed by a climatic depression with increasing *Artemisia* and Poaceae values. The second interstadial (Mißaue Interstadial II) is rich in *Pinus* with some *Betula*. Following Urban et al. (1991), this interstadial complex should be correlative to the Pritzwalk Interstadial. After a subsequent stadial with high values of Poaceae and *Artemisia* another interstadial, even though very weak, was found in the sequence of Schöningen (Urban et al., 1991). The pollen content of this interstadial just shows elevated *Pinus* values.

From southwestern Germany at Riedmatt, close to the German-Swiss borderline, a bi-partite interstadial of probably Rissian age was described by Bludau et al. (1994). The lower interstadial shows a boreal forest consisting of *Picea* and *Pinus* and some *Betula*. Afterwards, the forest stands were partly displaced by steppe vegetation with Poaceae, *Artemisia*, *Thalictrum*, and many other (heliophile) herbs. Furthermore, cold-adapted aquatic plants spread. A pioneer phase with *Betula* and *Juniperus* followed introducing a reforestation which culminated in boreal forest stands with *Picea* and *Pinus* again. Compared to the previous interstadial, *Pinus* slightly played a more pronounced role in the forest composition.

In northern Germany the Dömnitz (syn. Wacken) warm stage is known, which followed the Fuhne cold period. It was studied by Erd (1970, 1973) in the pollen record of Pritzwalk and by Menke (1968a) in the pollen sequence of Wacken. Unfortunately, the records do not cover the entire forest succession, but they reflect interglacial conditions due to the relatively frequent occurrences of *Carpinus*, *Quercus*, *Taxus*, and *Corylus*. Of special importance is the presence of *A. filiculoides* in both records. The Schöningen Interglacial is also assumed to be time-equivalent (Urban et al., 1991). U/Th dating at the open cast mine with uncorrected ages between 227 and 180 ka placed the Schöningen warm stage in MIS 7 (Urban, 1995). IR-RF dating at the sequence of Wacken assigned the warm stage to MIS 7e (Krbetschek and Stephan, 2010; cited in Stephan, 2014). Zagwijn (1973) studied a possibly time-equivalent sequence (= Hoogeven warm stage) in the Netherlands, which shows some similarities in the vegetation development compared Dömnitz and Wacken, but also some differences.

More recently, another thermomere, namely the Leck warm stage, was detected in Schleswig-Holstein, northern Germany (Stephan et al., 2011; Urban et al., 2011). Unfortunately, the investigated pollen records do not represent the whole interglacial, which is characterized by the dominance of *Alnus* accompanied by Quercetum mixtum with *Quercus*, *Tilia*, *Ulmus*, and *Corylus* during the climatic optimum followed by boreal forests with *Pinus*, *Betula*, *Picea*, and some *Larix*. U/Th dating failed because of open system behavior (Stephan et al., 2011; Urban et al., 2011). Nevertheless, this warm stage is tentatively assigned to MIS 7c (Stephan et al., 2011; Urban et al., 2011) due to the similar vegetational development as revealed in the pollen records of Nachtigall (Kleinmann et al., 2011) close to Höxter in North Rhine-Westphalia and Otto-Straße (Grüger, 1996; Grüger et al., 1994), urban area of Göttingen, in southern Lower Saxony. At Nachtigall U/Th dating yielded ages ranging from 227 + 9/−8 to 201 + 15/−13 ka pointing to MIS 7 (Kleinmann et al., 2011; Waas et al., 2011).

Concerning southern Germany, just a few records exist resembling the vegetation history of the Rissian stage. From the German Alpine foreland, close to Rosenheim, the outcrop of Höhenmoos revealed two sections with interglacial character separated by a cool phase (Herz et al., 2014). In the lower interglacial, the pollen assemblages show a succession from forests with *Quercus* and *Ulmus* (and without *Carpinus*) to forests consisting of *Abies* and *Picea* followed by an opening of the vegetation. The upper warm stage is characterized by thermophilous forests with *Quercus*, *Ulmus*, *Fraxinus*, *Carpinus*, and *Corylus* and the presence of *Celtis* pollen as well. In the course of the thermomere, *Abies* became dominant accompanied with findings of *Buxus* and *Fagus*. Later on, pollen assemblages dominated by *Picea*, *Abies*, and *Pinus* reflect the end phase of the warm stage. First luminescence dating with ages of 215, 200, and 190 ka tentatively point to a MIS 7 age.

Most recently, a long pollen record from the Neualbenreuth maar, northeastern Bavaria, was published by Stebich et al. (2020). There, three pre-Eemian warm stages with the Eemian Interglacial in stratigraphic superposition were found. These warm stages preceding the Eemian were placed into MIS 7e, 7c, and 7a with intervening stadial periods assigned to MIS 7d and 7b. The warm stages are of weak interglacial character compared to the Eemian. The pollen assemblages show deciduous forests as climax vegetation with *Quercus*, *Corylus* and delayed *Carpinus* together with conifer taxa consisting of *Picea* and *Pinus*.

From Switzerland, the long pollen record Meikirch is known, which was correlated by Welten (1982, 1988) with the Eemian and two different Holsteinian Interglacials. Based on luminescence dating and palynological comparisons with other profiles, Preusser et al. (2005) tentatively re-interpreted the Meikirch record to MIS 7.

According to luminescence dating, the Saalian glaciation, namely the Drenthe and the Warthe stage, took place in late MIS 6 (Lauer and Weiss, 2018; Lauer et al., 2020). The Saalian Late Glacial in northern Germany was first described by Menke and Ross (1967). According to their zonation, an unforested pollen zone A with dominating heliophilous taxa with lots of reworked pre-Quaternary elements can be separated from a following zone B with the presence of *Hippophaë* shrub stands and a subsequent zone C with shrub formations composed of *Betula*, *Juniperus*, and *Hippophaë*. Hermsdorf and Strahl (2008) refined this zonation as they subdivided zone C into a zone C1 corresponding to the zone C of Menke and Ross (1967) and a zone C2 with the occurrence of very open *Pinus* forests which is proved by macro remain findings at some sites.

During the Rissian Late Glacial a tundra vegetation prevailed in the German Alpine foreland mainly composed of heliophytic taxa, like Poaceae, Cyperaceae, *Artemisia*, Chenopodiaceae, *Helianthemum*, and *Selaginella*. Especially in the ending phase of the Late Glacial, shrub stands (*Ephedra*, *Hippophaë*, *Juniperus*, and *Salix*) grew at climatically favorable sites. At the sites Eurach (Jung, 1979) and Zeifen (Jung et al., 1972) the presence of *Betula nana* is witnessed by macrofossils.

6. Late Pleistocene

6.1 Eemian Interglacial

The forest development during the Eemian (syn. Riß/Würm Interglacial in southern Germany), lasting from c. 128,000 until 117,000 years before present and corresponding to MIS 5e, was rather uniform in northern Germany (Caspers et al., 2002). Many Eemian sites are known there (e.g., Behre and Lade, 1986; Caspers, 1997; Hahne et al., 1994b; Menke and Tynni, 1984; Müller, 1974b; Selle, 1962). There are two different labelings for the zonation of the Eemian in northwestern Germany (Table 2). Based on the zonation of Menke and Tynni (1984), the onset of the reforestation is marked with the spread of *Betula* (= Zone E I). Subsequently, *Pinus* reimmigrated and formed open woodlands together with *Betula* (= Zone E II). Due to the immigration of thermophilous deciduous trees (*Quercus* and *Ulmus* first) these pioneer woodlands were replaced resulting in mixed oak forests with accompanied *Pinus* (= Zone E III). During pollen zone E IVa, mixed oak forest rich in *Corylus* prevailed. Zone E IVb is also characterized by mixed oak forests, but with more frequent occurrences of *Tilia* and *Taxus* followed by the strong increase of *Carpinus* accompanied by a (slight)

Table 2 Zonation of the Eemian.

Pollen zones after Menke and Tynni (1984)	**Pollen zones after Selle (1962) and Müller (1974a, 1974b)**
VII	VIb
	VIa
VI	Vb
	Va
Vb	IVb
Va	IVa
IVb	IIIc
IVa	IIIb
	IIIa
III	IIb
II	IIa
I	I

In eastern Germany, another zonation exists (Erd, 1973; Hermsdorf and Strahl, 2008; Litt, 1994).

After Menke, B., Tynni, R., 1984. Das Eeminterglazial und das Weichselfrühglazial von Rederstall/Dithmarschen und ihre Bedeutung für die mitteleuropäische Jungpleistozän-Gliederung. Geol. Jb. A 76, 3–120 and Selle, W., 1962. Geologische und vegetationskundliche Untersuchungen an einigen wichtigen. Vorkommen des letzten Interglazials in Nordwestdeutschland. Geol. Jb. 79, 295–352.

spread of *Picea* (= Zone Va and Vb). The pollen zones E IVb, Va and Vb mark the climatic optimum of this interglacial (Caspers et al., 2002). In the course of a first climatic deterioration, *Carpinus* decreased and coniferous trees (*Pinus*, *Picea*, and some *Abies*) expanded during zone E VI. Zone E VII shows *Pinus* dominance with some *Betula* and *Picea*. At the end of the Eemian, these boreal forests demised. Additionally, a widespread soil acidification took place causing increased presence of *Calluna* and other acid-tolerant plants (Behre, 1989). In order to make the vegetational succession described above more clear, the reader is refered to Fig. 7. The duration of the Eemian, mentioned above, is in good agreement with the palynogical definition of this interglacial. Varves counts from Müller (1974b), Hahne et al. (1994b), and Caspers (1997), although partly incomplete and complemented by estimates of palynological and sedimentological means (Caspers et al., 2002), suggest a duration of 10,000–11,000 years. In the German

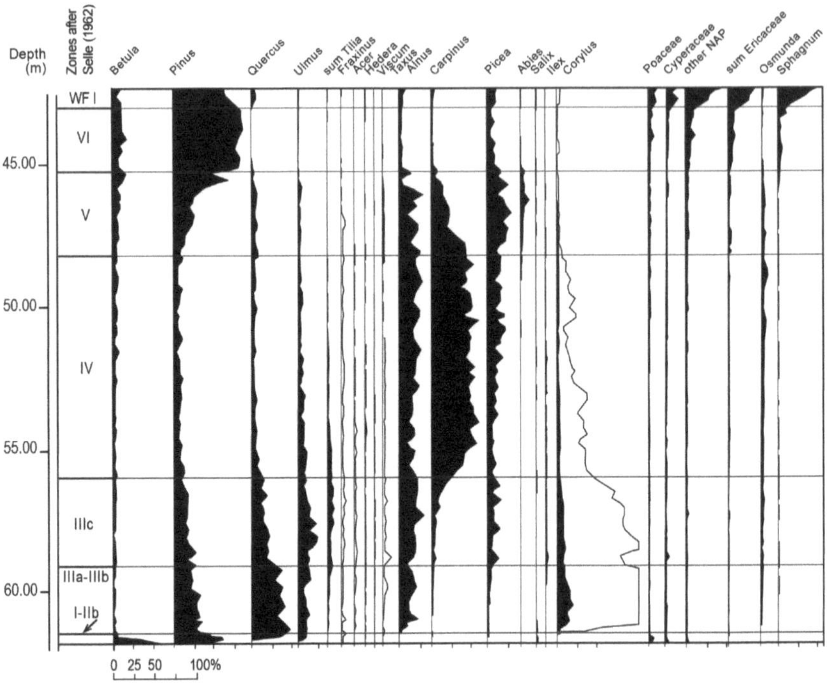

Fig. 7 Pollen diagram of the Eem Interglacial and the beginning of the Early Weichselian (= WF I) in the boring Quakenbrück GE 2. *Modified after Hahne, J., Kemle, S., Merkt, J., Meyer, K.-D., 1994b. Eem-, weichsel- und saalezeitliche Ablagerungen der Bohrung "Quakenbrück GE 2". Geol. Jb. A 134, 9–69.*

Alpine foreland, the duration of the terminal zone with *Pinus* and *Picea* was recorded from annually laminated sediments lasting at least 1650 years (Frenzel and Bludau, 1987).

In southern Germany, the reforestation at the beginning of the Riß/Würm-Interglacial resulted in open forests consisting of the pioneer trees *Betula* and *Pinus sylvestris*. The presence of these trees is witnessed by macro remains found, e.g., at Samerberg (Grüger, 1979a) and Zeifen (Jung et al., 1972). In the open forests several thermophilous trees immigrated when the climate became warmer. The following forest succession is well represented at Samerberg (Grüger, 1979a, 1979b; Fig. 8). First *Ulmus* re-immigrated into the Alpine foreland and spread, depending on the site conditions, in different extent. More or less at the same time, *Picea abies* immigrated together with *Ulmus* and also *Quercus*. Subsequently, *Quercus* expanded and became the dominating tree genus. Beside *Quercus*, other deciduous trees, like *Tilia*, *Fraxinus*, and *Ulmus*, grew in these forests. In the montane belt *P. abies*

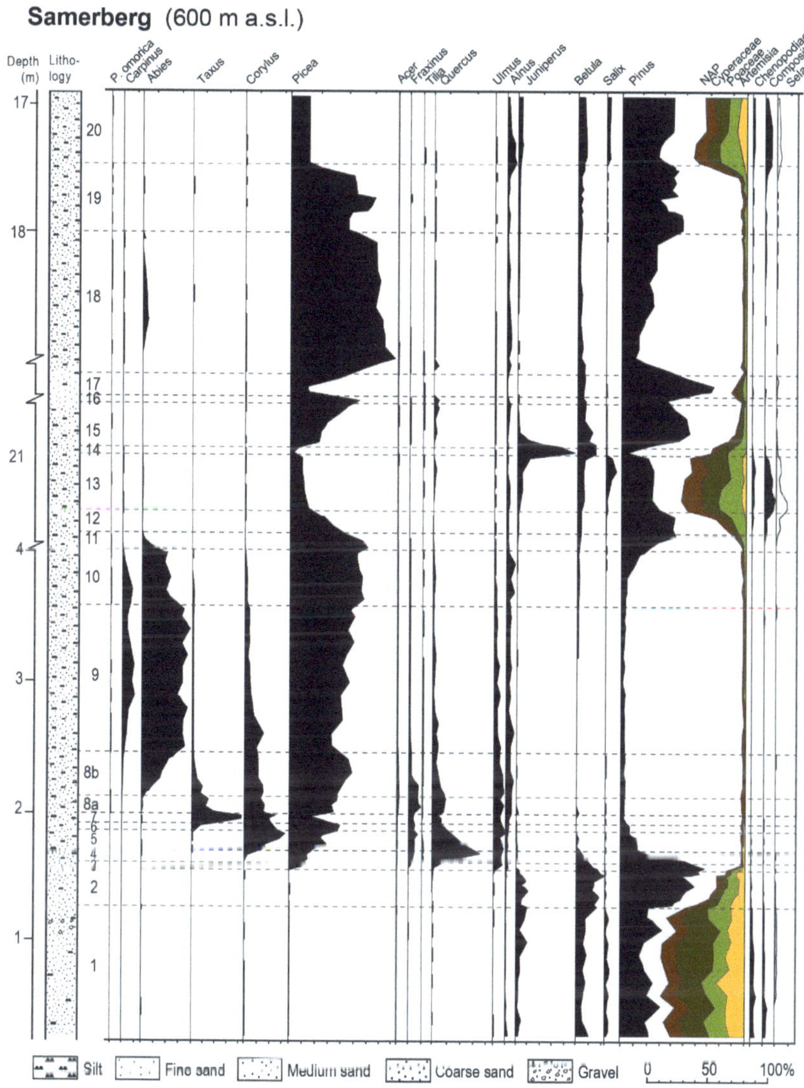

Fig. 8 Pollen diagram of the Eem Interglacial and part of the Early Würmian (first two stadials and first interstadial) in the outcrop and in the boring Samerberg I. *Modified after Grüger, E., 1979a. Spätriß, Riß/Würm und Frühwürm am Samerberg in Oberbayern - ein vegetationsgeschichtlicher Beitrag zur Gliederung des Jungpleistozäns. Geologica Bavarica 80, 5–64*

became important. Due to the expansion of thermophilous trees and *P. abies*, the pioneer trees *Betula* and *P. sylvestris* were outcompeted, but still present at poor soil sites. Afterwards *Corylus avellana* established in the understorey of the deciduous forests, followed by a significant expansion of *Taxus*.

Subsequently, the forest development can be assigned to two different types (Drescher-Schneider, 2000b), the 'Tieflandtyp'(lowland type) typically represented at Zeifen (Jung et al., 1972) and Krumbach (Frenzel, 1991) and the ‚Alpentyp' (alpine type) recorded at, for instance, Wurzacher Becken (Grüger and Schreiner, 1993), Pfefferbichl (Filzer, 1967), Eurach (Beug, 1979), Samerberg (Grüger, 1979a, 1979b), and Mondsee (Drescher-Schneider, 2000b, 2000c) in adjacent Austria. Regarding the Tieflandtyp (syn. Zeifen type), *Carpinus* became the dominant tree after succeeding an intercalated *Picea-Abies-Carpinus*-phase. At sites belonging to the Alpentyp (syn. Pfefferbichl type) *Carpinus* immigrated, but gained no further importance in the forest composition. The forests were dominated by *P. abies* and *Abies alba* there. The climatic deterioration towards the end of the Riß/Würm Interglacial led to more open forests, first still composed of *Picea* and *Abies*, later on just constisting of *Pinus* and *Picea*.

6.2 Early Weichselian and Early Würmian respectively

In northwestern Germany the first stadial, termed Herning Stadial (= MIS 5d), is characterized by a tundra vegetation (e.g., Behre and Lade, 1986) and can be divided into two subzones (e.g., Caspers, 1997; Caspers and Freund, 1997). The first subzone (= *Calluna* subzone) contains high *Calluna* values, whereas the second subzone (= Poaceae subzone) shows much less *Calluna* and dominating Poaceae together with maxima of, for instance, *Artemisia*, Chenopodiaceae, Plantaginaceae, and *Thalictrum*. The first stadial in the German Alpine foreland was, close to the Alps, treeless (Grüger, 1979a, 1979b). The open vegetation was mainly composed of Poaceae and Cyperaceae followed by *Artemisia*, but in much lesser frequency. Furthermore, other heliophytic elements, like Chenopodiaceae, *Helianthemum*, *Thalictrum*, *Botrychium*, and *Selaginella selaginoides* occurred with different proportions depending on the site conditions. At climatically favorable sites, patchy dwarf shrub stands were growing. In these stands *B. nana*, *Ephedra*, *Hippophaë rhamnoides*, *Juniperus*, and *Salix* were distributed. In regions more distant from the Alps, the occurrence of trees is assumed. According to Zagwijn (1989), a forest steppe existed in southern Germany. At Samerberg, located at the northern margin of the Alps, the first tree macro remains (*Betula* spec.) are witnessed around the end of the stadial (Grüger, 1979a).

The first interstadial, the so called Brørup Interstadial (= MIS 5c), is known from many sites in northern Germany (e.g., Averdieck, 1967,

1998; Behre and Lade, 1986; Caspers, 1997; Hahne et al., 1994b; Menke and Tynni, 1984). The vegetational succession started with a long *Betula* period culminating in open forests interrupted by a short climatic deteriotation (= Montaigu event). Afterwards, *Pinus*, *Larix*, and *Picea* immigrated during the *Betula* period leading to boreal forests dominated by conifers (= *Pinus* period) and accompanied by *Betula*. In most of the investigated profiles *Larix* spread before *Picea* (Averdieck, 1998; Behre, 1989; Behre and Lade, 1986) except of Oerel, where both trees expanded at the same time (Behre, 1989; Behre and Lade, 1986). The first interstadial in southern Germany, correlating with the Brørup in northern Germany, started with the spread of *Juniperus*, *Betula*, and usually *Pinus*. Subsequently, the pioneer woods were replaced by *Pinus* and *Picea* forming boreal forests. The Montaigu event, led to a reduction of *Picea* and the expansion of open vegetation. Additionally, *Pinus* benefited from this vegetational succession at favorable sites. As stated by Grüger (1989, 1995), the strongest vegetation response took place in the German Alpine foreland close to the Alps. With the onset of more favorable climatic conditions *Picea* became the dominant tree genus at sites above 600 m a.s.l. as well as at lower elevations close to the Alps. In the northern part of the Alpine foreland *Pinus* gained dominance (Stojakowits et al., 2016). In most of the pollen records studied so far *Abies* slightly spread for a short period of time. At climatically favorable sites even thermophilous trees were present, such as *Quercus* and *Carpinus*. The immigration of these trees took place from west to east as indicated in the pollen records (Müller and Sanchez Goñi, 2007), e.g., with a maximum of 16% at Füramoos (Müller et al., 2003) in the western part of the German Alpine foreland and 2% at Samerberg (Grüger, 1979a, 1979b) further in the east. Accentuating, this pattern is overlain by the increasing continentality towards the east. From the second half of the interstadial onwards, *Larix* played a minor role in the forest composition.

The second stadial in northern Germany (= Rederstall Stadial; MIS 5b) is charaterized by a tundra vegetation again. Compared to the Herning Stadial, this stadial revealed more elevated *Artemisia* values pointing to more continentality (Caspers and Freund, 2001). The second stadial led to the disappearence of forests close to the Alps. Following the interpretation of Welten (1981), timberline conditions were present at elevations between 300 and 700 m a.s.l. Compared to the first stadial, the higher proportions of *Artemisia* suggest a more continental climate during this stadial (Müller and Sanchez Goñi, 2007) as well as in northern Germany. Taxa like *Ephedra*, *Hippophaë rhamnoides*, *Juniperus*, and *Salix* formed dwarf shrub stands again. At Samerberg (Grüger, 1979a) macro remains of *Picea* (redeposited?), *Betula* spec. and *Betula humilis* were identified in the lower part of the stadial and much more frequent findings of *Pinus* (probably *Pinus mugo*), *Juniperus*, *B. nana*, *Salix* spec., and *Salix retusa* in the upper part.

The vegetation of the second interstadial in Northern Germany, termed Odderade (= MIS 5a), culminated in forests again (e.g., Behre, 1989; Behre and Lade, 1986). Reforestation began with a short *Betula* period followed by a long *Pinus* period. In contrast to the Brørup Interstadial, *Larix* and *Picea* spread in the course of the *Pinus* period with an expansion of *Picea* first (Averdieck, 1998; Behre, 1989; Behre and Lade, 1986) so that both interstadials can be easily distinguished, if they are more or less completely represented (Behre, 1989).

The start of the second interstadial in southern Germany is characterized by a pioneer forest phase with dominating *Betula* and *Pinus*. In the course of the ongoing warming within this interstadial, *Picea* spread. Depending on site conditions, differently composed boreal forests were growing in which *Larix* and presumably *Abies* were also present. At climatically favorable sites, thermophilous deciduous trees were present again, such as *Ulmus*, *Quercus*, *Carpinus*, and *Corylus*. These trees demand higher temperatures and re-occurred rather early indicating that their refugia in the preceding stadial were located not very far away, as indicated by, e.g., the pollen record of Sulzberg-Baden in northwestern Switzerland (Welten, 1981). The thermophilous trees already reached their maxima in the early part of this interstadial: 20% at Füramoos and 5% at Samerberg suggesting that the climatic optimum centered at this time (Müller and Sanchez Goñi, 2007). Additionally, a comparable vegetation gradient regarding the predominance of *Picea* and *Pinus* evolved. Close to the Alps, *Picea* prevailed, whereas *Pinus* was the dominant taxon in the northern part of the Alpine foreland (Stojakowits et al., 2016). In contrast to northern Germany, the occurrence of thermophilous trees point to a warmer character of the first two Würmian interstadials in the south (Behre, 1989). In the context of the pollen content, two taxa can be considerd as characteristic for the first two Würmian interstadials: on the one hand *Bruckenthalia spiculifolia* and on the other hand *Picea omorika*. The first named species was found several times in the Füramoos pollen record (Müller et al., 2003) as well as in two other yet unpublished pollen profiles from Hörlkofen, eastern Bavaria (Stojakowits et al., in prep.), and Blinder See, Baden-Württemberg (Knipping, pers. communication, 2016). This species is also known from sites in northern Germany, like Osterwanna, with both pollen and macrofossil findings (Behre, 1974) and from Oerel with discontinuous finds in the pollen content during the Brørup and Odderade and once during the Middle Weichselian interstadial Glinde. Furthermore, the cores of the boring Groß Todtshorn revealed pollen grains of *Bruckenthalia spiculifolia* belonging to the Brørup and Oderradde as well as macro remains during the Brørup (Caspers, 1997; Freund, 1997a). The *Picea omoricoides* pollen type has been found in several records of southern Germany, such as during both interstadials at Samerberg (Grüger, 1979a, 1979b), during the first interstadial at Pfefferbichl (Filzer, 1967), and at Hörlkofen (Stojakowits et al., in prep.). *P. omorika* was also registered during the first two interstadials in northern Germany (Behre et al., 2005). Regarding southern Germany, the fern *Osmunda* seems to be characteristic for the second interstadial. Continuous finds of its spores were reported from Mondsee (Drescher-Schneider, 2000b) and Füramoos (Müller, 2001; Müller et al., 2003). In the unpublished Hörlkofen pollen record and at Unterangerberg *Osmunda* is represented with comparativley high values peaking around 30% of the pollen content.

In order to make the vegetational succession described above more clear, the reader is referred to Figs. 8–10.

6.3 (Middle) Weichselian-Pleniglacial and Middle Würmian

The third stadial in northern Germany, termed Schalkholz Stadial after Menke and Tynni (1984), represents the onset of the Weichselian Pleniglacial (Behre, 1989; Behre and Lade, 1986). Just a few sites revealed pollen assemblages from this period. At Groß Todtshorn (Caspers, 1997), for instance, the beginning of the stadial is represented by a tundra vegetation with relatively frequent Poaceae and many other heliophytic taxa, like *Artemisia*, Caryophyllaceae, and *Thalictrum*, but also dwarf shrub stands composed of *B. nana*, *Juniperus*, *Salix*, *Calluna*, and *Empetrum* grew in the surroundings. Macrofossils of *B. nana* and *Salix* point to their local presence (Freund, 1997a). In the boring of Quakenbrück (Hahne et al., 1994b; Fig. 10), the Schalkholz Stadial was also found reflecting tundra conditions as well. The third stadial, which marks the end of the Early Würmian and the start of the Middle Würmian (Chaline and Jerz, 1984), had a very cool climate (= first cooling maximum within the Würmian) leading to a widspread deforestation in the Alpine foreland which can be seen in many pollen records, like Jammertal (Müller, 2000, 2001), Füramoos (Müller

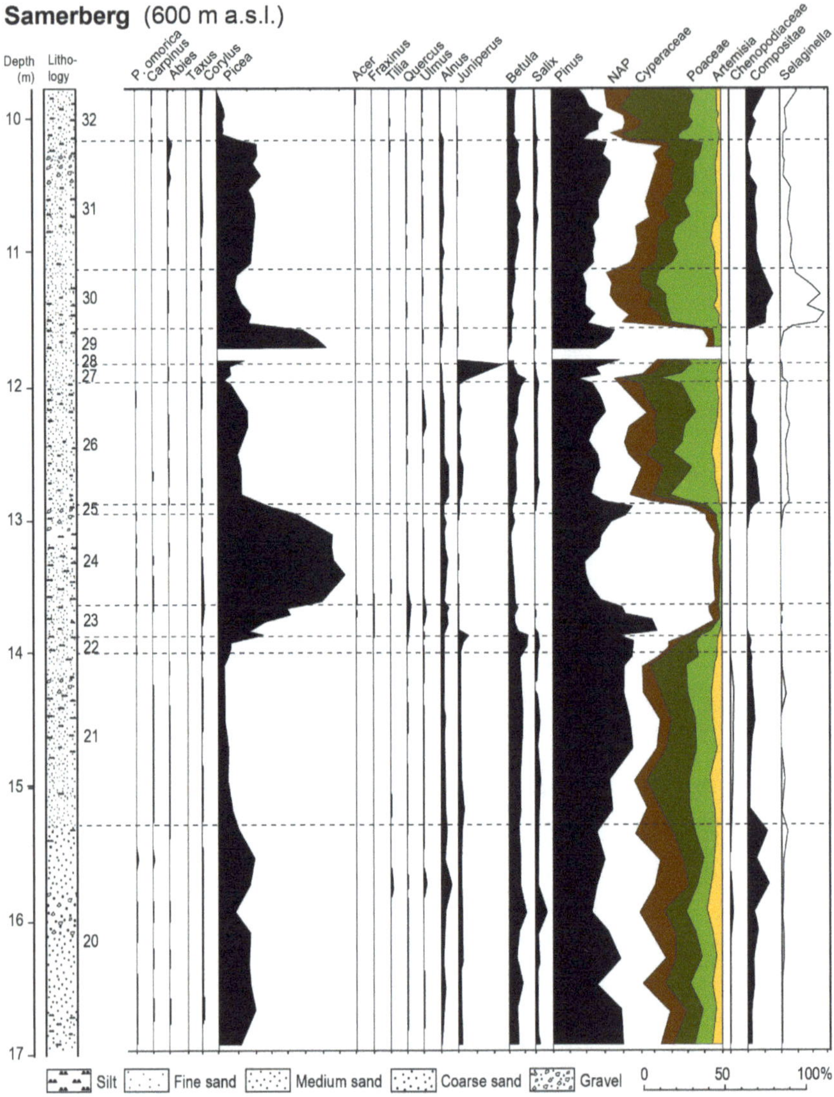

Fig. 9 Pollen diagram of the second stadial and interstadial as well of the third stadial and interstadial followed by the fourth stadial in the boring Samerberg I. *Modified after Grüger, E., 1979a. Spätriß, Riß/Würm und Frühwürm am Samerberg in Oberbayern - ein vegetationsgeschichtlicher Beitrag zur Gliederung des Jungpleistozäns. Geologica Bavarica 80, 5–64.*

et al., 2003), Hörlkofen (Stojakowits et al., in prep.), and Samerberg (Grüger, 1979a, 1979b; Fig. 9). According to the pollen and macro remain evidences at Samerberg (Grüger, 1979a), the vegetation was similar to the one of the second stadial.

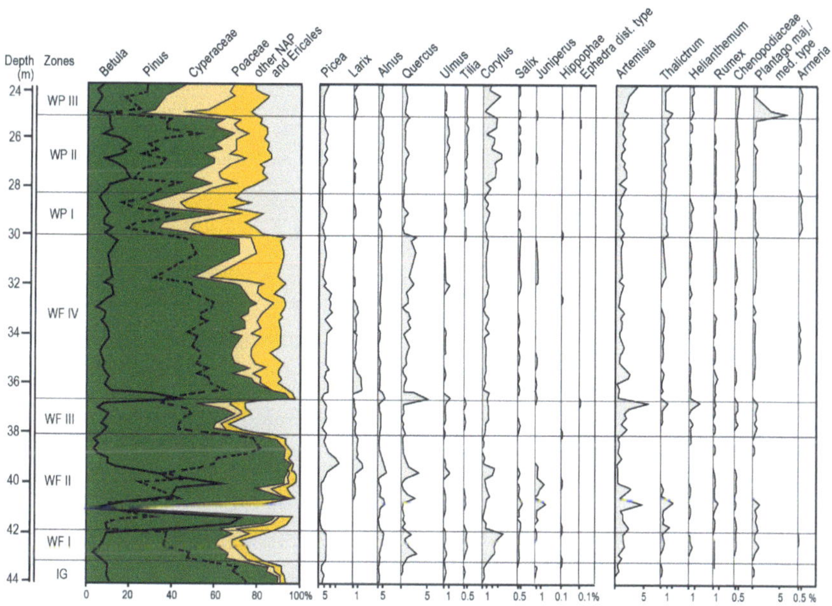

Fig. 10 Pollen diagram of the outgoing Eem Interglacial, the Early Weichselian (= WF I to WF IV), and the lower part of the Weichselian Pleniglacial (= WP I to III) in the boring Quakenbrück GE 2. *Modified after Hahne, J., Kemle, S., Merkt, J., Meyer, K.-D., 1994b. Eem-, weichsel- und saalezeitliche Ablagerungen der Bohrung "Quakenbrück GE 2". Geol. Jb. A 134, 9–69.*

The third interstadial in northern Germany, first found by Selle and Schneekloth (1965) at the site Oerel in superposition of the preceding Odderade Interstadial. It was termed Oerel Interstadial as a result of the site's detailed re-investigation by Behre and Lade (1986). Meanwhile, some other sites exist containing the Oerel Interstadial: Quakenbrück (Hahne et al., 1994b; Fig. 10), Groß Todtshorn (Caspers, 1997), and Neuhaus (Freund, 1997b). The pollen assemblages show a shrub tundra rich in Poaceae and other heliophytic taxa. Among the shrubs *Betula* ssp., *Juniperus*, *Salix*, *Calluna*, and *Empetrum* occurred. At Groß Todtshorn many macro remains of *B. nana* were found, but also seeds of *Juniperus communis* and one fruit scale belonging to *Betula humilis* (Freund, 1997a). At Oerel the presence of *B. nana* is also witnessed by macro remains (Behre, 1989; Behre and Lade, 1986) as well as the occurrence of *Empetrum nigrum* (Behre et al., 2005). Characteristic for the second part of the Oerel Interstadial are increasing pollen percentages of *Calluna* and *Empetrum* as well as elevated *Juniperus* proportions (Caspers and Freund, 2001). The likely equivalent third interstadial in southern Germany, termed Dürnten Interstadial

(Welten, 1981, 1982), began with a shrub phase mainly composed of *Betula*, *Hippophaë*, and *Juniperus*. The following expansion of *Pinus* and *Picea* resulted in an open reforestation (Fig. 9). In the western Alpine foreland, *Pinus* was dominant, in the eastern Alpine foreland *Picea* due to increasing continentality towards the east. *Larix* was relatively frequently present in these open forests.

The fourth stadial in northern Germany (= Ebersdorf-Stadial after Behre and Lade, 1986), represented in the pollen profiles of Groß Todtshorn (Caspers, 1997) and Quakenbrück (Hahne et al., 1994b; Fig. 10), is characterized by a tundra vegetation with dominating Cyperaceae, followed by Poaceae and many other heliophytes, such as *Armeria*, *Artemisia*, *Helianthemum*, Plantaginaceae, and *Thalictrum*. The onset of the fourth stadial led to a widespread deforestation in the Alpine foreland again (Fig. 10). The formerly existing open forest was substituted by formations rich in heliophytic taxa, like Poaceae, Chenopodiaceae, Cyperaceae, *Artemisia*, *Helianthemum*, and *Thalictrum*.

In northern Germany, a fourth interstadial located in superposition was found at Oerel and termed Glinde Interstadial (Behre and Lade, 1986; Fig. 11). As climax vegetation a shrub tundra with *Betula*, *Salix*, and *Empetrum* was reconstructed. Macrofossils of *B. nana* and *E. nigrum* confirm the pollen assemblages (Behre and Lade, 1986; Behre et al., 2005). The youngest available radiocarbon date has an age of $48{,}720 \pm 360$ BP (Behre and van der Pflicht, 1992). Nowadays this age can be re-calibrated with the IntCal20 curve (Reimer et al., 2020) resulting in an age of 52,700–50,120 cal. BP (re-calibrated with IntCal20; Reimer et al., 2020).

A possible time-equivalent of the Glinde was investigated in the opencast brown coal mine of Reichwalde, eastern Germany (Bohncke et al., 2008). There pollen and macro analyses revealed a low shrub tundra with *B. nana*, *Salix*, Poaceae, and heliophilous herbs.

Regarding younger time slices than the Glinde Interstadial, only short pollen sequences reflect the ongoing vegetation development in the further course of the Weichselian Pleniglacial. The majority of the investigated sites is located in the Netherlands, where three other interstadials of Pleni-Weichselian age were described. The oldest one is termed Moershoofd (Zagwijn, 1961; Zagwijn and Paepe, 1968). The pollen content points to a comparatively wet tundra vegetation with a striking predominance of Cyperaceae and little occurrences of dwarf shrubs (*Betula (nana)*, *Salix*, and *Ericales*) at climatically favorable sites (Teunissen and Teunissen-van Oorschot, 1974; van der Meer et al., 1984; Zagwijn, 1961).

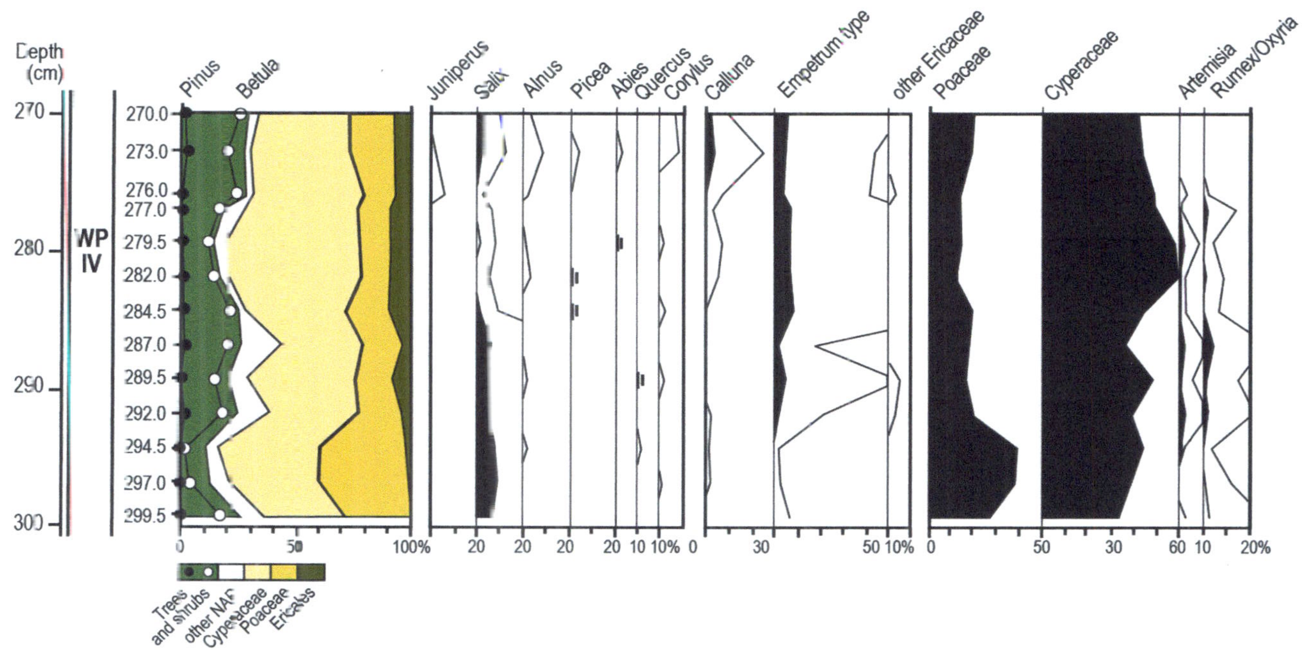

Fig. 11 Pollen diagram of the Glinde Interstadial from the Oerel sequence. *Modified from Behre, K.-E., Lade, U., 1986. Eine Folge von Eem und 4 Weichsel-Interstadialen in Oerel/Niedersachsen und ihr Vegetationsablauf. Eiszeitalter u. Gegenwart 36, 11–36.*

Radiocarbon datings with finite values yielded ages of 46,250 ± 1500 and 43,500 ± 1000 BP (van der Hammen et al., 1967; Zagwijn, 1961), 45,300 ± 2000 BP (Teunissen and Teunissen-van Oorschot, 1974), and 44,200 ± 1000 BP (van der Meer et al., 1984). According to van der Hammen (1971a, 1971b), Zagwijn (1974c), and Kolstrup and Wijmstra (1977), the pollen assemblages of the Moershoofd Complex contain lower *Betula* values with less pronounced peaks compared to the following Hengelo and Denekamp Interstadial. The Hengelo Interstadial was first described by van der Hammen et al. (1967) and further investigated by Zagwijn (1974c). The pollen content is dominated by Cyperaceae and other heliophytes, like Poaceae, *Artemisia*, Plantaginaceae, and *Thalictrum*. During the optimal phase of the Hengelo Interstadial, shrub stands developed consisting mainly of *Betula (nana)* as can be seen in a distinct pollen peak of the taxon (Zagwijn, 1974c), but also *Salix* and Ericales were present. Radiocarbon dates from localities assumed to be of Hengelo age centered around ca. 39,000 and 36,000 BP, e.g., van der Hammen et al. (1967) with 38,700 ± 400 BP, van der Hammen et al. (1967) and Vogel and Zagwijn (1967) with 37,500 ± 650, 37,000 ± 600, and 36,600 ± 600 BP, Kasse et al. (1995) with 37,750 ± 900, and Kolstrup and Wijmstra (1977) with 36,850 ± 620 and 38,420 ± 1400 BP. The youngest interstadial in the Weichselian interstadial described is the Denekamp Interstadial (van der Hammen (1971a, 1971b); van der Hammen et al., 1967) and characterized by tundra to shrub tundra vegetation dominated by Cyperaceae and relatively abundant Poaceae. *Betula* (*nana*) and *Salix* were also present forming patches of dwarf shrub stands (e.g., Kolstrup and Wijmstra, 1977). Radiocarbon dating yielded ages of 32,200 ± 500, 32,000 ± 400, 30,400 ± 450, 30,400 ± 450, 30,100 ± 300, 29,900 ± 460, 29,300 ± 300, and 28,860 ± 260 BP (van der Hammen et al., 1967). With exception of the Oerel Interstadial, however, the other interstadials described in the Middle Weichselian cannot be easily distinguished based on their pollen assemblages (e.g., Caspers and Freund, 2001). Additional radiocarbon data are needed in order to place these interstadials in the palyno- and chronostratigrapical framework. Nevertheless, although some records were sufficiently radiocarbon-dated, like Nochten in eastern Germany with more than 10 radiocarbon dates mainly from *Salix* twigs (Bos et al., 2001), the yielded radiocarbon ages do not fit to the age determinations from the Moershoofd complex, the Hengelo, and the Denekamp Interstadial. As stated by Caspers and Freund (1997), the dating techniques lack of a sufficient exact accuracy for detailed stratigraphical purposes, what especially holds true for old conventional radiocarbon dating methods.

The further palynostratigraphy of the Middle Würmian in southern Germany is unclear in detail, because most of the long pollen records end during the fourth Stadial, like Samerberg (Grüger, 1979a, 1979b). This also applies for adjacent parts of the Alpine foreland in Austria (e.g., Drescher-Schneider, 2000b) and Switzerland (Burga and Perret, 1998; Drescher-Schneider et al., 2007). The vegetation development in the ongoing Middle Würmian was presumably regionally differentiated as it can be derived from the situation during the first three Würmian interstadials. Several factors modified the vegetational succession, one of them being glacier advances. As an example, a Middle Würmian glacier advance in the Iller region was detected by Link and Preusser (2005), which remained just 10 km behind the moraines of the last glacial maximum (LGM). In the periglacial area of the Alpine foreland severe erosion on the one hand and sediment accumulation on the other hand (e.g., of Aeolian dust in the form of loess) took place. Such strong morphodynamic processes resulted in reworked pollen spectra in some records.

The Füramoos pollen record (Müller et al., 2003) revealed three interstadials in superposition of the Dürnten Interstadial. Based on radiocarbon dating, the lowermost interstadial should be correlative to the Glinde Interstadial introduced by Behre and Lade (1986). The pollen assemblages indicate an open shrub tundra with *B. nana*, *Juniperus*, and *Hippophaë*. However, tree species of *Betula* grew in the further surroundings at climatically favorable sites, too. The second interstadial could be time-equivalent to the Moershoofd Interstadial as indicated by radiocarbon dating with a similar climax vegetation distributed in the underlying interstadial. Close to the end of this pollen profile, another weak interstadial could be detected. Being aware of the uncertainties, Müller et al. (2003) did not correlate the detected interstadials with the chronology from northern Germany and the adjacent Netherlands. They just termed these three interstadials after a little village close to the Füramoos site (= Bellamont 1 to 3). In the re-investigation of the Füramoos site by Kern et al. (2022) (Fig. 12) a continuous record lasting from ca. 130,000 to 30,000 years before present showed that the two most prominent interstadials within MIS 3 correlate with the Greenland Interstadial 14 and 12. The climax vegetation culminated in open forests composed of *Betula*, *Pinus*, and some *Picea*.

The Nesseltalgraben record (Stojakowits et al., 2020), situated in the small, inner-Alpine Berchtesgaden basin at the northern rim of the Northern Calcareous Alps, revealed a pollen succession on top of a diamict covering the time span from approximately 60,000 to 30,000 years before present (Fig. 12). From around 59,000 to 52,000 cal. BP open forests prevailed at

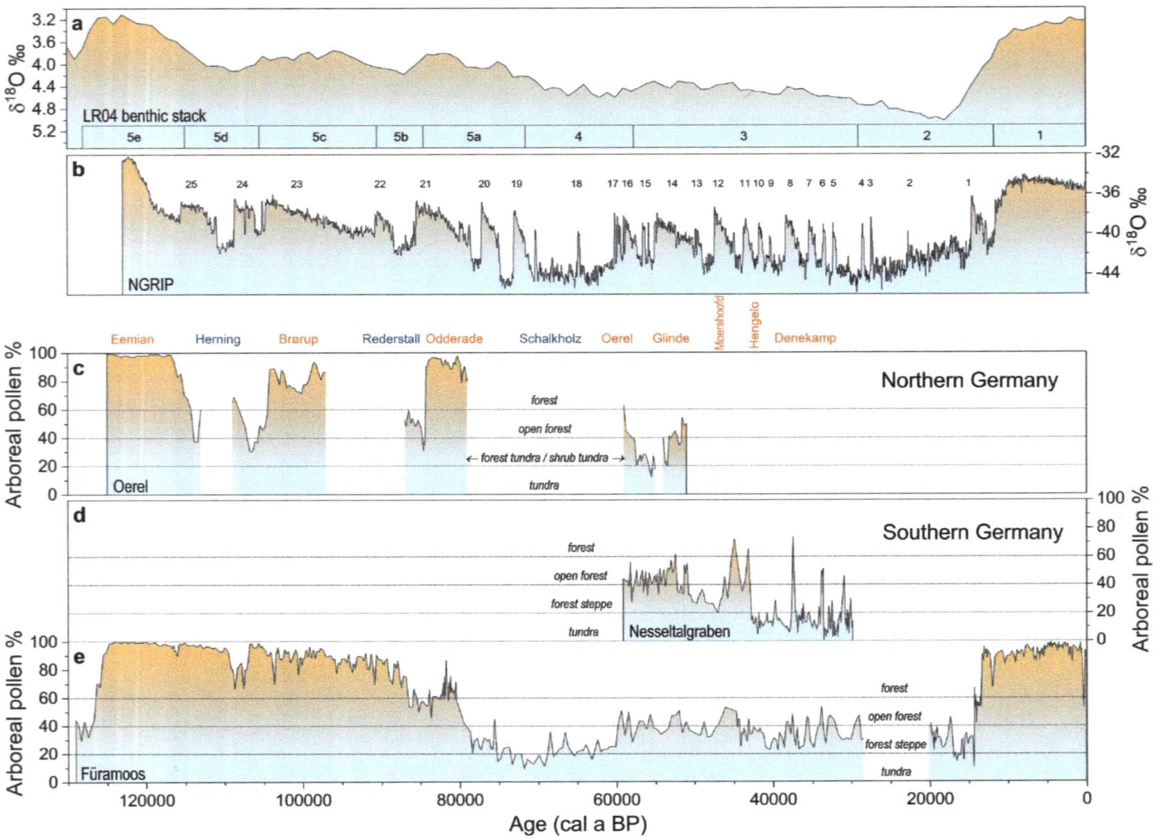

Fig. 12 See figure legend on opposite page.

climatically favorable sites in the valleys consisting of *P. sylvestris/mugo*, *Picea*, *Betula*, and some *Larix*, *Pinus cembra*, and *Alnus*. At higher altitudes tundra-like grasslands occurred. Subsequently, timberline oscillations took place repeatedly. At around 37,000 cal BP, *Pinus* stands substantially expanded followed by a spread of Alpine grasslands again. Around 34,000 cal. BP, *Pinus* and *Betula* slightly expanded at climatically favorable sites for the last time. Shortly after ca. 30,000 cal. BP, the onset of continuous glacifluvial sedimentation is dated (Mayr et al., 2019). Below the diamictic layer a short section is correlative to the first Early Würmian interstadial (Mayr et al., 2017).

From Breinetsried, a compressed peat was studied by Peschke (1983). Two radiocarbon dates with ages of 46,800–48,770 and 49,430–52,570 cal. BP (re-calibrated with IntCal20, Reimer et al., 2020) suggest that the peat was formed during the Moershoofd Interstadial. During the optimal phase of this interstadial, open forest patches grew at climatically favorable sites. The forest stands consisted primarily of *Pinus* accompanied by *Picea* and *Betula*. The presence of *Picea* is witnessed by wood findings incorporated in the compressed peat.

In the Wurzach basin (Grüger and Schreiner, 1993) another Middle Würmian interstadial was probably found. The pollen content showed a strong increase of *Picea* values and subordinated one of *Pinus*. Additionally, the percentages of *Betula* and *Hippophaë* are slightly elevated. However, a succession of different vegetation types to a forest vegetation is not visible.

From the Upper Rhine valley between Karlsruhe and Mannheim, an interstadial rich in *Pinus* with some *Picea* is known (Küttel et al., 1986).

Fig. 12 (A) LR04 benthic stack and MIS boundaries based on the dataset from Lisiecki and Raymo (2005), (B) Greenland ice core NGRIP (Rasmussen et al., 2014), (C) Oerel pollen record. In contrast to the original publication, Cyperaceae were excluded from the pollen sum. Pollen data were extracted from the European Pollen Database (EPD; http://www.europeanpollendatabase.net/) and the work of the data contributors and the EPD community is gratefully acknowledged. (D) Nesseltalgraben pollen record and (E) Füramoos pollen record. Pollen data were extracted from PANGAEA (Kern, 2021). *(C) After Behre, K.-E., Lade, U., 1986. Eine Folge von Eem und 4 Weichsel-Interstadialen in Oerel/Niedersachsen und ihr Vegetationsablauf. Eiszeitalter u. Gegenwart 36, 11–36. (D) After Stojakowits, P., Mayr, C., Lücke, A., Wissel, H., Hedenäs, L., Lempe, B., Friedmann, A., Diersche, V., 2020. Impact of climatic extremes on Alpine ecosystems during MIS 3. Quat. Sci. Rev. 239 (106) 333. (E) After Kern, O.A., Koutsodendris, A., Allstädt, F.J., Mächtle, B., Peteet, D.M., Kalaitzidis, S., Christanis, K., Pross, J., 2022. A near-continuous record of climate and ecosystem variability in Central Europe during the past 130 kyrs (Marine Isotope Stages 5–1) from Füramoos, southern Germany. Quat. Sci. Rev. 284 (107) 505.*

Radiocarbon dating from the top of the investigated peat section yielded an age of 35,640 ± 1000 BP confirmed by another date from molluscs in the overlying sand with 31,900 ± 1000 BP. This interstadial was assigned to be correlative to the Hengelo Interstadial in the Dutch palynostratigraphy. Close to Mannheim, fossil wood found in a gravel pit was dated (Löscher et al., 1980, 1983). Noteworthy is the occurrence of *A. alba* with a uncalibrated age of 31,270 ± 830 BP. Beside *A. alba*, wood of *P. sylvestris* and of one deciduous tree was dated resulting in ages of 30,860 ± 500 and 36,130 ± 1300 BP respectively. The latter two ages reflect a time-equivalent to the Denekamp-Interstadial in the Dutch palynostratigraphy. Another correlative of the Denekamp-Interstadial comes from the southern rim of the Franconian Alb with a radiocarbon age of 29,290 ± 160 (Brande, 1975, 1982). There, at the location Mauern, small patches of woodland (*Pinus* with *Betula*) existed at climatically favorable sites together with some shrubs in the surroundings, including *Juniperus* and *Salix*. The existence of trees is witnessed by a clear progressive succession from a *Salix* shrub stadium to a *Betula* stadium followed by a stable final phase with *P. sylvestris*-type and with single occurrences of *P. cembra*-type. The presence of *Betula* is testified by one fruit scale. Comparable pollen assemblages of this time were also found at Lake Bergsee, southern Black Forest (Duprat-Oualid et al., 2017).

The pollen analyses carried out so far on the Middle Würmian strata show a weak expansion of *Pinus*, *Betula*, and partly *Picea* in the two warmer phases after the Dürnten Interstadial. However, these taxa just formed open patches of woodland. Shrub stands consisted of *Juniperus*, *Hippophaë*, *Salix*, and *Betula*, which belonged to low and dwarf shrubs. The open conditions during these interstadials are witnessed, amongst others, by the occurrences of *Selaginella* and *Botrychium*. In contrast, the vegetation development of the time-equivalent Oerel Interstadial in northern Germany showed a shrub tundra as climax vegetation (Fig. 12). This also applies to the subsequent interstadials, like Glinde, pointing to a more favorable climate in the southern part of Germany during these warmer phases.

6.4 Pleniglacial and Late Würmian of MIS 2 age up to the last glacial maximum

In southern Germany, not much is known about the vegetation development during the Late Würmian before the LGM. The same also applies to the rest of Germany. Sensu Lang (1994) a 'steppe tundra' occurred. Close to the village Bellenberg in the northern Alpine foreland, an organic-rich section from a clay pit was studied by Peschke (published in Stojakowits et al., 2021) which revealed an open, tree-less tundra composed of Poaceae, Cyperaceae, *Artemisia*, *Helianthemum*, *Thalictrum*, Brassicaceae, and Caryophyllaceae. Furthermore, *Armeria* and other heliopytic taxa were present as well as likely some dwarf shrubs, such as *Salix* and *Ephedra*. The radiocarbon dates from this section are of early MIS 2 age. According to Lang (1994), Chenopodiaceae were frequent during early MIS 2, too. From the Bergsee record (Duprat-Oualid et al., 2017), similar pollen assemblages were found representing a grass tundra. In eastern Germany, the Nochten record contained plant assemblages which occurred shortly before the LGM according to radiocarbon and luminescence dating. A tundra rich in Poaceae, Cyperaceae, and mosses was reconstructed, but also some dwarf shrubs, like *Salix*, grew in the surroundings (Bos et al., 2001).

6.5 Late Glacial

The start of the Weichselian Late Glacial is reflected by a rise in the *Artemisia* values indicating a first warming phase and a replacement of the former distributed steppe tundra, given the name Meiendorf Interstadial (Menke, 1968b). Soon after, the *Salix* abundances increased, too, and the first presence of *Hippophaë* seems likely (Menke, 1968b) as well as a slight spread of *Juniperus* and *Betula* was registered (Litt and Stebich, 1999; Litt et al., 2001; Merkt and Müller, 1999). Towards the end of the Meiendorf Interstadial, the shrub vegetation diminished. Subsequently, with the onset of the Oldest Dryas a cold steppe tundra returned again. Heliophytes became more important. During the so-called Bølling Interstadial, a first substantial reforestation took place. Based on the pollen assemblages (e.g., Litt et al., 2001; Merkt and Müller, 1999), an open *Betula* forest developed preceded by a short shrub phase with lots of *Juniperus*. The presence of trees is witnessed by macrofossils, like in the Hämelsee record with many fruits and bud scales originated from *Betula pubescens*. At the end of the Bølling the woodlands composed of *Betula* trees declined. Throughout the Older Dryas very open Betula forests prevailed (Litt et al., 2001). Compared to the preceding Bølling, NAP gained more importance, especially Poaceae and *Artemisia*. Subsequently, the *Betula* forests became denser in the next interstadial (= Allerød). Later on, *Pinus* spread at the expense of *Betula* leading to a *Betula-Pinus* woodland. In the last third of the Allerød the Laacher

See Tephra with an age of 12,880 varve years was deposited (Leroy et al., 2000; Litt et al., 2001). The climatic deterioration of the Younger Dryas caused a strong rise of NAP (especially Poaceae and Cyperaceae as well as *Artemisia* and *Helianthemum*) resulting in the increase of open habitats and a severe thinning or locally even disappearance of the forests (e.g., Behre, 1967; Litt et al., 2001; Merkt and Müller, 1999). The Late Glacial division, obtained from the Eifel region, can also be applied to other regions of the northern German lowlands (e.g., Strahl, 2005).

The palynostratigraphically defined boundary between the Late Glacial and the Holocene was dated with varve counts of annually laminated lake sediments. In the Meerfelder Maar, Eifel region, the Younger Dryas–Preboreal boundary was dated to 11,590 varve years (Litt and Stebich, 1999; Litt et al., 2001). Composite varved sections from Belauer See, Plußsee, and Hämelsee in northern Germany placed the termination of the Younger Dryas at 11,560 varve years (Merkt and Müller, 1999). These dates are in good accordance with the boundary defined by dendrochronology in southern Germany with an age of 11,570 cal. BP (Friedrich et al., 2004).

After the retreat of the Würmian piedmont glaciation, a pioneer vegetation with little cover consisting of grasses and herbs established in the earliest phase of the Oldest Dryas in the German Alpine foreland. Beside Poaceae and Cyperaceae, other heliophytic taxa were present in this tundra-like vegetation, such as *Artemisia*, Chenopodiaceae, *Gypsophila repens*, *Helianthemum*, and *Thalictrum*. Depending on the site conditions, species-rich meadows and pioneer communities with predominantly tundra and steppe elements grew in the subsequent *Helianthemum* subzone. Furthermore, first single dwarf shrubs occurred at climatically favorable sites as suggested by macro remains of *B. nana* and *Salix* found at Simssee (Beug, 1976) and of *B. nana* at Schleinsee (Mielke and Müller, 1981). During the so called *B. nana* subzone, which corresponds to the youngest part of the Oldest Dryas (Ammann et al., 1994), dwarf shrub stands slightly expanded witnessed by macro remains at several sites, like Kulzer Moos in eastern Bavaria (Schmeidl, 1969), Lauterer Filze and Simssee in the eastern Alpine foreland (Beug, 1976; Schmeidl, 1971) and Schleinsee (Mielke and Müller, 1981) in the western part of the Alpine foreland north of Lake Constance. Nevertheless, the vegetation mosaic was still dominated by tundra- and steppe-like plant communities. The Bølling Interstadial, the first substantial warm phase during the Late Glacial, began with a shrub phase rich in *Juniperus* and *Hippophaë*, first described by Bertsch (1961).

Other shrub species were also present, like *B. nana* und several *Salix* species. Subsequently, the reforestation with *Betula* (*B. pubescens* and *B. pendula*) took place. In the course of the Bølling *Populus* immigrated and gained a presumably strong distribution in alluvial areas. At the end of the Bølling *P. sylvestris* immigrated and first *Pinus* specimens grew in the open *Betula* forests. The immigration of the conifer is also witnessed by needles, e.g., at Schleinsee (Mielke and Müller, 1981), and stomata findings, e.g., at Degersee (Lang, 1952). The short climatic deterioration of the Older Dryas was just detected at a few sites due to the short duration of this climatic oscillation. The Allerød is characterized by the predominance of open *Pinus* forests more or less accompanied by *Betula*, depending on site conditions. At some sites, like Ammersee (Kleinmann, 1995) and Mehlblockmoos (Stojakowits, 2014) in the eastern Alpine foreland and Degersee and Schleinsee (Kleinmann et al., 2015; Mielke and Müller, 1981) in the western Alpine foreland as well in the Black Forest (e.g., Lang, 1952; Lotter and Hölzer, 1989), the Laacher See Tephra with an age of 12,880 varve years (Leroy et al., 2000; Litt et al., 2001) was detected. The climatic deterioration of the Younger Dryas led to a weak opening of the prevailing forests, which is, for instance, indicated by increased values of *Juniperus* and heliophytic taxa, especially *Artemisia*, *Helianthemum*, and Poaceae. Areas more distanced to the Alps and situated below 400 m a.s.l. just show very weak signals (e.g., Brande, 1975; Knipping, 2001; Kortfunke, 1992), whereas in the low mountain ranges of southern Germany a forest opening is also witnessed (e.g., Hahne, 1991; Knipping, 1989; Stalling, 1987). At higher altitudes the timberline was depressed, such as in the Black Forest (e.g., Lotter and Hölzer, 1994).

7. Holocene

The Holocene vegetation history including its timing and human impact is well known in detail. Altogether, more than (two) thousand sites were investigated so far since the beginning in the 1920s. Consequently, a lot of details could be discussed regarding, for instance, the delayed immigration of trees from south to north, the vegetation gradient from south to north and west to east, different times for the start of human impact, and so on. However, this would exceed the scope of this book chapter which is intended to give an overview over the entire Quaternary palynostratigraphy in Germany. Therefore, the reader is referred to some selected articles and monographs from different sites and altitudes all over Germany (Table 3).

Table 3 Overview of some selected articles and monographs from different sites and altitudes all over Germany.

Author	Region
Behre (1976), Caspers (1993), Chen (1988), Dörfler et al. (2012), Golombek (1980, 1990), Jahns (2000, 2001, 2007), Jahns et al. (2013), Kramm (1978), Nelle and Dörfler (2008), Pott (1984), and Pott and Hüppe (2001)	Northern German lowlands
Litt et al. (2009)	Eifel
Beug (2016) and Voigt et al. (2008)	Harz
Beug (1957), Enters et al. (2008), Hahne (1991, 1992), Knipping (1989), Lagies (2006), Nelle (2002), Schmeidl (1969), Stalling (1987), and van der Knaap et al. (2019)	Low mountain ranges of Bavaria
Hölzer and Hölzer (1987), Lang (2005), Rösch (1989), and Rösch (2000)	Black Forest
Clark et al. (1989), Eusterhues et al. (2002), Kleinmann et al. (2015), Lechterbeck and Rösch (2020), Rösch (1993), Rösch and Lechterbeck (2016), Rösch and Wick (2019), Rösch and Marinova (2020), Rösch et al. (2021a), and Ryabogina et al. (2021)	Western Alpine foreland
Kleinmann (1995), Kortfunke (1992), Küster (1988, 1995), Rausch (1975), Rösch (2020), Rösch et al. (2021b), Stojakowits (2014), and Voigt (1996).	Eastern Alpine foreland
Bludau (1985), Mayer (1966), and Stojakowits et al. (2019)	Bavarian Alps

8. Conclusions and outlook

Based on the palynostratigraphical overview the following conclusions can be drawn in general:

1. The Late Pleistocene vegetation history including palynostratigraphy is comparatively well-known except of the still inadequately reconstructed vegetation development between 50,000 and 30,000 cal. BP, i.e., the Weichselian Pleniglacial and Middle Würmian in northern and southern Germany, respectively. Only a few pollen records provide complete vegetational successions for this period. This also applies to the time period between 30,000 cal. BP and the start of the Late Glacial.
2. The palynostratigraphy of the Saalian and the Rissian is poorly understood so far. More and longer records need to be discovered and precisely dated with modern methods including some of the pollen profiles mentioned.

3. Unlike other Middle Pleistocene interglacials, the Holstein vegetation history is fairly well known.
4. For the Cromerian Complex and the Early Pleistocene, apart the type localities, almost no time-correlative records from other regions of Germany have been found. Furthermore, it is unclear whether all warm stages within the Cromerian Complex and the Early Pleistocene have yet been discovered.

In summary, future research is needed to accurately date regional type localities belonging to the Cromerian Complex and the Early Pleistocene in order to place them within the stratigraphic framework provided, e.g., by the Marine Isotope Stages. Compared to northern Germany, the palynostratigraphy of southern Germany is much more incomplete and requires more detailed reconstructions in the future. Both northern and southern Germany still lack details for the palynostratigraphy in the period between ca. 50,000 and 20,000 cal. BP. Moreover, there are considerable discrepancies in the subdivision of the Quaternary in southern Germany, e.g., between the states of Baden-Würtemberg and Bavaria, which could eventually be solved with palynostratgraphic methods in the future.

References

Aguirre, E., Pasini, G., 1985. The Pliocene-Pleistocene boundary. Episodes 8, 116–120.

Ammann, B., Lotter, A.F., Eicher, U., Gaillard, M.-J., Wohlfarth, B., Haeberli, W., Lister, G., Maisch, M., Niessen, F., Schlüchter, C., 1994. The Würmian Late-glacial in lowland Switzerland. J. Quat. Sci. 9, 119–125.

Andersen, S.T., 1965. Interglacialer og interstadialer i Danmarks kvartaer. Medd. Dansk Geol. Foren. 15, 486–506.

Averdieck, F.-R., 1967. Die Vegetationsentwicklung des Eem-Interglazials und der Frühwürm-Interstadiale von Odderade/Schleswig-Holstein. Fundamenta B2, 101–125.

Averdieck, F.-R., 1998. Palynologische Neuuntersuchung der Typlokalität Odderade. Meyniana 50, 25–45.

Averdieck, F.-R., von der Brelie, G., 1963. Neue Beiträge zur pollenanalytischen Untersuchung des Interglazials von Bilshausen (Unter-Eichsfeld). Geol. Jb. 80, 437–446.

Baas, J., 1932. Eine frühdiluviale Flora im Mainzer Becken. Z. Botanik 25, 289–371.

Behre, K.-E., 1967. The Late Glacial and early postglacial history of vegetation and climate in northwestern Germany. Rev. Palaeobotan. Palynol. 4, 149–161.

Behre, K.-E., 1974. Die Vegetation des Spätpleistozan von Osterwanna/Niedersachsen. Geol. Jb. A 18, 3–48.

Behre, K.-E., 1976. Pollenanalytische Untersuchungen zur Vegetations- und Siedlungsgeschichte bei Flögeln und im Ahlenmoor (Elb-Weser-Winkel). Probleme der Küstenforschung 11, 101–118.

Behre, K.-E., 1989. Biostratigraphy of the last glacial period in Europe. Quat. Sci. Rev. 8, 25–44.

Behre, K.-E., 2004. Das mittelpleistozäne Interglazial von Surheide. Eiszeitalter u. Gegenwart, 36–47.

Behre, K.-E., Lade, U., 1986. Eine Folge von Eem und 4 Weichsel-Interstadialen in Oerel/ Niedersachsen und ihr Vegetationsablauf. Eiszeitalter u. Gegenwart 36, 11–36.

Behre, K.-E., van der Pflicht, J., 1992. Towards an absolute chronology for the last glacial period in Europe: radiocarbon dates from Oerel, northern Germany. Veget. Hist. Archaeobot. 1, 111–117.

Behre, K.-E., Hölzer, A., Lemdahl, G., 2005. Botanical macro-remains and insects from the Eemian and Weichselian site of Oerel (northwest Germany) and their evidence for the history of climate. Veget. Hist. Archaeobot. 14, 31–53.

Bertsch, K., 1924. Paläobotanische Untersuchungen im Reichermoos. Jahresh. Ver. Vaterländ. Naturk. in Württ. 80, 1–19.

Bertsch, A., 1961. Untersuchungen zur spätglazialen Vegetationsgeschichte Südwestdeutschlands. Flora 151, 243–280.

Beug, H.-J., 1957. Untersuchungen zur spätglazialen und frühpostglazialen Vegetationsgeschichte einiger Mittelgebirge (Fichtelgebirge, Harz und Rhön). Flora 145, 167–211.

Beug, H.-J., 1976. Die spätglaziale und frühpostglaziale Vegetationsgeschichte im Gebiet des ehemaligen Rosenheimer Sees Oberbayern. Botanisches Jahrbuch für Systematik 95, 373–400.

Beug, H.-J., 1979. Vegetationsgeschichtlich-pollenanalytische Untersuchungen am Riß/Würm-Interglazial von Eurach am Starnberger See/Obb. Geologica Bavarica 80, 91–106.

Beug, H.-J., 2016. Die spät- und nacheiszeitliche Vegetationsentwicklung am Nordrand der niedersächsischen und hessischen Mittelgebirge (Harz bis Weser). Pfeil-Verlag, München, pp. 1–124.

Bibus, E., Bludau, W., Ellwanger, D., Fromm, K., Kösel, M., Schreiner, A., 1996. On pre-Würm glacial and interglacial deposits oft he Rhine Glacier (South German Alpine foreland, Upper Swabia, Baden-Württemberg). In: Turner, C. (Ed.), The Early Middle Pleistocene in Europe. Balkema, Rotterdam, pp. 195–204.

Bittmann, F., Müller, H., 1996. The Kärlich Interglacial site and its correlation with the Bilshausen sequence. In: Turner, C. (Ed.), The Early Middle Pleistocene in Europe. Balkema, Rotterdam, pp. 187–193.

Bludau, W., 1985. Zur Paläoökologie des Ammergebirges im Spät- und Postglazial. Schäuble, Rheinfelden.

Bludau, W., 1995. Altpleistozäne Warmzeiten im Alpenvorland und im Oberrheingraben? Ein Beitrag der Palynologie zum "Uhlenberg-Problem". Geologica Bavarica 99, 119–133.

Bludau, W., Groschopf, R., Schreiner, A., 1994. Ein Riß-Interstadial bei Riedmatt am Hochrhein. Jber. Mitt. oberrhein. geol. Ver. N.F. 76, 295–323.

Boenigk, W., von der Brelie, G., Brunnacker, K., Koči, A., Schlickum, W.R., Strauch, F., 1974. Zur Pliozän-Pleistozän-Grenze im Bereich der Ville (Niederrheinische Bucht). Newsl. Stratigr. 3, 219–241.

Boenigk, W., von der Brelie, G., Brunnacker, K., Kempf, E.K., Koči, A., Schirmer, W., Stadler, G., Streit, R., Tobien, H., 1977. Jungtertiär und Quartär im Horloff-Graben/Vogelsberg. Geol. Abh. Hessen 75, 1–80.

Bohncke, S.J.P., Bos, J.A.A., Engels, S., Heiri, O., Kasse, C., 2008. Rapid climatic events as recorded in Middle Weichselian thermokarst lake sediments. Quat. Sci. Rev. 27, 162–174.

Bos, J.A.A., Bohncke, S.J.P., Kasse, C., Vandenberghe, J., 2001. Vegetation and Climate during the Weichselian Early Glacial and Pleniglacial in the Niederlausitz, eastern Germany—macrofossil and pollen evidence. J. Quat. Sci. 16, 269–289.

Brande, A., 1975. Vegetationsgeschichtliche und pollenstratigraphische Untersuchungen zum Paläolithikum von Mauern und Meilenhofen (Fränkische Alb). Quartär 26, 73–106.

Brande, A., 1982. Vegetation and landscape changes at the paleolithic site of Mauern (Bavaria). Early Man News 5/6. Newsletter for Human Paleoecology, 3–5.

Brunnacker, K., Boenigk, W., Koči, A., Tillmanns, W., 1976. Die Matuyama/Brunhes-Grenze am Rhein und an der Donau. N. Jb. Geol. Paläont. (Abh.) 151, 358–378.

Burga, C., Perret, R., 1998. Vegetation und Klima der Schweiz seit dem jüngeren Eiszeitalter. Ott, Thun.

Caspers, G., 1993. Vegetationsgeschichtliche Untersuchungen zur Flußauenentwicklung an der Mittelweser im Spätglazial und Holozän. Abhandlungen aus dem Westfälischen Museum für Naturkunde 55, 3–101.

Caspers, G., 1997. Die eem- und weichselzeitliche Hohlform von Gros Todtshorn (Kr. Harburg; Niedersachsen) – Geologische und palynologische Untersuchungen zu Vegetations- und Klimaverlauf der letzten Kaltzeit. Schriftenreihe der Deutschen Geologischen Gesellschaft 4, 7–59.

Caspers, G., Freund, H., 1997. Die Vegetations- und Klimaentwicklung des Weichsel-Früh- und -Hochglazials im nördlichen Mitteleuropa. Schriftenreihe der Deutschen Geologischen Gesellschaft 4, 201–249.

Caspers, G., Freund, H., 2001. Vegetation and climate in the Early- and Pleni-Weichselian in northern central Europe. J. Quat. Sci. 16, 31–48.

Caspers, G., Jordan, H., Merkt, J., Meyer, K.-D., Müller, H., Streif, H., 1995. III. Niedersachsen. In: Benda, L. (Ed.), Das Quartär Deutschlands. Schweizerbart, Stuttgart, pp. 23–58.

Caspers, G., Merkt, J., Müller, H., 2002. The Eemian Interglaciation in Northwestern Germany. Quat. Res. 58, 49–52.

Chaline, J., Jerz, H., 1984. Arbeitsergebnisse der Subkommission für Europäische Quartärstratigraphie – Stratotypen des Würm-Glazials (Berichte der SEQS 6). Eiszeitalter u. Gegenwart 34, 185–206.

Chanda, S., 1962. Untersuchungen zur pliozänen und pleistozänen Floren- und Vegetationsgeschichte im Leinetal und im südwestlichen Harzvorland (Untereichsfeld). Geol. Jb. 79, 783–844.

Chen, S.-H., 1988. Neue Untersuchungen über die spät- und postglaziale Vegetationsgeschichte im Gebiet zwischen Harz und Leine (BRD). Flora 181, 147–177.

Clark, J.S., Merkt, J., Müller, H., 1989. Post-glacial fire, vegetation, and human history on the Northern Alpine Forelands, South-Western Germany. J. Ecol. 77, 897–925.

von der Brelie, G., 1955. Die pollenstratigraphische Gliederung des Pleistozäns in Nordwestdeutschland. 2. Die Pollenstratigraphie im jüngeren Pleistozän. Eiszeitalter u. Gegenwart 6, 25–38.

Diehl, M., 2007. Palynologie und Sedimentologie der Interglazialprofile Döttingen, Bonstorf, Munster und Bilshausen. PhD thesis, Johannes Gutenberg-Universität, Mainz, p. 310.

Diehl, M., Sirocko, F., 2007. A new Holsteinian pollen record from the dry maar at Döttingen (Eifel). In: Sirocko, F., Claussen, M., Sanchez Goñi, M.F., Litt, T. (Eds.), The Climate of Past Interglacials. Developments in Quaternary Science 7. Elsevier, Amsterdam, pp. 397–416.

Doppler, G., Kroemer, E., Rögner, K., Wallner, J., Jerz, H., Grottenthaler, W., 2011. Quaternary stratigraphy of Southern Bavaria. E&G Quat. Sci. (Eiszeitalter u. Gegenwart) 60, 329–365.

Dörfler, W., Feeser, I., van den Boogard, C., Dreibrodt, S., Erlenkeuser, H., Kleinmann, A., Merkt, J., Wiethold, J., 2012. A high-quality annually laminated sequence from Lake Belau, Northern Germany: revised chronology and its implications for palynological and tephrochronological studies. Holocene 22, 1413–1426.

Drescher-Schneider, R., 2000a. 2. Halt: Kiesgrube Thalgut.—Pollen- und grossrestanalytische Untersuchungen. In: Kelly, M., Linde, U., Schlüchter, C. (Eds.), DEUQUA 2000—Exkursionsführer. Bern, pp. 128–136.
Drescher-Schneider, R., 2000b. The Riss-Würm interglacial from West to East in the Alps: an overview of the vegetational succession and climatic development. Geologie en Mijnbouw/Netherlands J. Geosci. 79, 233–239.
Drescher-Schneider, R., 2000c. Die Vegetations- und Klimaentwicklung im Riß/Würm-Interglazial und im Früh- und Mittelwürm in der Umgebung von Mondsee. Ergebnisse der pollenanalytischen Untersuchungen. Mitt. Komm. Quartärforsch. 12, 39–92.
Drescher-Schneider, R., Jacquat, C., Schoch, M., 2007. Palaeobotanical investigations at the mammoth site of Niederweningen (Kanton Zürich), Switzerland. Quat. Int. 164–165, 113–129.
Duprat-Oualid, F., Ruis, D., Bégeot, C., Magny, M., Millet, L., Wulf, S., Appelt, O., 2017. Vegetation response to abrupt climate changes in Western Europe from 45 to 14.7 k cal a BP: the Bergsee lacustrine record (Black Forest, Germany). J. Quat. Sci. 32, 1008–1021.
Eißmann, L., Litt, T., Wansa, S., 1995. Elsterian and Saalian deposits in their type area in central Germany. In: Ehlers, J., Kozarski, S., Gibbard, P. (Eds.), Glacial Deposits in North West Europe. Balkema, Rotterdam, pp. 439–464.
Enters, D., Dörfler, W., Zolitschka, B., 2008. Historical soil erosion and land-use change during the last two millennia recorded in lake sediments of Frickenhauser See, northern Bavaria, central Germany. Holocene 18, 243–254.
Erd, K., 1970. Pollenanalytical classification of the Middle Pleistocene in the German Democratic Republic. Palaeogeogr. Palaeoclimatol. Palaeoecol. 8, 129–145.
Erd, K., 1973. Pollenanalytische Gliederung des Pleistozäns der Deutschen Demokratischen Republik. Z. Geol. Wiss. 1, 1087–1103.
Erd, K., Palme, H., Präger, J., 1987. Holsteininterglaziale Ablagerungen von Rossendorf bei Dresden. Z. Geol. Wiss. 15, 281–295.
Eusterhues, K., Lechterbeck, J., Schneider, J., Wolf-Brozio, U., 2002. Late- and Post-Glacial evolution of Lake Steisslingen (I). Sedimentary history, palynological record and inorganic geochemical indicators. Palaeogeogr. Palaeoclimatol. Palaeoecol. 187, 341–371.
Filzer, P., 1967. Das Interglazial Riß-Würm vom Pfefferbichl bei Buching im Allgäu. Vorzeit 16, 9–24.
Filzer, P., German, R., 1968. Das Eem und die Würmeiszeit im Profil der wissenschaftlichen Kernbohrung Wurzacher Becken 1 (DFG). Heidelberger Geographische Arbeiten 20, 79–94.
Filzer, P., Scheuenpflug, L., 1970. Ein frühpleistozänes Pollenprofil aus dem nördlichen Alpenvorland. Eiszeitalter u. Gegenwart 21, 22–32.
Firbas, F., 1928. Über die Flora und das interglaziale Alter des Helgoländer Süßwassertöcks. Senckenbergiana 10, 185–195.
Firbas, F., 1949. Spät- und nacheiszeitliche Waldgeschichte Mitteleuropas nördlich der Alpen. Erster Band: Allgemeine Waldgeschichte. Fischer, Jena.
Firbas, F., 1952. Spät- und nacheiszeitliche Waldgeschichte Mitteleuropas nördlich der Alpen. Zweiter Band: Waldgeschichte der einzelnen Landschaften. Fischer, Jena.
Florschütz, F., 1938. Die beiden Azolla-Arten des niederländischen Pleistozäns. Rec. Trav. Bot. Néerl. 35, 932–945.
Frenzel, B., 1973a. 2. Some remarks on the Pleistocene vegetation. Eiszeitalter u. Gegenwart 23 (24), 281–292.
Frenzel, B., 1973b. 3. On the Pleistocene vegetation history. Eiszeitalter u. Gegenwart 23 (24), 321–332.
Frenzel, B., 1978. Das Problem der Riß/Würm-Warmzeit im deutschen Alpenvorland. In: Frenzel, B. (Ed.), Führer Exkurs. IGCP 73/1/24 "Quat. Glac. Northern Hemisphere". Stuttgart, pp. 103–114.

Frenzel, B., 1991. Über einen frühen letzteiszeitlichen Vorstoß des Rheingletschers in das deutsche Alpenvorland. In: Frenzel, B. (Ed.), Klimageschichtliche Probleme der letzten 130000 Jahre. Paläoklimaforschung 1. Fischer, Stuttgart, pp. 377–400.

Frenzel, B., Bludau, W., 1987. On the duration of the interglacial to glacial transition at the end of the Eemian Interglacial (deep sea stage 5 E): botanical and sedimentological evidence. In: Berger, W.H., Labeyrie, L.D. (Eds.), Abrupt Climate Change. Reidel, Dordrecht, pp. 151–162.

Freund, H., 1997a. Die Vegetation im Weichsel-Früh- und -Hochglazial von Groß Todtshorn (Kr. Harburg; Niedersachsen) anhand botanischer Makroreste. Schriftenreihe der Deutschen Geologischen Gesellschaft 4, 61–86.

Freund, H., 1997b. Eem- und weichselzeitliche Ablagerungen der Bohrung Neuenhaus-Veldhausen (Kr. Grafschaft Bentheim; Niedersachsen). Schriftenreihe der Deutschen Geologischen Gesellschaft 4, 137–155.

Friedrich, M., Remmele, S., Kromer, B., Hofmann, J., Spurk, M., Kaiser, K.F., Orcel, C., Küppers, M., 2004. The 12,460-year Hohenheim oak and pine tree-ring chronology from Central Europe — a unique annual record for radiocarbon calibration and paleoenvironment reconstructions. Radiocarbon 46, 1111–1122.

Fromm, K., 1994. Paläomagnetische Bestimmungen an Quartärsedimenten in Nordwest-Deutschland. Geol. Jb. A 134, 229–257.

German, R., Dehm, R., Ernst, W., Filzer, P., Käss, W., Müller, G., Witt, W., 1965. Ergebnisse der wissenschaftlichen Kern-Bohrung Ur-Federsee 1. Oberrhein. Geol. Abh. 14, 97–139.

German, R., Borneff, J., Brunnacker, K., Dehm, R., Filzer, P., Käss, W., Kunte, H., Müller, G., Witt, W., 1967. Ergebnisse der wissenschaftlichen Kern-Bohrung Ur-Federsee 2. Oberrhein. Geol. Abh. 16, 45–110.

Geyh, M.A., Müller, H., 2005. Numerical ^{230}Th/U dating and a palynological review of the Holsteinian/Hoxnian Interglacial. Quat. Sci. Rev. 24, 1861–1872.

Gibbard, P.L., Head, M.J., 2010. The newly-ratified definition of the Quaternary System/ Period and redefinition of the Pleistocene Series/Epoch, and comparison of proposals advanced prior to formal ratification. Episodes 33, 152–158.

Gistl, R., 1928. Die letzte Interglazialzeit der Lüneburger Heide pollenanalytisch betrachtet. Bot. Arch. 21, 648–710.

Golombek, E.B., 1980. Pollenanalytische Untersuchungen zur spät- und postglazialen Vegetationsgeschichte im Drömling (Ostniedersachsen). Ber. naturhist. Ges. Hannover 123, 79–157.

Golombek, E.B., 1990. Palynologische Untersuchungen im Altwarmbüchener Moor bei Hannover. Ber. naturhist. Ges. Hannover 132, 31–45.

Göttlich, K., Werner, J., 1974. Vorrißzeitliche Interglazialvorkommen in der Altmoräne des östlichen Rheingletschergebiets. Geol. Jb. A 18, 49–79.

Grüger, E., 1967. Vegetationsgeschichtliche Untersuchungen an cromerzeitlichen Ablagerungen im nördlichen Randgebiet der deutschen Mittelgebirge. Eiszeitalter u. Gegenwart 18, 204–235.

Grüger, E., 1979a. Spätriß, Riß/Würm und Frühwürm am Samerberg in Oberbayern - ein vegetationsgeschichtlicher Beitrag zur Gliederung des Jungpleistozäns. Geologica Bavarica 80, 5–64.

Grüger, E., 1979b. Die Seeablagerungen vom Samerberg/obb. und ihre Stellung im Jungpleistozän. Eiszeitalter u. Gegenwart 29, 23–34.

Grüger, E., 1983. Untersuchungen zur Gliederung und Vegetationsgeschichte des Mittelpleistozäns am Samerberg in Oberbayern. Geologica Bavarica 84, 21–40.

Grüger, E., 1989. Palynostratigraphy of the Last Interglacial/Glacial cycle in Germany. Quat. Int. 3 (4), 69–79.

Grüger, E., 1995. Correlation of Middle-European Late-Pleistocene pollen sequences of the Pfefferbichl and Zeifen types. Meded. Rijks Geol. Dienst 52, 97–104.

Grüger, E., 1996. Palynostratigraphy of the Middle Pleistocene sequence from Göttingen, Otto-Straße. In: Turner, C. (Ed.), The Early Middle Pleistocene in Europe. Balkema, Rotterdam, pp. 173–180.

Grüger, E., Schreiner, A., 1993. Riß/Würm- und würmzeitliche Ablagerungen im Wurzacher Becken (Rheingletschergebiet). N. Jb. Geol. Paläont. (Abh.) 189, 81–117.

Grüger, E., Jordan, H., Meischner, D., Schlie, P., 1994. Mittelpleistozäne Warmzeiten in Göttingen, Bohrungen Ottostraße und Akazienweg. Geol. Jb. A 134, 167–210.

Hahne, J., 1991. Untersuchungen zur spät- und postglazialen Vegetationsgeschichte im nördlichen Bayern (Rhön, Grabfeld, Lange Berge). Flora 185, 17–32.

Hahne, J., 1992. Untersuchungen zur spät- und postglazialen Vegetationsgeschichte im nordöstlichen Bayern (Bayerisches Vogtland, Fichtelgebirge, Steinwald). Flora 187, 169–200.

Hahne, J., 1996. The interglacial site of Hunteburg near Quakenbrück. In: Turner, C. (Ed.), The Early Middle Pleistocene in Europe. Balkema, Rotterdam, pp. 181–186.

Hahne, J., Mengeling, H., Merkt, J., Gramann, F., 1994a. Die Hunteburg-Warmzeit ("Cromer-Komplex") und Ablagerungen der Elster-, Saale- und Weichsel-Kaltzeit in der Forschungsbohrung Hunteburg GE 58 bei Osnabrück. Geol. Jb. A 134, 117–166.

Hahne, J., Kemle, S., Merkt, J., Meyer, K.-D., 1994b. Eem-, weichsel- und saalezeitliche Ablagerungen der Bohrung "Quakenbrück GE 2". Geol. Jb. A 134, 9–69.

Hahne, J., Ellwanger, D., Stritzke, R., 2008. Evidence for a Waalian thermomer pollen record from the borehole Heidelberg UniNord, Upper Rhine Graben, Baden-Württemberg. Eiszeitalter u. Gegenwart (E&G Quaternary Science Journal) 57, 403–410.

Hahne, J., Ellwanger, D., Franz, M., Stritzke, R., Wielandt-Schuster, U., 2012. Pollenanalytische Untersuchungsergebnisse aus dem baden-württembergischen Rheinsystem Oberrheingraben, Hochrhein und Oberschwaben – eine Zusammenfassung des aktuellen Kenntnisstandes. LGRB Informationen 26, 119–154.

Hallik, R., 1960. Die Vegetationsentwicklung der Holstein-Warmzeit in Nordwest-Deutschland und die Altersstellung der Kieselgurlager der südlichen Lüneburger Heide. Z. dt. geol. Ges. 112, 326–333.

Hermsdorf, N., Strahl, J., 2008. Karte der Eem-Vorkommen des Landes Brandenburg. Brandenburg. geowiss. Beitr. 15, 23–55.

Herz, M., Knipping, M., Kromer, E., 2014. The Rosenheim Basin: Würmian and Pre-Würmian deposits and the Höhenmoos interglacial (MIS 7). In: Kerschner, H., Krainer, K., Spötl, C. (Eds.), From the Foreland to the Central Alps. Field Trips to Selected Sites of Quaternary Research in the Tyrolean and Bavarian Alps. Deuqua Excursions, Geozon, Berlin, pp. 6–17.

Heumann, G., Litt, T., 2002. Stratigraphy and palaeoecology of the late Pliocene and early Pleistocene in the open-cast mine Hambach (Lower Rhine Basin). Netherlands J. Geosci. Geologie en Mijnbouw 81, 193–199.

Hölzer, A., Hölzer, A., 1987. Paläoökologische Moor-Untersuchungen an der Hornisgrinde im Nordschwarzwald. Carolinea 45, 43–50.

Homann, M., Lepper, J., 1994. Das Cromer-Profil von Sohlingen (Süd-Niedersachsen). Geol. Jb. A 134, 211–228.

Jahns, S., 2000. Late-glacial and Holocene woodland dynamics and land-use history of the Lower Oder valley, north-eastern Germany, based on two, AMS ^{14}C-dated, pollen profiles. Veget. Hist. Archaeobot. 9, 111–123.

Jahns, S., 2001. On the Late Pleistocene and Holocene history of vegetation and human impact in the Ücker valley, north-eastern Germany. Veget. Hist. Archaobot. 10, 97–104.

Jahns, S., 2007. Palynological investigations into the Late Pleistocene and Holocene history of vegetation and settlement at the Löddigsee, Mecklenburg, Germany. Veget. Hist. Archaeobot. 16, 157–169.

Jahns, S., Christiansen, J., Kierleis, W., Sudhaus, D., 2013. On the Holocene vegetation history of Brandenburg and Berlin. In: Kadrow, S., Wlodarzcak, P. (Eds.), Environment and Subsistence—Forty Years after Janusz Kruk's "Settlement Studies…" Studien zur Archäologie in Ostmitteleuropa/Studia nad Pradziejami Europy Środkowej 11, Habelt, Bonn, pp. 1–20.

Jerz, H., Linke, G., 1987. Arbeitsergebnisse der Subkommission für Europäische Quartärstratigraphie: Typusregion des Holstein-Interglazials. Eiszeitalter u. Gegenwart 37, 145–148.

Jessen, K., Milthers, V., 1928. Stratigraphical and palaeontological studies of interglacial fresh-water deposits in Jutland and north-west Germany. Danm. Geol. Unders. II (48), 1–379.

Jonas, F., 1937. Das Quakenbrücker Interglazial. Bot. Zentralbl. 57, 343–366.

Jung, W., 1979. Die pflanzlichen Großreste des Riß/Würm-Interglazials von Eurach am Starnberger See/Obb. Geologica Bavarica 80, 107–113.

Jung, W., Beug, H.J., Dehm, R., 1972. Das Riß/Würm-Interglazial von Zeifen, Landkreis Laufen a. d. Salzach. Bayer. Akad. Wissensch., Math.-Nat. Kl., Abh., N.F. 151, 1–131.

Kasse, C., Bohncke, S.J.P., Vandenberghe, J., 1995. Fluvial periglacial environments, climate and vegetation during the middle Weichselian in the northern Netherlands, with special reference to the Hengelo Interstadial. Meded. Rijks Geol. Dienst N.S. 52, 387–414.

Kern, O.A., 2021. Percentages of pollen data from late MIS 6 to MIS 1 from Füramoos. PANAGAEA, Southern Germany, https://doi.org/10.1594/PANGAEA.934305.

Kern, O.A., Koutsodendris, A., Allstädt, F.J., Mächtle, B., Peteet, D.M., Kalaitzidis, S., Christanis, K., Pross, J., 2022. A near-continuous record of climate and ecosystem variability in Central Europe during the past 130 kyrs (Marine Isotope Stages 5–1) from Füramoos, southern Germany. Quat. Sci. Rev. 284 (107), 505.

Kleinmann, A., 1995. Seespiegelschwankungen am Ammersee. Ein Beitrag zur spät- und postglazialen Klimageschichte Bayerns. Geologica Bavarica 99, 253–367.

Kleinmann, A., Müller, H., Lepper, J., Waas, D., 2011. Nachtigall: a continental sediment and pollen sequence of the Saalian Complex in NW-Germany and its relationship to the MIS-framework. Quat. Int. 241, 97–110.

Kleinmann, A., Merkt, J., Müller, H., 2015. Sedimente des Degersees: Ein Umweltarchiv—Sedimentologie und Palynologie. Materialhefte zur Archaologie in Baden-Württemberg 102, 409–471.

Knipping, M., 1989. Zur Vegetationsgeschichte des Oberpfälzer Waldes. Dissertationes Botanicae 140, 1–209.

Knipping, M., 2001. Pollenanalytische Untersuchungen an einem Profil aus dem Ottmaringer Tal (Südliche Frankenalb). Quartär 51 (52), 211–227.

Knipping, M., 2008. Early and Middle Pleistocene pollen assemblages of deep core drillings in the northern Upper Rhine Graben. Netherlands J. Geosci./Geologic en Mijnbouw 87, 51–65.

Kolstrup, E., Wijmstra, T., 1977. A palynological investigation of the Moershoofd, Hengelo and Denekamp Interstadials in the Netherlands. Geol. Mijnb. 56, 85–102.

Kortfunke, C., 1992. Über die spät- und postglaziale Vegetationsgeschichte des Donaumooses und seiner Umgebung. Dissertationes Botanicae 184, 1–177.

Koutsodendris, A., Müller, U., Pross, J., Brauer, A., Kotthoff, U., Lotter, A.F., 2010. Vegetation dynamics and climate variability during the Holsteinian interglacial based on a pollen record from Dethlingen (northern Germany). Quat. Sci. Rev. 29, 3298–3307.

Kramm, E., 1978. Pollenanalytische Hochmooruntersuchungen zur Floren- und Siedlungsgeschichte zwischen Ems und Hase. Abhandlungen aus dem Westfälischen Museum für Naturkunde 40, 3–49.

Kühl, N., Gobet, E., 2010. Climatic evolution of the Middle Pleistocene warm period of Bilshausen, Germany, compared to the Holocene. Quat. Sci. Rev. 29, 3736–3749.
Küster, H., 1988. Vom Werden einer Kulturlandschaft. Vegetationsgeschichtliche Studien am Auerberg (Südbayern). Quellen und Forschungen zur prähistorischen und provinzialrömischen Archäologie 3, 1–163.
Küster, H., 1995. Postglaziale Vegetationsgeschichte Südbayerns: Geobotanische Studien zur prähistorischen Landschaftskunde. Akademie Verlag, Berlin.
Küttel, M., Löscher, M., Hölzer, A., 1986. Ergebnisse paläobotanischer Untersuchungen zur Stratigraphie und Paläoökologie des Würms im Oberrheingraben zwischen Karlsruhe und Mannheim. Eiszeitalter u. Gegenwart 36, 75–88.
Lagies, M., 2006. Neue pollenanalytische Forschungen in Spessart und Odenwald — eine Zusammenfassung. Carolinea 63, 113–134.
Lang, G., 1952. Zur späteiszeitlichen Vegetations- und Florengeschichte Südwestdeutschlands. Flora 139, 243–294.
Lang, G., 1994. Quartäre Vegetationsgeschichte Europas. Gustav Fischer, Stuttgart.
Lang, G., 2005. Seen und Moore des Schwarzwaldes als Zeugen spätglazialen und holozänen Vegetationswandels. Stratigraphische, pollenanalytische und großrestanalytische Untersuchungen. Andrias 16, 3–160.
Lauer, T., Weiss, M., 2018. Timing of the Saalian- and Elsterian glacial cycles and the implications for Middle – Pleistocene hominin presence in central Europe. Sci. Rep. 8, 5111.
Lauer, T., Weiss, M., Bernhardt, W., Heinrich, S., Rappsilber, I., Stahlschmidt, M.C., von Suchodoletz, H., Wansa, S., 2020. The Middle Pleistocene fluvial sequence at Uichteritz, central Germany: chronological framework, paleoenvironmental history and early human presence during MIS 11. Geomorphology 354 (107), 016.
Lechterbeck, J., Rösch, M., 2020. Böhringer See, western Lake Constance (Germany): an 8500 year record of vegetation change. Grana 60, 119–131.
Leroy, S.A.G., Zolitschka, B., Negendank, J.F.W., Seret, G., 2000. Palynological analyses in the laminated sediment of Lake Holzmaar (Eifel, Germany): duration of Lateglacial and Preboreal biozones. Boreas 29, 52–71.
Link, A., Preusser, F., 2005. Hinweise auf eine Vergletscherung des Kemptener Beckens (Südwest-Bayern) im Mittleren Würm. Eiszeitalter u. Gegenwart 55, 64–87.
Linke, G., Hallik, R., 1993. Die pollenanalytischen Ergebnisse der Bohrungen Hamburg-Dockenhuden (qho 4), Wedel (qho 2) und Hamburg-Billbrook. Geol. Jb. A 138, 169–184.
Lisiecki, L.E., Raymo, M.E., 2005. A Pliocene-Pleistocene stack of 57 globally distributed benthic $\delta^{18}O$ records. Paleoceanography 20, PA1003.
Litt, T., 1994. Paläoökologie, Paläobotanik und Stratigraphie des Jungquartärs im nordmitteleuropäischen Tiefland. Dissertationes Botanicae 227, 1–185.
Litt, T., Stebich, M., 1999. Bio- and chronostratigraphy of the lateglacial in the Eifel region, Germany. Quat. Int. 61, 5–16.
Litt, T., Brauer, A., Gozlar, T., Merkt, J., Balaga, K., Müller, H., Ralska-Jasiewiczowa, M., Stebich, M., Negendank, J.F.W., 2001. Correlation and synchronisation of Lateglacial continental sequences in northern central Europe based on annually laminated lacustrine sediments. Quat. Sci. Rev. 20, 1233–1249.
Litt, T., Ellwanger, D., Villinger, E., Wansa, S., 2005. Das Quartär in der Stratigraphischen Tabelle von Deutschland. Newsl. Stratigr. 41, 385–399.
Litt, T., Behre, K.-E., Meyer, K.-D., Stephan, H.-J., Wansa, S., 2007. Stratigraphische Begriffe für das Quartär des norddeutschen Vereisungsgebietes. Eiszeitalter u. Gegenwart (E & G Quaternary Science Journal) 56, 7–65.
Litt, T., Schölzel, C., Kühl, N., Brauer, A., 2009. Vegetation and climate history in the Westeifel Volcanic Field (Germany) during the past 11,000 years based on annually laminated lacustrine maar sediments. Boreas 38, 679–690.

Löscher, M., Becker, B., Bruns, M., Hieronymus, U., Mäusbacher, R., Münnich, M., Münzing, K., Schedler, J., 1980. Neue Ergebnisse über das Jungquartär im Neckarschwemmfächer bei Heidelberg. Eiszeitalter u. Gegenwart 30, 89–100.

Löscher, M., Cordes-Hieronymus, U., Schloss, S., 1983. Holozäne und jungpleistozäne Sedimente im Oberrheingraben bei Heidelberg. Geol. Jb. A71, 61–72.

Lotter, A.F., Hölzer, A., 1989. Spätglaziale Umweltverhältnisse im Südschwarzwald: Erste Ergebnisse paläolimnologischer und paläoökologischer Untersuchungen an Seesedimenten des Hirschenmoores. Carolinea 47, 7–14.

Lotter, A.F., Hölzer, A., 1994. A high-resolution Late-Glacial and early Holocene environmental history of Rotmeer, southern Black Forest. Dissertationes Botanicae 234, 365–388.

Lüttig, G., 1965. The Bilshausen type section. Geol. Soc. Am. 80, 159–178. Special Paper.

Lüttig, G., Rein, U., 1954. Das Cromer- (Günz/Mindel-) Interglazial von Bilshausen (Unter-Eichsfeld). Geol. Jb. 70, 159–166.

Mayer, H., 1966. Waldgeschichte des Berchtesgadener Landes (Salzburger Kalkalpen). Beihefte zum Forstwissenschaftlichen Centralblatt 22, 1–42.

Mayr, C., Brandlmeier, B., Diersche, V., Stojakowits, P., Kirscher, U., Matzke-Karasz, R., Bachtadse, V., Eigler, M., Haas, U., Lempe, B., Reimer, P.J., Spötl, C., 2017. Nesseltalgraben, a new reference section of the last glacial period in southern Germany. J. Paleolimnol. 58, 213–229.

Mayr, C., Stojakowits, P., Lempe, B., Blaauw, M., Diersche, V., Grohganz, M., López Correa, M., Ohlendorf, C., Reimer, P., Zolitschka, B., 2019. High-resolution geochemical record of environmental changes during MIS 3 from the northern Alps (Nesseltalgraben, Germany). Quat. Sci. Rev. 218, 122–136.

Menke, B., 1968a. Beiträge zur Biostratigraphie des Mittelpleistozäns in Norddeutschland (Pollenanalytische Untersuchungen aus Westholstein). Meyniana 18, 35–42.

Menke, B., 1968b. Das Spätglazial von Glüsing. Ein Beitrag zur Kenntnis der spätglazialen Vegetationsgeschichte in Westholstein. Eiszeitalter u. Gegenwart 19, 73–84.

Menke, B., 1969. Vegetationsgeschichtliche Untersuchungen an altpleistozänen Ablagerungen aus Lieth bei Elmshorn. Eiszeitalter u. Gegenwart 20, 76–83.

Menke, B., 1970. Ergebnisse der Pollenanalyse zur Pleistozän-Stratigraphie und zur Pliozän-Pleistozän-Grenze in Schleswig-Holstein. Eiszeitalter u. Gegenwart 21, 5–21.

Menke, B., 1975. Vegetationsgeschichte und Florenstratigraphie Nordwestdeutschlands im Pliozän und Frühquartär. Mit einem Beitrag zur Biostratigraphie des Weichselfrühglazials. Geol. Jb. A 26, 3–151.

Menke, B., Behre, K.-E., 1973. 2. History of vegetation and biostratigraphy. Eiszeitalter u. Gegenwart 23 (24), 251–267.

Menke, B., Ross, P.H., 1967. Der erste Fund von Kieselgur in Schleswig-Holstein bei Brokenlande, südlich von Neumünster. Eiszeitalter u. Gegenwart 18, 113–126.

Menke, B., Tynni, R., 1984. Das Eeminterglazial und das Weichselfrühglazial von Rederstall/Dithmarschen und ihre Bedeutung für die mitteleuropäische Jungpleistozän-Gliederung. Geol. Jb. A 76, 3–120.

Merkt, J., Müller, H., 1999. Varve chronology and palynology of the Lateglacial in Northwest Germany from lacustrine sediments of Hämelsee in Lower Saxony. Quat. Int. 61, 41–59.

Meyer, K.-J., 1974. Pollenanalytische Untersuchungen und Jahresschichtenzählungen an der holsteinzeitlichen Kieselgur von Hetendorf. Geol. Jb. A21, 87–125.

Mielke, K., Müller, H., 1981. Palynologie. In: Bender, F. (Ed.), Angewandte Geowissenschaften 1. Enke Stuttgart, pp. 393–407.

Müller, H., 1965. Eine pollenanalytische Neubearbeitung des Interglazialprofils von Bilshausen (Unter-Eichsfeld). Geol. Jb. 83, 327–352.

Müller, H., 1974a. Pollenanalytische Untersuchungen und Jahresschichtenzählungen an der holsteinzeitlichen Kieselgur von Munster-Breloh. Geol. Jb. A 21, 107–140.
Müller, H., 1974b. Pollenanalytische Untersuchungen und Jahresschichtenzählungen an der eemzeitlichen Kieselgur von Bispingen/Luhe. Geol. Jb. A 21, 149–169.
Müller, H., 1986. Altquartäre Sedimente im Deckgebirge des Salzstockes Gorleben. Z. dt. geol. Ges. 137, 85–95.
Müller, H., 1992. Climatic changes during and at the end of the interglacials of the Cromerian Complex. In: Kukla, G.J., Went, E. (Eds.), Start of a Glacial. NATO ASI Series I 3. Springer, Berlin, pp. 51–69.
Müller, U., 2000. A Late-Pleistocene pollen sequence from the Jammertal, south-western Germany with particular reference to location and altitude as factors determining Eemian forest composition. Veget. Hist. Archaeobot. 9, 125–131.
Müller, U., 2001. Die Vegetations- und Klimaentwicklung im jüngeren Quartär anhand ausgewählter Profile aus dem südwestdeutschen Alpenvorland.—Tübinger Geowiss. Arb. Reihe D 07, 1–118.
Müller, H., Höfle, H.-C., 1994. Die Holstein-Interglazialvorkommen bei Bossel westlich von Stade und Wanhöden nördlich Bremerhaven. Geol. Jb. A 134, 71–116.
Müller, U., Sanchez Goñi, M., 2007. Vegetation dynamics in southern Germany during marine isotope stage 5 (~ 130 to 70 kyr ago). In: Sirocko, F., Claasen, M., Sanchez Goñi, M. (Eds.), The Climate of Past Interglacials. Developments in Quaternary Sciences 7. Elsevier, Amsterdam, pp. 277–287.
Müller, U., Pross, J., Bibus, E., 2003. Vegetation response to rapid climate change in Central Europe during the past 140,000 yr based on evidence from the Füramoos pollen record. Quat. Res. 59, 235–245.
Nelle, O., 2002. Zur holozänen Vegetations- und Waldnutzungsgeschichte des Vorderen Bayerischen Waldes anhand von Pollen- und Holzkohleanalysen. Hoppea 63, 161–361.
Nelle, O., Dörfler, W., 2008. A summary of the late- and post-glacial vegetation history of Schleswig-Holstein. Mitt. Arbeitsgem. Geobot. Schleswig-Holstein Hamb. 65, 45–68.
Peschke, P., 1983. Palynologische Untersuchungen interstadialer Schieferkohlen aus dem schwäbisch-oberbayerischen Alpenvorland. Geologica Bavarica 84, 69–99.
Pott, R., 1984. Pollenanalytische Untersuchungen zur Vegetations- und Siedlungsgeschichte im Gebiet der Borkenberge bei Haltern in Westfalen. Abhandlungen aus dem Westfälischen Museum für Naturkunde 46, 3–28.
Pott, R., Hüppe, J., 2001. Flussauen- und Vegetationsentwicklung an der mittleren Ems. Zur Geschichte eines Flusses in Nordwestdeutschland. Abhandlungen aus dem Westfälischen Museum für Naturkunde 63, 5–119.
Preusser, F., Drescher-Schneider, R., Fiebig, M., Schlüchter, C., 2005. Re-interpretation of the Meikirch pollen record, Swiss Alpine Foreland, and implications for Middle Pleistocene chronostratigraphy. J. Quat. Sci. 20, 607–620.
Proborukmi, M.S., Urban, B., Frechen, M., Grube, A., Rolf, C., 2017. Late Pliocene-Pleistocene record of Garding-2 research drill core, Northwest Germany. Z. Dt. Ges. Geowiss. (German J. Geol.) 168, 141–167.
Rasmussen, S.O., Bigler, M., Blockley, S.P., Blunier, T., Buchardt, S.L., Clausen, H.B., Cvijanovic, I., Dahl-Jensen, D., Johnsen, S.J., Fischer, H., Gkinis, V., Guillevic, M., Hoek, W.Z., Lowe, J.J., Pedro, J.B., Popp, T., Seierstadt, I.K., Steffensen, J.P., Svensson, A.M., Vallelonga, P., Vinther, B.M., Walker, M.J.C., Wheatley, J.J., Winstrup, M., 2014. A stratigraphic framework for abrupt climatic changes during the Last Glacial period based on three synchronized Greenland ice-core records: refining and extending the INTIMATE event stratigraphy. Quat. Sci. Rev. 106, 14–28.
Rausch, K.-A., 1975. Untersuchungen zur spät- und nacheiszeitlichen Vegetationsgeschichte im Gebiet des ehemaligen Inn-Chiemseegletschers. Flora 164, 235–282.

Reich, H., 1953. Die Vegetationsentwicklung der Interglaziale von Großweil-Ohlstadt und Pfefferbichl im Bayerischen Alpenvorland. Flora 140, 386–443.
Reimer, P.J., Austin, W.E.N., Bard, E., Bayliss, A., Blackwell, P.G., Ramsey, C.B., Butzin, M., Cheng, H., Edwards, R.L., Friedrich, M., Grootes, P.M., Guilderson, T.P., Hajdas, I., Heaton, T.J., Hogg, A.G., Hughen, K.A., Kromer, B., Manning, S.W., Muscheler, R., Palmer, J.G., Pearson, C., van der Plicht, J., Reimer, R.W., Richards, D.A., Scott, E.M., Southon, J.R., Turney, C.S.M., Wacker, L., Adolphi, F., Büntgen, U., Capano, M., Fahrni, S.M., Fogtmann-Schulz, A., Friedrich, R., Köhler, P., Kudsk, S., Miyake, F., Olsen, J., Reinig, F., Sakamoto, M., Sookdeo, A., Talamo, S., 2020. The IntCal20 northern hemisphere radiocarbon age calibration curve (0–55 cal kBP). Radiocarbon 62, 725–757.
Rein, U., 1955. Die pollenstratigraphische Gliederung des Pleistozäns in Nordwestdeutschland. 1. Die Pollenstratigraphie im älteren Pleistozän. Eiszeitalter u. Gegenwart 6, 16–24.
Reitner, J., 2022. The imprint of Quaternary processes on the Austrian landscape. In: Embleton-Hamann, C. (Ed.), Landscapes and Landforms of Austria, World Geomorphological Landscapes. Springer Nature, Switzerland, pp. 47–72.
Rösch, M., 1989. Pollenprofil Breitnau-Neuhof: Zum zeitlichen Verlauf der holozänen Vegetationsentwicklung im südlichen Schwarzwald. Carolinea 47, 15–24.
Rösch, M., 1993. Prehistoric land use as recorded in a lake shore core at Lake Constance. Veget. Hist. Archaeobot. 2, 213–232.
Rösch, M., 2000. Long-term human impact as registered in an upland pollen profile from the southern Black Forest, south-western Germany. Veget. Hist. Archaeobot. 9, 205–218.
Rösch, M., 2020. 48. Mires around the yew forest of Paterzell (Upper Bavaria). Grana 60, 155–157.
Rösch, M., Lechterbeck, J., 2016. Seven Millennia of human impact as reflected in a high resolution pollen profile from the profundal sediments of Litzelsee, Lake Constance region, Germany. Veget. Hist. Archaeobot. 25, 339–358.
Rösch, M., Marinova, E., 2020. 51. Zeller See. Grana 60, 243–245.
Rösch, M., Wick, L., 2019. 43. Buchensee (Lake Constance region, Germany). Grana 58, 308–310.
Rösch, M., Stojakowits, P., Friedmann, A., 2021a. Does site elevation determine the start and intensity of human impact? Pollen evidence from southern Germany. Veget. Hist. Archaeobot. 30, 255–268.
Rösch, M., Friedmann, A., Rieckhoff, S., Stojakowits, P., Sudhaus, D., 2021b. A late Würmian and Holocene pollen profile from Tüttensee, Upper Bavaria, evidence of 15 millennia of vegetation history in the Chiemsee glacier region. Acta Palaeobot. 61, 136–147.
Ryabogina, N., Marinova, E., Rösch, M., 2021. 56. Gnadensee. Grana 60, 477–479.
Schedler, J., 1979. Neue pollenanalytische Untersuchungen am Schieferkohlevorkommen des Uhlenberges bei Dinkelscherben (Schwaben). Geologica Bavarica 80, 165–182.
Schedler, J., 1981. Vegetationsgeschichtliche Untersuchungen an altpleistozänen Ablagerungen in Südwestdeutschland. Dissertationes Botanicae 58, 1–158.
Scheuenpflug, L., 1979. Der Uhlenberg in der östlichen Iller-Lech-Platte. Geologica Bavarica 80, 159–164.
Schmeidl, H., 1969. Beitrag zur spätglazialen Vegetations- und postglazialen Waldentwicklung im südlichen Oberpfalzer Wald, in: Rückert, G. (1969), Erläuterungen zur Bodenkarte von Bayern 1:25000, Blatt Nr. 6640 Neunburg vorm Wald. Bayerisches Geologisches Landesamt, München, pp. 103–113.
Schmeidl, H., 1971. Ein Beitrag zur spätglazialen Vegetations- und Waldentwicklung im westlichen Salzachgletschergebiet. Eiszeitalter u. Gegenwart 22 (1971), 110–126.

Selle, W., 1962. Geologische und vegetationskundliche Untersuchungen an einigen wichtigen. Vorkommen des letzten Interglazials in Nordwestdeutschland. Geol. Jb. 79, 295–352.

Selle, W., Schneekloth, H., 1965. Ergebnisse der Kernbohrung Oerel, Krs. Bremervörde: drei Interstadiale über Ablagerungen des Eem-Interglazials. Z. dt. geol. Ges. 115, 109–117.

Stalling, H., 1987. Untersuchungen zur spät- und postglazialen Vegetationsgeschichte im Bayerischen Wald. Dissertationes Botanicae 105 (1987), 1–201.

Stark, P., 1923. Zur Entwicklungsgeschichte der badischen Bodenseemoore. Ber. Deut. Bot. Ges. 41, 361–373.

Stark, P., 1924. Pollenanalytische Untersuchungen an zwei Schwarzwaldmooren. Zeitschrift f. Bot. 16, 593–618.

Stebich, M., Höfer, D., Mingram, J., Nowaczyk, N., Rohrmüller, J., Mrlina, J., Kämpf, H., 2020. A contribution towards the palynostratigraphical classification of the Middle Pleistocene in Central Europe: the pollen record of the Neualbenreuth Maar, northeastern Bavaria (Germany). Quat. Sci. Rev. 250 (106), 681.

Stephan, H.-J., 2014. Climato-stratigraphic subdivision of the Pleistocene in Schleswig-Holstein, Germany and adjoining areas. E&G Quaternary Science Journal 63, 3–18.

Stephan, H.-J., Urban, B., Lüttig, G., Menke, B., Sierralta, M., 2011. Palynologische, petrographische und geochronologische Untersuchungen an Ablagerungen der Leck-Warmzeit (spätes Mittelpleistozän) und begleitender Sedimente. Geol. Jb. A160, 3–80.

STG 2016 (German Stratigraphic Commission; editing, coordination and layout), Menning, M., Hendrich, A., 2016. (1) Table plain 100 × 141 cm, (2) Table folded A4. In: Stratigraphic Table of Germany 2016. Potsdam (German Research Centre for Geosciences) in Germany.

Stojakowits, P., 2014. Pollenanalytische Untersuchungen zur Rekonstruktion der Vegetationsgeschichte im südlichen Iller-Wertach-Jungmoränengebiet seit dem Spätglazial. Phd Thesis, University of Augsburg, Germany.

Stojakowits, P., Peters, M., Friedmann, A., 2016. Vegetationsentwicklung und –wandel im süddeutschen Alpenvorland während des Würm. Ber. d. Reinhold-Tüxen-Ges. 28, 123–134.

Stojakowits, P., Korch, O., Grashey-Jansen, S., Friedmann, A., 2019. 45. Zugspitzplatt, Wetterstein Mountains, Germany. Grana 58, 396–398.

Stojakowits, P., Mayr, C., Lücke, A., Wissel, H., Hedenäs, L., Lempe, B., Friedmann, A., Diersche, V., 2020. Impact of climatic extremes on Alpine ecosystems during MIS 3. Quat. Sci. Rev. 239 (106), 333.

Stojakowits, P., Mayr, C., Ivy-Ochs, S., Preusser, F., Reitner, J., Spötl, C., 2021. Environments at the MIS 3/2 transition in the northern Alps and their foreland. Quat. Int. 581–582, 99–113.

Strahl, J., 2005. Zur Pollenstratigraphie des Weichsel-Spätglazials in Brandenburg. Brandenburg. geowiss. Beiträge 12, 87–112.

Strahl, J., Zwirner, R., 2002. Lithologische und pollenanalytische Untersuchungen an holstein- bis fuhnezeitlichen Ablagerungen im Gebiet Dömitz-Lenzen (NW-Brandenburg). Brandenburgische Geowiss. Beitr. 9, 33–43.

Teunissen, D., Teunissen-van Oorschot, H., 1974. Eine interstadiale Torfschicht bei Nijmegen (Niederlande) und deren Bedeutung für die Erklärung der dortigen Landschaftsmorphologie. Geol. Mijnb. 53, 393–400.

Urban, B., 1978a. Vegetationsgeschichtliche Untersuchungen zur Gliederung des Altquartärs der Niederrheinischen Bucht. Sonderveröff. Geol. Inst. Univ. Köln 34, 1–165.

Urban, B., 1978b. The Interglacial of Frechen I/Rheinland—a section of the Tiglian A-type. Geol. Mijnb. 57, 401–406.

Urban, B., 1995. Palynological evidence of younger Middle Pleistocene Interglacials (Holsteinian, Reinsdorf and Schoningen) in the Schoningen open cast lignite mine (eastern Lower Saxony, Germany). Meded. Rijks Geol. Dienst N.S. 52, 175–186.

Urban, B., 2007. Stratigraphische Begriffe für das Quartär des Periglazialraums in Deutschland. Eiszeitalter u. Gegenwart (E & G Quaternary Science Journal) 56, 84–95.

Urban, B., Thieme, H., Elsner, H., 1988. Biostratigraphische, quartärgeologische und urgeschichtliche Befunde aus dem Tagebau "Schöningen". Ldkr. Helmstedt. Z. dt. geol. Ges. 139, 123–154.

Urban, B., Lenhard, R., Mania, D., Albrecht, B., 1991. Mittelpleistozän im Tagebau Schöningen. Ldkr. Helmstedt. Z. dt. geol. Ges. 142, 351–372.

Urban, B., Sierralta, M., Frechen, M., 2011. New evidence for vegetation development and timing of Upper Middle Pleistocene interglacials in Northern Germany and tentative correlations. Quat. Int. 241, 125–142.

van der Hammen, T., 1971a. The upper Quaternary stratigraphy of the Dinkel Valley. Mededelingen Rijks Geologische Dienst N.S. 22, 59–72.

van der Hammen, T., 1971b. The Denekamp, Hengelo and Moershoofd Interstadials. Mededelingen Rijks Geologische Dienst N.S. 22, 81–85.

van der Hammen, T., Maarleveld, G.C., Vogel, J.C., Zagwijn, W.H., 1967. Stratigraphy, climatic succession and radiocarbon dating of the last glacial in the Netherlands. Geol. Mijnb. 45, 33–35.

van der Knaap, W.O., van Leeuwen, J.F.N., Fahse, L., Szidat, S., Studer, T., Baumann, J., Heurich, M., Tinner, W., 2019. Vegetation and disturbance history of the Bavarian Forest National Park. Germany. Veget. Hist. Archaeobot. 29, 277–295.

van der Meer, J.J.M., Slotboom, R.T., de Vires-Bruynsteen, I.M.E., 1984. Lithology and palynology of Weichselian alluvial fan deposits near Eerbeek, The Netherlands. Boreas 13, 393–402.

Vogel, J.C., Zagwijn, W.H., 1967. Groningen radiocarbon dates VI. Radiocarbon 9, 63–106.

Voigt, R., 1996. Paläolimnologische und vegetationsgeschichtliche Untersuchungen an Sedimenten aus Fuschlsee und Chiemsee (Salzburg und Bayern). Dissertationes Botanicae 270, 1–303.

Voigt, R., Grüger, E., Baier, J., Meischner, D., 2008. Seasonal variability of Holocene climate: a palaeolimnological study on varved sediments in Lake Jues (Harz Mountains, Germany). J. Paleolimnol. 40, 1021–1052.

Waas, D., Kleinmann, A., Lepper, J., 2011. Uranium-series dating of fen peat horizons from pit Nachtigall in northern Germany. Quat. Int. 241, 111–124.

Weber, C.A., 1893. Über die diluviale Vegetation von Klinge in Brandenburg und ihre Herkunft. Engl. Bot. Jb. 17. Beiblatt 1, 1–20.

Weber, C.A., 1896. Die fossile Flora von Honerdingen und das nordwestdeutsche Diluvium. Abh. Naturwiss. Ver. Bremen 13, 413–468.

Welten, M., 1981. Verdrängung und Vernichtung der anspruchsvollen Gehölze am Beginn der letzten Eiszeit und die Korrelation der Frühwürm-Interstadiale in Mittel- und Nordeuropa. Eiszeitalter u. Gegenwart 31, 187–202.

Welten, M., 1982. Pollenanalytische Untersuchungen im Jüngeren Quartär des nördlichen Alpenvorlandes der Schweiz. Beitr. geol. Karte der Schweiz N.F. 156, 1–174.

Welten, M., 1988. Neue pollenanalytische Ergebnisse über das Jüngere Quartär der Schweiz (Mittel- und Jungpleistozän). Beitr. geol. Karte der Schweiz N.F. 162, 1–40.

Zachos, J., Pagani, M., Sloan, L., Thomas, E., Billups, K., 2001. Trends, rhythms, and aberrations in globale climate 65 Ma to present. Science 292, 686–693.

Zagwijn, W.H., 1957. Vegetation, climate and time correlations in the Early Pleistocene of Europe. Geol. Mijnb. 19, 233–244.

Zagwijn, W.H., 1960. Aspects of the Pliocene and early Pleistocene vegetation in the Netherlands. Meded. Geol. Stichting Ser. C-III-1(5), 1–78.

Zagwijn, W.H., 1961. Vegetation, climate and radiocarbon datings in the Late Pleistocene of the Netherlands — Part I. Eemian and Early Weichselian. Meded. Geol. Sticht. N.S. 14, 15–45.

Zagwijn, W.H., 1963. Pollenanalytical investigations in the Tiglian of the Netherlands. Meded. Geol. Sticht. N.S. 16, 49–71.

Zagwijn, W.H., 1973. Pollenanalytic studies of Holsteinian and Saalian beds in the Northern part of the Netherlands. Meded. Rijks Geol. Dienst N.S. 24, 139–156.

Zagwijn, W.H., 1974a. The Pliocene-Pleistocene boundary in western and southern Europe. Boreas 3, 75–97.

Zagwijn, W.H., 1974b. Bemerkungen zur stratigraphischen Gliederung der plio-pleistozänen Schichten des niederländischen Grenzgebietes zwischen Venlo und Brüggen. Z. dt. geol. Ges. 125, 2–16.

Zagwijn, W.H., 1974c. Vegetation, climate and radiocarbon datings in the Late Pleistocene of the Netherlands, Part II: Middle Weichselian. Meded. Rijks Geol. Dienst N.S. 25, 101–111.

Zagwijn, W.H., 1989. Vegetation and climate during the warmer intervals in the Late Pleistocene of Western and Central Europe. Quat. Int. 3/4, 57–67.

Zagwijn, W.H., 1996. The Cromerian Complex Stage of the Netherlands and correlation with other areas in Europe. In: Turner, C. (Ed.), The Early Middle Pleistocene in Europe. Balkema, Rotterdam, pp. 145–172.

Zagwijn, W.H., de Jong, J., 1984. Die Interglaziale von Bavel und Leerdam und ihre stratigraphische Stellung im niederländischen Früh-Pleistozän. Med. Rijks Geol. Dienst. 37, 155–169.

Zagwijn, W.H., Paepe, R., 1968. Die Stratigraphie der weichselzeitlichen Ablagerungen der Niederlande und Belgiens. Eiszeitalter u. Gegenwart 19, 129–146.

Zagwijn, W.H., Zonneveld, J.I.S., 1956. The interglacial of Westerhoven. Geol. en Mijnb. N.S. 18, 37–46.

Zagwijn, W.H., van Montfrans, H.M., Zandstra, J.G., 1971. Subdivision of the "Cromerian" in the Netherlands, pollen-analysis, paleomagnetism and sedimentary petrology. Geol. Mijnb. 50, 41–58.

CHAPTER FOUR

The latest European record of *Chasmaporthetes lunensis lunensis* (Del Campana, 1914) from Schernfeld (Bavaria, Germany) in terms of the changes in the European carnivore paleoguilds

Adrian Marciszak[a,*], Wiktoria Gornig[a], Aleksandra Kropczyk[a], and Gertrud E. Rössner[b,c]

[a]Department of Paleozoology, University of Wrocław, Wrocław, Poland
[b]Staatliche Naturwissenschaftliche Sammlungen Bayerns, Bayerische Staatssammlung für Paläontologie und Geologie, München, Germany
[c]Department für Geo- und Umweltwissenschaften, Palaeontologie und Geobiologie, Ludwig Maximilians Universität München, München, Germany
*Corresponding author: e-mail address: adrian.marciszak@uwr.edu.pl

Contents

Abstract

The hyaena *Chasmaporthetes lunensis* lacks the robust and hyper-robust dentition of its bone-cracking relatives. Its gracile skeleton reflects an adaption to fast running and chasing down prey. The species was widespread on the northern hemisphere throughout the Pliocene and the Early Pleistocene ca. 4.5–1.5 myr, and, hence, an important component of the Holarctic fauna as a cursorial carnivore. By the earliest Pleistocene the relatively long-lived and seemingly fairly stable carnivore guild of the European Pliocene turned over, and canids of genera *Lycaon* and *Canis* replaced *Chasmaporthetes*. Its last known European occurrence is that from the German site Schernfeld with five isolated teeth documenting one large individual.

Stratigraphy & Timescales, Volume 7
ISSN 2468-5178
https://doi.org/10.1016/bs.sats.2022.09.003

1. Introduction

Together with Machairodontinae Gill, 1872, Hyaenidae Gray, 1821 were the main Neogene Afro-Eurasian to North American carnivores. Their relative abundance, their widespread, geographical range and their role in the consumption of carcasses make them an important component of Neogene and Quaternary paleofaunas. Hyaenidae were also considered as being of great stratigraphic interest in biochronological and palaeoecological analyses (Kurtén and Werdelin, 1988; Sen and Saraç, 2018; Turner et al., 2008). In this context, the genus *Chasmaporthetes* Hay, 1921, reached an extraordinarily wide Afro-Eurasian distribution, and even entered North America as the only hyena (Antón et al., 2006; Berta, 1981). *Chasmaporthetes* exhibit a set of morphological features indicative of an adaptation to a more active hunting behavior as compared to other Hyaenidae (Kurtén and Werdelin, 1988). Among these characters are a broad rostrum, a less crushing but more trenchant post-canine dentition than that of scavenging hyenas, and long, slender limbs adapted for high speed running (Antón et al., 2006). Although widely distributed, remains of *Chasmaporthetes* are nowhere abundant, and the latest species, *C. lunensis* disappeared already in the Early Pleistocene. The latest occurrences are reported from the German site Schernfeld (Qiu, 1987; Turner, 1992, 1995; Turner et al., 2008) and the Ukrainian locality Taurida Cave (Lavrov et al., 2022).

Although the classification of the Schernfeld remains as *C. lunensis* has been suggested by various authors e.g., Qiu (1987), Turner (1992, 1995), Turner et al. (2008), they have not been adequately revised and published. Thus, the main goal of this chapter is to examine these remains, along with the occurrence of *C. lunensis* in time and space. Additionally, hypotheses are put forward regarding the possible reasons for the disappearance of this species.

2. Material and methods

All material is stored at the SNSB—Bayerische Staatssammlung für Paläontologie und Geologie in Munich (Germany). Teeth were provisionally described by Dehm (1962: 36) as *Hyaena perrieri* Croizet and Jobert, 1828. The material counts five isolated teeth (Fig. 1), all steaming from the same adult individual:

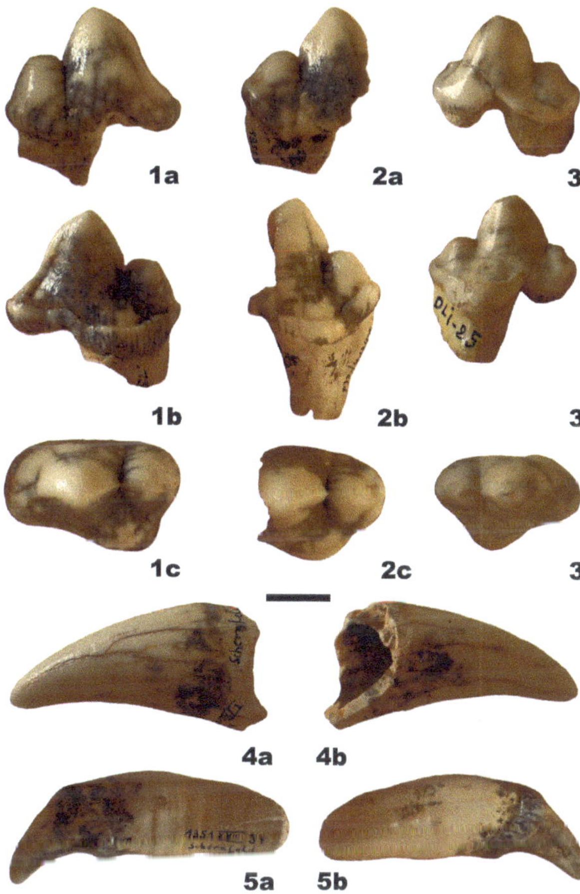

Fig. 1 Teeth of *Chasmaporthetes lunensis lunensis* from German site Schernfeld: 1—left P3 (SNSB-BSPG 1951 XXVI 34), 2—right p4 (SNSB-BSPG 1951 XXVI 34), 3—left P2 (SNSB-BSPG 1951 XXVI 34), 4—left C1 (SNSB-BSPG 1951 XXVI 34), 5—left I3 (SNSB-BSPG 1951 XXVI 34). Scale bar equals 10 mm. a—buccal view, b—lingual view, c—occlusal view.

(1) left complete I3 (SNSB-BSPG 1951 XXVI 34);
(2) crown of the left C1 (SNSB-BSPG 1951 XXVI 34);
(3) left P2 without mesial root and with a fragment of the distal root (SNSB-BSPG 1951 XXVI 34);
(4) left P3 without mesial root (SNSB-BSPG 1951 XXVI 35);
(5) distal crown half of the right p4 with the distal root (SNSB-BSPG 1951 XXVI 34).

Measurements were taken point to point, with an electronic caliper, to the nearest 0.01 mm. Measurements and dental terminology follows Schmid

(1940). For comparison of the Schernfeld teeth, we used material of *C. lunensis*, *P. perrieri*, *P. brevirostris* and *C. crocuta* from Eurasian localities, according to Schütt (1971), Kurtén and Poulianos (1977, 1981), Qiu (1987), Kurtén and Werdelin (1988), Cardoso (1993), Turner and Antón (1996), Turner (2001), Argant (2004), Antón et al. (2006), Baryshnikov and Tsoukala (2010), Vinuesa et al. (2014), Sen and Saraç (2018) and unpublished data of *C. Crocuta* from various Polish localities, measured by AM.

3. Locality

Schernfeld (48°54′17″N, 11°6′23″E, ca. 544 a. s. l.) is a paleontological site, located NW of Eichstätt in Middle Franconia and 65 km S of Nuremberg in the Jurassic limestone of the Franconian Alb (Dehm, 1962; Kowalski, 2001). The site was found during limestone quarrying in October 1951 in a vertical karstic fissure filled with red-brown loamy fossiliferous sands containing numerous teeth and bones. Among large carnivores, as a first canine of *Megantereon* was recognized. Complete skeletal remains were not found. Most of the fossil material is well preserved in light colors (Carls and Rabeder, 1988; Dehm, 1962). As a result of excavations, 55 animal taxa were identified:

(1) four gastropod species: *Pupilla muscorum* (Linnaeus, 1758); *Vallonia costata* (O. F. Müller, 1774); *Limax* sp.; *Helicopsis striata* (O. F. Müller, 1774);

(2) two squamate species: *Natrix* sp., *Ophisaurus* sp.;

(3) 49 mammal species: *Talpa fossilis* Petényi, 1864; *Talpa minor* Freudenberg, 1914; *Beremendia fissidens* (Petényi, 1864); *Petenyia hungarica* Kormos, 1934; *Sorex* aff. *Araneus* Linnaeus, 1758; *Sorex praealpinus* Heller, 1930; *Sorex runtonensis* Hinton, 1911; *Sorex minutus* Linnaeus, 1766; *Erinaceus europaeus* Linnaeus, 1758; *Rhinolophus* cf. *ferrumequinum* (Schreber, 1774); *Myotis* sp.; *Myotis* cf. *kormosi* Heller, 1936; *Myotis* cf. *exilis* Heller, 1936; *Apodemus* sp.; *Apodemus atavus* Heller, 1936; *Apodemus* cf. *sylvaticus* (Linnaeus, 1758); *Borsodia newtoni* (F. Major, 1902); *Clethrionomys kretzoii* (Kowalski, 1958); *Glirulus pusillus* Heller, 1936; *Lemmus kowalskii* Carls and Rabeder, 1988; *Mimomys ostramosensis* Jánossy and van der Meulen, 1975; *Mimomys pitymoides* Jánossy and van der Meulen, 1975; *Mimomys tornensis* Jánossy and van der Meulen, 1975; *Muscardinus* sp.; *Myoxus sackdillingensis* (Heller, 1930); *Petauria helleri* Dehm, 1962; *Pliomys* *episcopalis* Méhely, 1914; *Pliomys schernfeldensis* Carls and Rabeder, 1988; *Ungaromys dehmi* Carls and Rabeder, 1988; *Hypolagus brachygnathus* Kormos, 1934; *Ochotona* sp.; *Lycaon falconeri* (Forsyth Major, 1877); *Canis etruscus* Forsyth Major, 1877; *Vulpes praeglacialis* (Kormos, 1932); *Ursus etruscus* Cuvier, 1823; *Gulo schlosseri* Kormos, 1914; *Meles meles atavus* (Kormos, 1914); *Martes vetus* Kretzoi, 1942; *Mustela palerminea* (Petényi, 1864); *Mustela praenivalis* Kormos, 1934; *Panthera gombaszoegensis toscana* (Schaub, 1949); *Megantereon cultridens* Cuvier 1824; *Acinonyx pardinensis pleistocaenicus* (Zdansky, 1925); *Felis lunensis* Martelli, 1906; *Chasmaporthetes lunensis lunensis* (Del Campana, 1914); *Equus stenonis* Cocchi, 1867; *Cervus* sp., *Cervus nestii* (Azzaroli, 1947); Bovidae (large species), Elephantidae (Anderson, 1970; Carls and Rabeder, 1988; Dehm, 1962; Heller, 1963, 1967; Kościów and Nadachowski, 2002; Kowalski, 2001; Marciszak et al., 2021a; Musil, 1992; Rabeder, 1986a,b).

4. Results

Since the discovery in 1951, the hyena material from Schernfeld was not revisited. Dehm (1962) only briefly mentioned this material, and referred to *Hyaena perrieri* Croizet and Jobert, 1828. He described the I3 as relatively strong, in contrast the C was mentioned as a weak tooth. In the p2 morphology, he pointed to an elongated, occlusal outline with the prominent and strong distal part of the crown. According to Dehm (1962), the P3 fragment shows a strong distal protrusion bordered by the strong cingulum. In addition, Dehm (1962) compared the Schernfeld teeth with material of *P. perrieri* from Saint-Vallier, described by Viret (1954). He wrote that their morphologies generally coincide with those of the *P. perrieri* teeth. But he also noted that some features like the long P2 and in general the especially elongated and prominent distal crown parts of the premolars do not match those of *P. perrieri*. Nevertheless, he assessed the differences not sufficient enough to erect a new species. Instead, Qiu (1987) was the first to notice that the remains from Schernfeld do not belong to *P. perrieri*. He noticed that the dimensions of the teeth were too large for *P. perrieri* and that the P2 from Schernfeld and the P2 from Olivola were very similar in morphology. Both teeth have a crown slightly oriented lingually with a strongly developed distal part of the crown. Metrically and morphologically the teeth of the Schernfeld hyena correspond well with the equivalent teeth of *C. lunensis*. Later authors, e.g., Turner (1992, 1995), Turner et al. (2008), do not revisit the material and based their conclusions on Qiu (1987).

The I3 (L=13.7 mm, B=10.3 mm) is a canine-like tooth with strong lingual and distal cingula and a crest along its distal margin on its crown. The long and massive root is laterally flattened and covers approx. 60% of the total length of the tooth. It narrows slightly toward the end. The distal flange is of semilunar shape and characterized by the presence of a small fossa lunaris which is surrounded by the mesial and distal edge. The C crown is curved, moderately robust, and strongly flattened bucco-lingually. The crown is proportionally short, with smooth distal edges, which are accompanied by hollows. The tooth bears a thin and well-defined distal crest, running from its apex to the base.

The C from Schernfeld is large (L=17.5 mm, B=12.1 mm), and its size places in the upper range of variability of *C. lunensis* (L C=16.6 mm, 14.5–18.4 mm, n=17; B C=12.5 mm, 11.0–13.9 mm, n=16). Contrary to the size, it is quite narrow (B/L=69.6), and the massiveness index places it in the lower range of variation (B/L=74.7, 68.6–83.3, n=16). The C of *P. perrieri* is slightly larger (L C=17.7 mm, 15.7–20.1 mm, n=13; B C=13.2 mm, 12.0–15.0 mm, n=11), but the massiveness index is similar (B/L=73.7, 66.7–79.4, n=10). *C. crocuta* (L C=17.2 mm, 14.0–26.0 mm, n=130; B C=13.1 mm, 10.0–21.0 mm, n=134) and especially *P. brevirostris* (L C=21.8 mm, 19.4–27.0 mm, n=28; B C=17.2 mm, 14.5–21.0 mm, n=30) are characterized by larger C. But they are only slightly more massive, and the mean of the B/L index for *P. brevirostris* is 78.6 (67.9–92.4, n=28) and the mean for *C. crocuta* is 75.8 (66.5–86.5, n=130).

The P2 (L=21.5 mm, L pa=13.1 mm, H pa=16.6 mm, Ba=11.2 mm, Bp=13.2 mm) crown has an irregular, somewhat triangular occlusal outline. Mesial and distal margins are rounded, while the buccal margin is almost straight. Only on the bucco-mesial edge occurs a gentle convexity. The mesial half of the lingual margin is straight, and on the distal half occurs as a considerable, median convexity, which reaches almost the end of the crown. The paracone, is an elongated, oval-shaped and moderately high cusp, located almost exactly at the median of the crown. The protocone is smaller and lower than the hypocone, and shifted strongly mesio-lingually. It is oval-shaped and closely appressed to the paracone. The higher and more rounded hypocone is placed almost exactly at the median of the crown, just distal of the paracone. The distinction between the protocone, paracone and hypocone, as well as between the hypocone and the weak, distal cingulum is moderate.

The dimensions of P2 from Schernfeld are among the largest of *C. lunensis* (L P2=18.6mm, 16.4–21.2mm, n=33; B P2=11.2mm, 8.1–12.3mm, n=33). The P2 of *P. perrieri* is distinctly smaller (L P2=17.6mm, 16.2–18.8mm, n=28; B P2=11.8mm, 9.7–16.6mm, n=25). The mean of the length (L P2=17.6mm, 14.5–22.2mm, n=141) and breadth of P2 (B P2=13.1mm, 10.4–16.2mm, n=142) of *C. crocuta* is slightly smaller than that of *C. lunensis*. But *C. crocuta* is characterized by considerable size variability of P2, and the ranges of variability of both species overlap. The P2 of *P. brevirostris* is distinctly larger (L P2=21.1mm, 17.5–23.9mm, n=47; B P2=14.4mm, 11.4–19.0mm, n=46). The P2 from Schernfeld is large and moderately massive, with the breadth to the length ratio B/L=61.5, almost the same as the mean B/L=60.5 (44.8–67.2, n=32) of *C. lunensis*. All three other hyaenids have wider P2, where the B/L mean in *P. perrieri* is 66.7 (57.6–80.7, n=24), and in *P. brevirostris* the mean is 67.7 (54.1–78.2, n=44). The P2 is especially robust in *C. crocuta*, where B/L mean is 74.1 (52.2–83.8, n=140).

In the lateral view, the left P3 (L=25.7mm, L pr=14.5mm, H pr=19.3mm, Ba=14.3mm, Bp=16.7mm) is a compact and moderately high tooth. The oval-shaped protoconid is proportionally short and high, pushed slightly mesio-medially. The mesial margin is blunt, the distal edge is irregular, more or less rounded. The buccal margin is straight, similar to the mesial half of the lingual margin. On the distal half of the lingual edge occurs a considerable convexity. A thick and prominent ridge runs from the protoconid apex in mesio-lingual direction. The ridge ends with the rudimentary paraconid, or at least a cuspid-like swelling, occurs. A second ridge runs distally from the protoconid apex. It is much thinner and less prominent and terminates at the base of the cusp. The relatively large and rounded hypoconid is situated almost exactly in the middle of the crown, in the same line and just distal of the paraconid. There is a clear distinction between the paraconid and the hypoconid in form of a deep, thin and V-shaped valley. Almost the whole crown is collared by a thick and strong cingulum, which is weakly developed on the buccal margin.

The dimensions of P3 from Schernfeld (L=25.7mm) ranks the tooth among the largest P3 of *C. lunensis* (L P3=22.4mm, 19.4–25.9mm, n=43; B P3=13.8mm, 11.5–16.7mm, n=41). *P. perrieri* possesses similarly large P3 (L P3 = 23.5 mm, 19.4–25.8 mm, n = 35; B P3 = 16.3 mm, 12.1–18.7 mm, n = 31), and the ranges of variation of the two species are nearly identical. The P3 of *C. crocuta* (L P3 = 24.1 mm, 17.7–28.1 mm, n = 144) is comparable in length, but on average is wider (Bp P3 = 17.7 mm, 13.2–20.6 mm, n = 147). The P3 of *P. brevirostris* is distinctly larger and broader (L P3 = 27.1 mm, 24.4–30.0 mm, n = 58; B P3 = 19.1 mm, 16.7–21.6 mm, n = 53). Among these four hyenas, *C. lunensis* has the narrower P3, with the B/L mean of 60.6 (52.4–69.6, m = 41), and in the Schernfeld hyena this ratio is 65.6. The B/L of *P. perrieri* is 70.1 (59.6–75.8, n = 31), in *P. brevirostris* is 72.7 (61.2–80.9, n = 53), and in *C. crocuta* is 75.7 (57.7–83.6, n = 142).

The preserved distal half of a p4 (H pa = 19.27 mm, Bp = 15.56 mm) has an irregular, somewhat triangular occlusal outline. The buccal margin is gently arched, while the lingual one is considerably convex. This convexity is developed into a prominent, median bulge. The distal margin is rounded. The paracone is a rounded and high cusp, located almost exactly medially on the crown. The hypocone is located just distal of the paracone and slightly disto-lingually. The hypocone is a large and low cuspid, well separated from the paracone by a deep, narrow and V-shaped valley. The thick and strong cingulum, weaker developed on the buccal margin, collared the whole the rest of the crown.

Metrical values and massiveness indexes showed, that the five isolated teeth from Schernfeld are different from *C. crocuta* and to *P. brevirostris*. The only hyena species, of which the analyzed teeth may stem instead of *C. lunensis*, is *P. perrieri*. However, there are features which excluded also this hyena. The size of P2, clearly above the range of *P. perrieri*, narrower P2 and P3, with a strongly developed distal part of the crowns and weaker paracone. Additionally, in the occlusal outline the P2-P3 row in *C. lunensis* is more arched, with moderately convex, buccal margins. Some of these features were already highlighted by Schütt (1971), Koufos (1993), and Sotnikova (2004). The occlusal outline and the buccal margin of the P2-P3 row of *P. perrieri* is almost straight. Hence, we confirm the earlier assignment of the Schernfeld material to *C. lunensis* in agreement with previous authors, e.g., Qiu (1987).

5. Discussion

The tooth specimens of *C. lunensis* from Schernfeld belonged to a large individual, comparable in size with the individual from Olivola, and with a dentition with less developed a bone-cracking capability, when compared with *P. perrieri*, *C. crocuta*, or *P. brevirostris*. Its teeth, C, P2-P3 were large, but proportionally narrow. However, we do not re-open the discussion on the taxonomy of *C. lunensis*, especially because several species have been erected on scanty material (Sotnikova, 2004). The fragmentary status and the highly unevenly represented Schernfeld material undermine the approach of a cladistic analysis, so we do not attempt to further analyze the taxonomic position of these remains (Tseng et al., 2013). Taking into account the evolutionary lineage in the broader concept of this species (Kurtén and Werdelin, 1988; Sotnikova, 1994), we attributed the Schernfeld material to *Chasmaporthetes lunensis lunensis* (Del Campana, 1914) sensu Qiu (1987) and Sotnikova (2004). These results support the interpretation that *C. lunensis* was one of the long living and most widespread Palearctic carnivore among large, Eurasian forms (Tseng et al., 2013).

The oldest identified members of the genus *Chasmaporthetes* are known from the late Miocene Eurasian deposits (8.2–6.6 myr) of Greece, Chad, and China (de Bonis and Koufos, 1994; Koufos, 1987, 2000, 2022; Kurtén and Werdelin, 1988; Qiu, 1987; Sotnikova, 2004; Tseng et al., 2013; Werdelin and Solounias, 1990, 1991; Zdansky, 1924, 1935). The fragmentary nature of these remains and thus a thorough comparison of their evolutionary positions is not possible at this point (Tseng et al., 2013). During the early Pliocene, *Chasmaporthetes* had spread to North America, where a single species *Chasmaporthetes ossifragus* Hay, 1921 survived till ca. 800 kyr (Pérez-Claros et al., 2021). In Europe 25 localities (Fig. 2), dated on the Pliocene and the Early Pleistocene, roughly between 4.9 and 1.5 myr, documented the presence of *C. lunensis* (and some related forms like the island form *Chasmaporthetes melei* Rook et al., 2004) from Sardinia (Abbazzi et al., 2004; Rook et al., 2004). The oldest occurrence of this species, dated on 4.9–4.7 myr, is documented from Maramena (Schmidt Kittler, 1995) and Serrat d'en Vacquer (Perpignan, Kurtén and Werdelin, 1988). The specimen from Serrat d'en Vacquer, which together with slightly younger (4.1–3.8 myr) record from Dermedzhi (3.1–2.9 myr; Moldavia; Khomenko, 1932; Kurtén and Werdelin, 1988), was assigned to *Chasmaporthetes borissiaki* (Khomenko, 1932), is recently regarded synonymous with *C. lunensis* (Rook et al., 2004). Later this species was recorded from the Turkish site Çalta 1 (4.0–3.6 myr; Ginsburg, 1998; Sen and Saraç, 2018), Spanish Layna (3.9–3.8 myr; Crusafont and Aguirre, 1971; Kurtén and Werdelin, 1988) and Las Higueruelas (3.3–3.2 myr; Arribas et al., 2010; Madurell-Malapeira et al., 2014; Mazo et al., 2003),

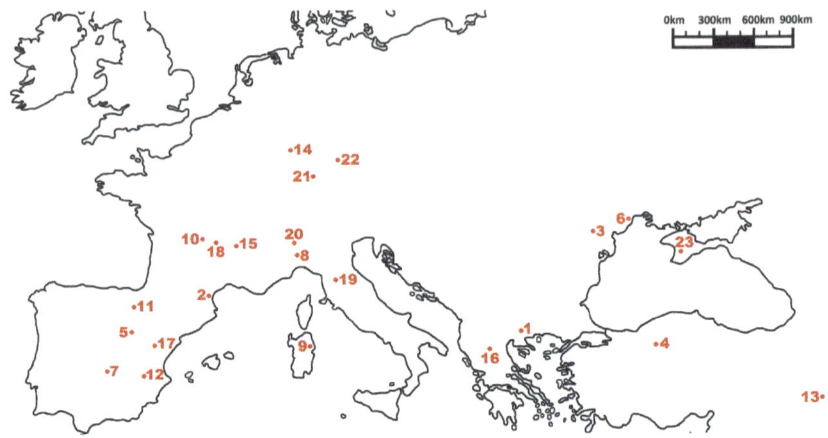

Fig. 2 Distribution of Pliocene and Early Pleistocene sites with *Chasmaporthetes lunensis* in Europe: 1—Maramena (Greece), 2—Serrat d'en Vacquer (France), 3—Dermedzhi (Moldavia), 4—Çalta 1 (Turkey), 5—Layna (Spain), 6—Odessa Catacombs (Ukraine), 7—Las Higueruelas (Spain), 8—Arondelli-Triversa (Italy), 9—Monte Tuttavista (as *C. melei*, Italy), 10—Perrier-Etouaires, Perrier-Roccaneyra, Perrier-Pardines (France), 11—Villaroya (Spain), 12—El Rincón (Spain), 13—Gülyazi (Turkey), 14—Neuleiningen 11 (Germany), 15—Saint Vallier (France), 16—Dafnero (Greece), 17—La Puebla de Valverde (Spain), 18—Sénèze (France), 19—Inferno (Upper Valdarno, Italy), 20—Olivola (Italy), 21—Erpfingen (Germany), 22—Schernfeld (Germany), 23—Taurida Cave (Ukraine).

Ukrainian locality Odessa Catacombs (3.6–3.2 myr; Sotnikova, 1989, 1994, 2004, 2006) and from Italy Arondelli-Triversa (3.3–3.1 myr; de Bruijn et al., 1992; Turner et al., 2008).

At the Pliocene/Pleistocene boundary and during the Early Pleistocene, species from genus *Chasmaporthetes* reaches its maximum range. Numerous records, like French sites Perrier-Etouaires (2.6–2.5 myr; Kurtén and Werdelin, 1988; Viret, 1954), Perrier-Roccaneyra (2.6–2.5 myr; Kurtén and Werdelin, 1988; Schaub, 1941; Stehlin, 1904), Perrier-Pardines (2.4–2.3 myr; Kurtén and Werdelin, 1988), Saint Vallier (2.2–2.1 myr; Argant, 2004; de Bruijn et al., 1992; Kurtén and Werdelin, 1988; Viret, 1954), Spanish localities Villaroya (2.6–2.5 myr; Kurtén and Werdelin, 1988; Villalta, 1952), El Rincón (2.6–2.5 myr; Turner et al., 2008), La Puebla de Valverde (2.1–2.0 myr; Antón et al., 2006; Crusafont Pairó et al., 1964; Heintz, 1978; Kurtén and Crusafont Pairó, 1977; Kurtén and Werdelin, 1988; Tseng et al., 2011), and single findings from Turkey (Gülyazi; 2.5–2.4 myr; Schütt, 1971), Germany (Neuleiningen 11; 2.4–2.2 myr; Heidtke, 1979) and Greece (Dafnero; 2.2–2.1 myr; Koufos, 1993), documented presence of the species between 2.7 and 2.0 myr. Also numerous Asiatic sites of Early Pleistocene age like Beregovaya, Kuruksay, Shamar and various Chinese localities yielded remains of *Chasmaporthetes* (Galiano and Frailey, 1977; Kurtén and Werdelin, 1988; Qiu, 1987; Sharapov, 1986; Sotnikova, 1974, 1976, 1980, 1989, 1994; Tseng et al., 2013; Turner et al., 2008).

Records younger than 2.0 myr are rarer, but *C. lunensis* is still known from France (Sénèze; 2.0–1.9 myr; Delson et al., 2006; Schaub, 1943), Italy (Inferno/Upper Valdarno and Olivola, both dated on 2.0–1.9 myr; Del Campana, 1914; Ferretti, 1999; Ficcarelli and Torre, 1967a,b) and Germany (Erpfinger Höhle, 1.8–1.7 myr; Lehmann, 1953, Lehmann and Bleich, 1957; Rathgeber, 1997, 2004; Rathgeber and Ufrecht, 2002). Finally, *C. lunensis* got extinct somewhere between 1.7 and 1.5 myr. The two latest occurrences are known from Schernfeld (1.8 1.6 myr; Dehm, 1962; Qiu, 1987; Turner, 1992, 1995; Turner et al., 2008) and Taurida Cave (1.8–1.6 myr; Lavrov et al., 2022).

During the Pliocene the European palaeocommunities were dominated by Machairodontinae (*Homotherium latidens, M. cultridens*), largely specialized as flesh-eaters, and two large hyenas (Turner, 2009; Werdelin and Turner, 1996). One of the most important features of the Pliocene carnivore guild is the almost complete absence of the canids, and certainly of larger taxa, and thus the apparent absence of a cursorial, pack-hunting species. However, the two hyenas of the time, *P. perrieri* and *C. lunensis*, were capable to fill that niche. *P. perrieri* was less robust than the *P. brevirostris* and *C. crocuta*, but was still a large hyena and specialized in scavenging and capable of operating in groups. In turn, *C. lunensis* was regarded as an active carnivore and a social hunter (Turner et al., 2008). Studies on its cranium confirm a skull morphology suggesting broad functional similarity with that of *Lycaon pictus* (Canidae) (Antón et al., 2006). Hence, *C. lunensis* was a hyena well-adapted to running and chasing down prey. Moreover, *C. lunensis* lacks the robust and hyper-robust dentition of the bone-cracking hyenas. From the latest Pliocene into the Early Pleistocene Eurasia faced a turnover of the larger carnivore guild in both the structural and the taxonomic makeup (Turner, 1995; Turner and Antón, 1996, 1999). This is the final stage in the long, evolutionary chain of hyaenas, where the late Miocene and Pliocene played the ecological role of cursorial hunters was later taken over by the canids (Turner, 2009; Turner et al., 2008).

However, we need to stress caution in drawing conclusions about diet on the basis on anatomy alone. So, scavenging in *C. lunensis* must not be excluded. There is a high likeliness that it was capable of obtaining a consistent living from scavenging when necessary. Yet, the major amount of food was taken from hunting. This might have forced *C. lunensis* in competition with speedy hunters like *Acinonyx pardinensis* (Croizet and Jobert, 1828). Being of similar size as the extant *Crocuta crocuta* (Erxleben, 1777), even if more slenderly built, social hunting in *C. lunensis* might have been advantageous during encounters with the larger, but solitary *A. pardinensis*. However, compared with other carnivorans, *C. lunensis* has clearly kept hyaenid characters, presumably because of a free, ecological niche and possibly to minimize competition with co-occurring *P. perrieri*. Both species were of similar body size, but the dentition of *P. perrieri* had greater bone-cracking adaptions. In co-occurring sites, Antón et al. (2006) found *P. perrieri* remains to be more abundant than those of *C. lunensis*.

This scenario worked well during the Pliocene, until the latest part of the Pliocene and the earliest Pleistocene, when new large carnivores appeared in Europe, like *Panthera gombaszoegensis* (Kretzoi, 1938) and *P. brevirostris* (Fig. 3). During the Early Pleistocene the relatively long-lived and seemingly fairly stable carnivore guild of the European later Pliocene was changing. *P. brevirostris* replaced *P. perrieri* and medium-sized larger canids replaced

Fig. 3 Biochronology of late Pliocene and Pleistocene large canids (genus *Lycaon* and *Canis*) and large hyenas (*Chasmaporthetes lunensis, Pachycrocuta brevirostris, Pliocrocuta perrieri*, and *Crocuta crocuta* ssp.). Artwork by W. Gornig.

C. lunensis (Turner, 2009; Turner et al., 2008; Turner and Antón, 1996). The scale of the impact of competition is difficult to reconstruct, but certainly competition came up. It seems, that the appearance of large, social dogs of the genus *Lycaon* and, to a lesser extent, *Canis*, also may have played a significant role in the disappearance of *C. lunensis*. The first appearance of *Canis* in Europe, based on the Vialette record, on the 3.1 myr in a form already comparable in size with *C. etruscus* (Fig. 3; Lacombat, 2006; Lacombat et al., 2008; Sotnikova and Rook, 2010). So far, conviction prevails, that the large, hypercarnivorous *Lycaon* arrived much later, ca. 2.2–2.1 myr (Madurell Malapeira et al., 2013; Martinez-Navarro and Rook, 2003). However, recent find from Roccaneyra (France), dated on 2.6 myr showed, that they were present in Europe much earlier than previously through (Bartolini-Lucenti and Spassov, 2022). In the environment previously dominated by large hyenas, have emerged social, highly intelligent, easily adaptable and aggressive canids from genus *Canis* and *Lycaon*, dominating these ecological niches for the next 2 myr. Robust *L. falconeri* and latter *L. lycaonoides* were as large or larger than the extant largest wolves (Marciszak et al., 2021b; Sotnikova, 2001). Their size and social behavior were advantageous in the co-occurrence with *Canis*, but also with *C. lunensis*. The appearance of *P. brevirostris* certainly had a major impact on both hyenas, *P. perrieri* and C. *lunensis*. However, the impact was certainly larger on *P. perrieri*, taking into account the similar adaptation to scavenging. It is highly likely that C. *lunensis*, as an active social hunter, suffered mainly from the retrieval of prey, just as *L. pictus* often loses its prey to *C. crocuta* attacks. Although, C. *lunensis* has no counterpart among hyenas in modern theriofaunas and is closest to *L. pictus*, many common analogies can be found. Contemporary *L. pictus*, with sufficiently numerous packs, copes well in competition with *C. crocuta* and the brown hyena *Parahyaena brunnea* (Thunberg, 1820). Even the extant, relatively small packs of *L. pictus* are able to successfully compete in the environments dominated by *C. crocuta*. However, as notes Eaton (1979: 11): "These data originate from recent populations in which pack size of dogs appears much smaller than in the early part of this century and before. (…) Thus, historically, wild dogs were probably dominant to spotted hyenas." It should be noted, however, that our assumptions of adaptations toward a cursorial and active hunting mode of life for *C. lunensis*, that it was necessarily in competition with extremely cursorial hunters like *L. falconeri* and *C. etruscus*.

It is known, that the decline of dog-like hyenas started 7–6.5 myr during a period of climate change, exacerbated by canids crossing the Bering land bridge from North America to Eurasia. In the opposite direction, only one hyena, *C. ossifragus*, managed to enter North America. In that context, *C. ossifragus* evolved into a cheetah-like sprinter (Tseng et al., 2013; Turner, 2009; Turner et al., 2008) and survived for some time in niches previously occupied by canids only, but eventually at ca. 800–750 kyr disappeared (Tseng et al., 2013). A similar scenario, but much earlier, happened in Eurasia. The earliest record of *L. falconeri* from Roccaneyra (2.6 myr) testifies to the fact that this huge and massive *Lycaon* species had already adapted there (Bartolini-Lucenti and Spassov, 2022). As a result of a combined set of factors, climate change, the alteration of the fauna of potential preys and increasing competition, *C. lunensis* became relaced in a relatively short time. The final extinction of this species took place in about 1.6–1.5 myr. *C. lunensis*, which was widespread on four continents from the end of the Miocene through the entire Pliocene, was relegated from an active and dominant carnivoran species to a survival-struggling guild element that disappeared as one of the first.

We find also here another analogy, which was repeated less than a 1 myr later in the case of *A. pardinensis*. Both species had large body dimensions but were relatively slim. *C. lunensis* was comparable in size to modern *C. crocuta*, but was much lightly built, probably weighing at least 30% less. As research shows, a massive body structure is one of the factors that guarantee an advantage in aggressive encounters with other carnivorans of similar size. Examples include the *Panthera spelaea fossilis* (von Reichenau, 1906) and *H. latidens* (Antón et al., 2005; Marciszak et al., 2021b), *P. gombaszoegensis* and *A. pardinensis* in the Pleistocene or the extant *C. lupus* and *Hyaena hyaena* (Linnaeus, 1758). Despite the fact that *A. pardinensis* was the largest felid of the early Pleistocene of Europe next to *P. gombaszoegensis* and right after *H. latidens*, its position in the paleohierarchy was relatively low (García and Virgós, 2007; Hemmer, 2004, 2011). This might be one of the reasons of their extinction, in combination with a number of other factors, of course. We suggest, that also *C. lunensis* was not powerful enough to successfully competed with the pair of effective cursorial hunters like *L. falconeri* and *C. etruscus*. As a result, this hyena disappeared as one of the first large carnivores. Contrary to *C. lunensis*, immense and extremely robust *P. brevirostris* was powerful enough to successfully compete not only with those canids, but also with *H. latidens* and *P. gombaszoegensis* (Iannucci et al., 2021; Liu et al., 2021; Palmqvist et al., 2011). Schernfeld remains of *C. lunensis,* together with the probably contemporaneous record from Taurida Cave, documented the last occurrence of this species.

Acknowledgments

The work was supported by an internal grant from the Faculty of Biological Sciences, University of Wrocław, entitled "The Middle Pleistocene Revolution—How the Modern Theriofauna of Eurasia was developed," as part of the programme "Inicjatywa Doskonałości—Uczelnia Badawcza (IDUB)," grant No. BPIDUB.4610.6.2021.KP.A.

References

Abbazzi, L., Angelone, C., Arca, M., Barisone, G., Bedetti, C., Delfino, M., Kotsakis, T., Marcolini, F., Palombo, M.R., Pavia, M., Piras, P., Rook, L., Torre, D., Tuveri, C., Valli, A.M.F., 2004. Plio-Pleistocene fossil vertebrates of Monte Tuttavista (Orosei, Eastern Sardinia, Italy), an overview. Riv. Ital. Paleontol. Stratigr. 110 (3), 681–706.

Anderson, E., 1970. Quaternary evolution of the genus *Martes* (Carnivora, Mustelidae). Acta Zool. Fenn. 130, 1–132.

Antón, M., Turner, A., Salesa, M.J., Morales, J., 2006. A complete skull of *Chasmaporthetes lunensis* (Carnivora, Hyaenidae) from the Spanish Pliocene site of La Puebla de Valverde (Teruel). Estud. Geol. 62 (1), 375–388.

Argant, A., 2004. Les Carnivores du gisement Pliocène final de Saint-Vallier (Drôme, France). Geobios 37 (1), 133–182.

Arribas, A., Garrido, G., García-Solano, J.A., 2010. Los hiénidos del Plioceno-Pleistoceno español: géneros y especies, distribución temporal y espacial. Zona Arqueológica 13, 34–48.

Bartolini-Lucenti, S., Spassov, N., 2022. Cave canem! The earliest *Canis* (*Xenocyon*) (Canidae, Mammalia) of Europe: taxonomic affinities and paleoecology of the fossil wild dogs. Quat. Sci. Rev. 276, 107315.

Baryshnikov, G.F., Tsoukala, E., 2010. New analysis of the Pleistocene carnivores from Petralona Cave (Macedonia, Greece) based on the collection of the Thessaloniki Aristotle University. Geobios 43, 389–402.

Berta, A., 1981. The Plio-Pleistocene hyaena *Chasmaporthetes ossifragus* from Florida. J. Vertebr. Paleontol. 1, 341–356.

Cardoso, J.L., 1993. Contribuição para o conhecimento dos grandes mamíferos do Plistocénico superior de Portugal. PhD Thesis, Universidade Nova de Lisboa. Câmara Municipal de Oeiras, Oeiras.

Carls, N., Rabeder, G., 1988. Die Arvicoliden (Rodentia, Mammalia) aus dem Ältestpleistozän von Schernfeld (Bayern). Beitr. Paläont. 14, 123–237.

Crusafont, M., Aguirre, E., 1971. *Euryboas lunensis* et *Hyaena donnezani* associés en Espagne dans le gisement d'âge pliocène terminal de Layna (Soria). C. R. Hebd. Seances Acad. Sci. 273, 2476–2478.

Crusafont Pairó, M., Hartenberger, J.L., Heintz, E., 1964. Un nouveau gisement de mammifères fossiles d'âge villafranchien à La Puebla de Valverde (Province de Teruel, Espagne). C. R. Acad. Sci. Paris 258, 2869–2871.

de Bonis, L., Koufos, G., 1994. Some Hyaenidae from the late Miocene of Macedonia (Greece) and a contribution to the phylogeny of the hunting hyaenas. Münchner Geowiss. Abh. B. 26, 81–96.

de Bruijn, H., Daams, R., Daxner-Höck, G., Fahlbusch, V., Ginsburg, L., Mein, P., Marles, J., Heinzmann, E., Mayhew, D.F., van der Meulen, A.J., Schmidt-Kittler, N., Antunes, M., Telles, M., 1992. Report of the regional committee on Mediterranean Neogene stratigraphy (RCMNS) Working Group on Fossil Mammals. Newsl. Stratigr. 26 (2–3), 65–118.

Dehm, R., 1962. Altpleistozäne Säuger von Schernfeld bei Eichstätt in Bayern. Mitt. Bayer. Staatssamml. Paläontol. Hist. Geol. 2, 17–61.

Del Campana, D., 1914. La *Lycyaena lunensis* n. sp. dell' ossario Pliocenico di Olivola (Val di Magra). Palaeontogr. Ital. 20, 87–103.

Delson, E., Faure, M., Guérin, C., Aprile, L., Argant, J., Blackwell, B.A.B., Debard, E., Harcourt-Smith, W., Martin-Suarez, E., Monguillon, A., Parenti, F., Pastre, J.-F., Sen, S., Skinner, A.R., Swisher, I.I.I.C.C., Valli, A.M.F., 2006. Franco-American renewed research at the late Villafranchian locality of Senèze (Haute-Loire, France). Cour. Forsch. Inst. 256, 275–290.

Eaton, R.L., 1979. Interference competition among carnivores: a model for the evolution of social behaviour. Carnivore 2, 9–16.

Ferretti, M.P., 1999. Tooth enamel structure in the hyaenid *Chasmaporthetes lunensis lunensis* from the late Pliocene of Italy, with implications for feeding behaviour. J. Vertebr. Paleontol. 19 (4), 767–770.

Ficcarelli, G., Torre, D., 1967a. Una mandibola di *Euryboas lunensis* (del Campana) nel giacimento villafranchiano di Olivola (val di Magra). Atti Soc Toscana Sci. Nat. Mem. B 74A, 193–198.

Ficcarelli, G., Torre, D., 1967b. A jaw of *Euryboas lunensis* from the Villafranchian beds of Olivola, Magra valley. Atti Soc. Toscana Sci. Nat. Mem. B 74 (1), 193–196.

Galiano, H., Frailey, D., 1977. *Chasmaporthetes Kani*, new species from China, with remarks on phylogenetic relationships of genera within the Hyaenidae (Mammalia, Carnívora). Am. Mus. Novit. 2632, 1–16.

García, N., Virgós, E., 2007. Evolution of community composition in several carnivore palaeoguilds from the European Pleistocene: the role of interspecific competition. Lethaia 40 (1), 33–44.

Ginsburg, L., 1998. Carnivores. In: Sen, S. (Ed.), Pliocene Vertebrate Locality of Çalta. vol. 20. Geodiversitas, Ankara, Turkey, pp. 379–396. 3.

Heidtke, U., 1979. Eine Großsäuger-Fauna aus dem älteren Pleistozän der Pfalz (Spaltenfüllung Neuleiningen 11). Mitt. Pollichia 67, 135–141.

Heintz, E., 1978. La faune villafranchienne de la Puebla de Valverde, Teruel, Espagne. Composition qualitative et quantitative. Geol. Mediter. 5 (2), 277–279.

Heller, F., 1963. Eine altquartäre Wirbeltierfauna des unteren Cromerium aus der nördlichen Frankenalb. Neues Jahrb. Geol. Palaontol. Abh. 118 (1), 1–20.

Heller, F., 1967. Die Wühlmäuse (Arvicolidae Gray, 1821) der altpleistozänen Säugetierfauna von Schernfeld bei Eichstätt in Bayern. Mitt. Bayer. Staatsslg. Paläont. Hist. Geol. 7, 201–203.

Hemmer, H., 2004. Notes on the ecological role of European cats (Mammalia: Felidae) of the last two million years. Paleontologia 2, 214–232. Miscelanea en homenaje a Emiliano Aquirre.

Hemmer, H., 2011. The story of the cave lion—*Panthera leo spelaea* (Goldfuss, 1810)—a review. Quaternaire, Hors-série 4, 201–208.

Iannucci, A., Mecozzi, B., Sardella, R., Iurino, D.A., 2021. The extinction of the giant hyena *Pachycrocuta brevirostris* and a reappraisal of the Epivillafranchian and Galerian Hyaenidae in Europe: faunal turnover during the early-middle Pleistocene transition. Quat. Sci. Rev. 272, 107240.

Khomenko, I.P., 1932. *Hyaena borissiaki* n. sp. From the Roussillon Fauna of Bessarabia. vol. 1 Trav. Inst. Pal. Acad. Sci. SSSR, pp. 81–134.

Kościów, R., Nadachowski, A., 2002. Type populations of some *Mimomys* species (Arvicolidae, Rodentia) at the Pliocene/Pleistocene boundary in Central Europe. Folia Zool. 51, 93–104.

Koufos, G.D., 1987. *Chasmaporthetes bonisi*, a new hyaenid (Carnivora, Mammalia) from the late Miocene of Macedonia (Greece). Bull. Soc. Géol. Fr. 8/3 (5), 913–920.

Koufos, G.D., 1993. Late Pliocene carnivores from western Macedonia (Greece). Paläont. Zeit. 67 (3–4), 357–376.

Koufos, G.D., 2000. Revision of the late Miocene carnivores from the Axios valley, Macedonia, Greece. Münchner Geowiss. Abh. B. 39, 51–92.

Koufos, G.D., 2022. The fossil record of hyaenids (Mammalia: Carnivora: Hyaenidae) in Greece. In: Vlachos, E. (Ed.), Fossil Vertebrates of Greece. Laurasiatherians, Artiodactyles, Perissodactyles, Carnivorans, and Island Endemics, vol. 2. Springer Cham, Basel, pp. 555–576.

Kowalski, K., 2001. Pleistocene rodents of Europe. Folia Quatern. 72, 1–389.

Kurtén, B., Crusafont Pairó, M., 1977. Villafranchian carnivores (Mammalia) from La Puebla de Valverde (Teruel, Spain). Comment. Biol., 85 Soc. Sci. Fenn., pp. 1–39.

Kurtén, B., Poulianos, N., 1977. New stratigraphic and faunal material from Petralona cave with special reference in the Carnivora. Anthropos 4, 47–130.

Kurtén, B., Poulianos, A.N., 1981. Remarks on the middle and lower Pleistocene vertebrate fauna in the Petralona cave. Anthropos 8, 57–72.

Kurtén, B., Werdelin, L., 1988. A review of the genus *Chasmaporthetes* Hay, 1921 (Carnivora, Hyaenidae). J. Vertebr. Paleontol. 8, 46–66.

Lacombat, F., 2006. New data on the early Villafranchian fauna from Vialette (haute-Loire, France). Collection of the Musee Crozatier (Le Puy-en-Velay, haute-Loire). Stratigraphy, paleontology and paleoenvironment of Pliocene-Pleistocene of Transbaikalia and interregional correlations. In: International Symposium, Ulan-Ude, 2006, Volume of Abstracts, p. 56.

Lacombat, F., Abbazzi, L., Ferretti, M.P., Martínez-Navarro, B., Moullé, P.E., Palombo, M.R., Rook, L., Turner, A., Valli, A.M.F., 2008. New data on the Early Villafranchian fauna from Vialette (Haute-Loire, France) based on the collection of the Crozatier Museum (Le Puy-en-Velay, Haute-Loire, France). Quat. Int. 179, 64–71.

Lavrov, A.V., Khantemirov, D., Gimranov, D., Lopatin, A.V., 2022. The first record of the hyena *Chasmaporthetes lunensis* (Del Campana, 1914) in Crimea (Tavrida cave, Early Pleistocene). In: Rozanov, A.Y., Petrov, O.V., Bogdanova, T.N., Bugrova, E.M., Vuks, V.Y., Gavrilova, V.A., Yanina, T.A. (Eds.), Paleontology and Stratigraphy: Current State and Ways of Development. Materials of the 68th Session of the Paleontological Society Dedicated to the 100th Anniversary of the Birth of Alexander Ivanovich Zhamoida. Palaeontological Society of the Russian Academy of Science, Saint Petersburg, pp. 222–223.

Lehmann, U., 1953. Eine Villafranchiano-Fauna von der Erpfinger Höhle (Schwäbische Alb). Neues Jb. Geol. Paläontol. Abh. 1953, 437–464.

Lehmann, U., Bleich, K., 1957. Weitere Fossilfunde aus dem ältesten Pleistozän der Erpfinger Höhle (Schwäbische Alb). Mitt. Geol. Hamburg 26, 60–99.

Liu, J., Liu, J., Zhang, H., Wagner, J., Jiangzuo, Q., Song, Y., Liu, S., Wang, Y., Jin, C., 2021. The giant short-faced hyena *Pachycrocuta brevirostris* (Mammalia, Carnivora, Hyaenidae) from Northeast Asia: a reinterpretation of subspecies differentiation and intercontinental dispersal. Quat. Int. 577, 29–51.

Madurell Malapeira, J., Rook, L., Martínez-Navarro, B., Alba, D.M., Garrido, J.A., Moya-Sola, S., 2013. The latest European painted dog. J. Vertebr. Paleontol. 33 (5), 1244–1249.

Madurell-Malapeira, J., Ros-Montoya, S., Espigares, M.P., Alba, D.M., Aurell-Garrido, J., 2014. Villafranchian large mammals from the Iberian Peninsula: paleobiogeography, paleoecology and dispersal events. J. Iber. Geol. 40 (1), 167–178.

Marciszak, A., Ambros, D., Hilpert, B., 2021a. Mustelids from Sackdilling cave (Bavaria, Germany) and their biostratigraphic significance. Geobios 68, 83–107.

Marciszak, A., Kropczyk, A., Lipecki, G., 2021b. The first record of *Cuon alpinus* (Pallas, 1811) from Poland and the possible impact of other large canids on the evolution of the species. J. Quat. Sci. 36 (6), 1101–1121.

Martinez-Navarro, B., Rook, L., 2003. Gradual evolution in the hunting dog lineage. C. R. Palevol 2, 695–702.

Mazo, A., Made, J., Der, V., Arribas, A., Sánchez, A., 2003. Hace 3 Millones de Años. Fundación de Cultura y Deportes de Castilla La Mancha, Junta de Comunidades de Castilla-La Mancha, Ciudad Real.

Musil, R., 1992. Die Pferde aus der oberpliozänen Spaltenfüllung Schernfeld bei Eichstätt. vol. 32 Mitt. Bayer. Staatssamml. Paläontol. Hist. Geol, pp. 115–162.

Palmqvist, P., Martínez-Navarro, B., Pérez-Claros, J.A., Torregrosa, V., Figueirido, B., Jiménez-Arenas, J.M., Patrocinio Espigares, M., Ros-Montoya, S., De Renzi, M., 2011. The giant hyena *Pachycrocuta brevirostris*: modelling the bone-cracking behavior of an extinct carnivore. Quat. Int. 243, 61–79.

Pérez-Claros, J.A., Coca-Ortega, C., Werdelin, L., 2021. How many hyenas in North America? A quantitative perspective. J. Vertebr. Paleontol. 41 (3), e1979988.

Qiu, Z., 1987. Die Hyaeniden aus dem Ruscinium und Villafranchium Chinas. In: Münchner Geowiss. Abh. vol. 9. Verlag Friedrich Pfeil, pp. 1–108.

Rabeder, G., 1986a. Herkunft und frühe Evolution der Gattung *Microtus* (Arvicolidae, Rodentia). Zeit. Säugetier. 51 (6), 350–367.

Rabeder, G., 1986b. Paläontologische Evolutionsforschung. vol. 71 Wissenschaftliche Nachrichten, pp. 7–10.

Rathgeber, T., 1997. Bärenhöhle bei Erpfingen. Beiträge zur Höhlen- und Karstkunde in Südwestdeutschland. Karstlandschaften in Südwestdeutschland, Exkursion vom Neckarland über die Schwäbische Alb zum Hochrhein. vol. 39, pp. 64–67.

Rathgeber, T., 2004. Die quartären Säugetier-Faunen der Bären- und Karlshöhle bei Erpfingen im Überblick. Laichinger Höhlenfreund 38 (2), 107–144.

Rathgeber, T., Ufrecht, W., 2002. Bären- und Karlshöhle bei Sonnenbühl-Erpfingen. Abhand. Karst- Höhlenk. 34, 86–90.

Rook, L., Ferretti, M.P., Arca, M., Tuveri, C., 2004. *Chasmaporthetes melei* n. sp., an endemic hyaenid (Carnivora, Mammalia) from the Monte Tuttavista fissure fillings (late Pliocene to early Pleistocene, Sardinia, Italy). Riv. Ital. Paleontol. Stratigr. 110, 707–714.

Schaub, S., 1941. Ein neues Hyaenidengenus von der Montagne de Perrier. Eclogae Geol. Helv. 34, 279–286.

Schaub, S., 1943. Die oberpliocaene Säugetierfauna von Sénèze und ihre verbreitungsgeschichtliche Stellung. Eclogae Geol. Helv. 36 (2), 270–289.

Schmid, E., 1940. Variationsstatistische Untersuchungen am Gebiß pleistozäner und rezenter Leoparden und anderer Feliden. Zeit. Säuget. 15, 1–179.

Schmidt-Kittler, N., 1995. Carnivora, Mammalia. In: Schmidt-Kittler, N. (Ed.), The Vertebrate Locality Maramena (Macedonia, Greece) at the Turolian/Ruscinian Boundary (Neogene). Münchner Geowiss. Abh. A, vol. 28. Verlag Friedrich Pfeil, pp. 75–86.

Schütt, G., 1971. Hyaenidenfunde aus dem Villafranchium der Türkei. In: Richter, K. (Ed.), Sonderheft anläßlich des 70. Geburtstages von Otto Sickenberg. vol. 10. Mitt. Geol. Inst. Univer. Hannover, pp. 119–140.

Sen, S., Saraç, G., 2018. Hyaenidae (Carnivora, Mammalia) from late Miocene and Pliocene of Çalta (Ankara, Turkey). Rev. Paléobiol. 37 (2), 561–575.

Sharapov, S.S., 1986. The Kuruksay Assemblage of the Late-Pliocene Mammals of the Afghan-Tadzhik Depression. Donish, Dushanbe., p. 272.

Sotnikova, M.V., 1974. Carnivorans of the Eopleistocene of southern Tadzikistan. Bull. Mosk. Obsh. Ispyt. Prirody, Otdel Geol. 49 (5), 143–144.

Sotnikova, M.V., 1976. The upper-Pliocene Carnivora of Central Asia. Izv. AN SSSR Ser. Geolog. 11, 133–137.

Sotnikova, M.V., 1980. The late-Pliocene Mustelidae from the locality Shamar (Mongolian People's Republic). Bull. Kom. Izuch. Chetv. Per. 50, 138–145.

Sotnikova, M.V., 1989. Late Pliocene early Pleistocene Carnivora Stratigraphic Significance. vol. 440 Union SSSR Acad. Scien. Tran., pp. 1–121.

Sotnikova, M.V., 1994. The genus *Chasmaporthetes* Hay, 1921 from the Pliocene of Russia, Ukraine, Mongolia and Tadzhikistan. In: Tatarinov, L.P. (Ed.), Paleotheriology. Nauka, Moscow, pp. 113–139.

Sotnikova, M.V., 2001. Remains of Canidae from the lower Pleistocene site of Untermassfeld. In: Kahlke, R.-D. (Ed.), Das Pleistozän von Untermassfeld bei Meiningen (Thüringen). vol. 2. Monog. Röm.-Germ. Zentral, pp. 607–632.

Sotnikova, M.V., 2004. New data on Pliocene carnivores of the Odessa catacombs fauna. In: Gozhik, P.F. (Ed.), Problems of Phanerozoic Stratigraphy of Ukraine. INGN ANAU, Kiev, pp. 199–202.

Sotnikova, M.V., 2006. Pliocene-Early Pleistocene carnivore assemblages of Transbaikalian area, Russia. In: Alexeeva, N.V., Erbajeva, M.A., Mironov, A.G. (Eds.), Stratigraphy, Paleontology and Paleoenvironment of Pliocene-Pleistocene of Transbaikalia and Interregional Correlations: Book of Abstracts, Ulan-Ude. Siberian Branch of the Russian Academy of Sciences, pp. 84–85.

Sotnikova, M.V., Rook, L., 2010. Dispersal of the Canini (Mammalia, Canidae: Caninae) across Eurasia during the late Miocene to early Pleistocene. Quat. Int. 212 (2), 86–97.

Stehlin, H.G., 1904. Une faune à *Hipparion* à Perrier. Bull. Soc. Géol. France 4 (4), 433–444.

Tseng, Z.H., Antón, M., Salesa, M.J., 2011. The evolution of the bone-cracking model in carnivorans: cranial functional morphology of the Plio-Pleistocene cursorial hyaenid *Chasmaporthetes lunensis* (Mammalia: Carnivora). Paleobiology 37 (1), 140–156.

Tseng, Z.H., Li, Q., Wang, X., 2013. A new cursorial hyena from Tibet, and analysis of biostratigraphy, paleozoogeography, and dental morphology of *Chasmaporthetes* (Mammalia, Carnivora). J. Vertebr. Paleontol. 33 (6), 1457–1471.

Turner, A., 1992. Villafranchian-Galerian larger carnivores of Europe, dispersions and extinctions. In: von Koenigswald, W., Werdelin, L. (Eds.), Mammalian Migration and Dispersal Events in the European Quaternary. vol. 153. Cour. Forsch. Senck, pp. 153–160.

Turner, A., 1995. The Villafranchian large carnivore guild: geographic distribution and structural evolution. Il Quaternario 8 (2), 349–356.

Turner, A., 2001. Remains of *Pachycrocuta brevirostris* (Mammalia, Hyaenidae) from the lower Pleistocene site of Untermassfeld near Meiningen. In: Kahlke, R.-D. (Ed.), Das Pleistozän von Untermassfeld bei Meiningen (Thüringen). vol. 40. Monog. Röm.-Germ. Zentral, pp. 673–690. 3.

Turner, A., 2009. The evolution of the guild of large Carnivora of the British Isles during the middle and late Pleistocene. J. Quat. Sci. 24, 991–1005.

Turner, A., Antón, M., 1996. The giant hyaena, *Pachycrocuta brevirostris* (Mammalian, Carnivora, Hyaenidae). Geobios 29, 455–468.

Turner, A., Antón, M., 1999. Climate and evolution: implications of some extinction patterns in African and European machairodontine cats of the Plio-Pleistocene. Estud. Geol. 54, 209–230.

Turner, A., Antón, M., Werdelin, L., 2008. Taxonomy and evolutionary patterns in the fossil Hyaenidae of Europe. Geobios 41 (5), 677–687.

Villalta, J.F., 1952. Contribución al conocimiento de la fauna de mamíferos fósiles del Plioceno de Villarroya (Logroño). Bol. Inst. Geol. Min. Esp. 64, 1–203.

Vinuesa, V., Madurell-Malapeira, J., Ansón, M., Alba, D.M., 2014. New cranial remains of *Pliocrocuta perrieri* (Carnivora, Hyaenidae) from the Villafranchian of the Iberian Peninsula. Boll. Soc. Paleontol. Ital. 53 (1), 39–47.

Viret, J., 1954. Le Loess à Bancs Durcis de Saint-Vallier (Drôme) et sa faune de Mammifères villafranchiens. Nouv. Arch. Mus. Hist. Nat. 4, 1–200.

Werdelin, L., Solounias, N., 1990. Studies of fossil hyaenids: the genus *Adcrocuta* Kretzoi and the interrelationships of some hyaenid taxa. Zool. J. Linn. Soc. 98, 363–386.

Werdelin, L., Solounias, N., 1991. The Hyaenidae: taxonomy, systematics and evolution. Fossils Strata 30, 1–104.

Werdelin, L., Turner, A., 1996. Turnover in the guild of larger carnivores in Eurasia across the Miocene-Pliocene boundary. Acta Zool. Cracov. 39, 585–592.

Zdansky, O., 1924. Jungtertiäre Carnivoren Chinas. Paleontol. Sin. C 2 (1), 1–149.

Zdansky, O., 1935. *Equus* und andere Perissodactyla. Paleontol. Sin. C 6 (5), 1–54.

CHAPTER FIVE

The Quaternary vertebrate fauna of cave deposits of the Podillia-Bukovynian Karst-Speleological Area (Western Ukraine)

Bogdan Ridush*
Department of Physical Geography, Geomorphology and Paleogeography, Yuriy Fedkovych Chernivtsi National University, Chernivtsi, Ukraine
*Corresponding author: e-mail address: b.ridush@chnu.edu.ua

Contents

Stratigraphy & Timescales, Volume 7
ISSN 2468-5178
https://doi.org/10.1016/bs.sats.2022.10.002

Abstract

The Podillia-Bukovinian Karst-Speleological Area is rich in numerous caves of karst, suffosion, and weathering origin. The gypsum karst is the most widespread in the area. It is famous for the World's largest maze-caves in gypsum. Some of these caves contain Quaternary sediments with vertebrate fauna that can be used for biostratigraphy. The first faunal remains were found as early as 1886. Up to now, at list 30 cave sites with the Quaternary vertebrate fauna are known. Most of the sites are in horizontal caves, which results in specific bone and sediment accumulations. The formation of most of the bone accumulations is related to the inhabitation of caves by the *troglophilic* species, especially carnivores, and only a few inhabited by humans. The oldest among the cave sites is Chortkiv, dated to the late Odesian Fauna (Kryzhanivka Stage, MIS 41–62). The best-studied cave sections with cave fauna are Bukovynka, Kryshtaleva (Kryvchanska), Pryima I, Chortovi Skeli (Lviv VIII). Several caves are promising for further research, such as Martynivka, Tovtry, and others. A very small part of the osteological material was published in detail. Only 15 radiocarbon data are known for the cave sites in the region. Many finds in the caves were made in the 1950s–1960s, sometimes just by accident. Part of the bone collections were lost. The material stored in museums should be revised. They still is a large potential for the future research of the Quaternary cave deposits of the Podillia-Bukovinian Karst-Speleological Area (Western Ukraine).

1. Introduction

The study of cave sediments for paleogeographical reconstructions is one of the essential areas in paleogeography. Conditions of continental sediment accumulation and preservation in caves differ from terrestrial ones. First, due to their intrazonal nature, caves (primarily subhorizontal) attracted the animal population (including hominids), becoming concentrators of their vital activity traces. Vertical cavities, karst pits, and shafts acted as natural traps and accumulated rich faunal material, as well as eolian and fluvial sediments enriched with palynological material.

Due to the significant temperature inertia of the bed rocks, the caves have relatively stable microclimatic (thermo- and hygro-constant) conditions, which contributes to the better preservation of organic residues. In some cases, when the cavities accumulated snow, they formed perennial snowfields and glaciers, which could store even whole tree trunks and other organic remains. Specific for karst massifs are water-chemogenic sediments (flowstones, speleothems), represented mainly by calcium carbonate, which, on the one hand, additionally preserve other sediments, and on the other hand, are themselves a source of radioisotope records.

Geochemically, cave sediments are dominated by an alkaline environment, which contributes to the preservation of calcium in osteological remains. In addition, calcium is part of hydroxyapatite, which, in turn, is a "shield" to protect the fossil DNA from decay. The preservation of DNA is also facilitated by the already mentioned constant and relatively low temperature.

Cave deposits, as a special type of continental deposits, have certain features compared to terrestrial types, which determine the specificity of their study and interpretation.

An important circumstance in the study of the paleogeography of certain regions is that terrestrial Neogene-Quaternary deposits are absent or largely destroyed in many areas of karst formation. In such cases, cave sediments will be almost the only source of information about the paleoenvironmental conditions of the Quaternary period in this particular area.

However, even in the presence of a complete section of terrestrial Quaternary sediments, cave sediments can be a significant source of information, due to the specificity of subterranean sediment genesis, diagenesis, and epigenesis of these deposits.

Sometime ago, we made the preliminary brief reviews of "bear caves" of Ukraine, including the territory to be described (Ridush, 2009, 2014). During the time that passed, new information was obtained, and several previous data were clarified. The purpose of this study is to review the state-of-art in the investigation of cave sediments of the Podillia-Bukovinian Karst-Speleological Area, and to estimate the palaeogeographic potential of these sediments in the karst areas of the continental part of Ukraine.

2. Literature review

The karst and caves in the region were researched by a dozen of scholars in the past. The greatest input in this field was done by B. Ivanov,

V. Dubliansky, A. Klimchouk, V. Rogozhnilov, V. Andreichouk, V. Korzhyk, I. Turchinov and others (Andreychouk and Klimchouk, 2017; Ivanov, 1956, 1961, 1965; Klimchouk, 2012; Klimchouk and Andrejchuk, 2002; Klimchouk and Andreychouk, 2017; Klimchouk and Rogozhnikov, 1982; Klimchuk et al., 2008; Klimchouk et al., 2009; Korzhyk, 2007; Ridush and Kuprich, 2003).

For decades, the studies of cave deposits were provided by paleozoologists: G. Bachynsky, K. Tatarinov, I. Marysova (Bachynskyi, 1965, 1967, 1970; Bachynskyi and Tatarinov, 1966; Bachynskyi et al., 1964; Marysova, 1962; Tatarinov, 1965, 1966b, 2000; Tatarinov and Bachynskyi, 1968). Archeological studiesof the cave Paleolithic were provided by Matskevyi (1998, 2001, 2005). During the time between 1958–1970, the sites of fossil vertebrates were found in 33 out of 600 surveyed caves in Ukraine. Now we know about 30 Quaternary paleontological cave sites only in the Podillia-Bukovynian area.

In recent years the complex studies of cave sediments in the area are carried out by an international team. The palynological studies are provided by N. Gerasimenko and Y. Avdieyenko (Avdieienko, 2018, 2019; Avdieienko et al., 2018; Gerasimenko et al., 2014a, 2014b). Research on the physical properties of cave deposits is conducted by Bondar et al. (2010) and Bondar and Ridush (2015).

Starting from the 1990s, paleozoological research in the caves of Ukraine was carried out by the author in close cooperation with international cooperation (Croitor et al., 2014; Marciszak et al., 2015; Ridush, 1998, 2004b, 2004c, 2012, 2014; Ridush et al., 2012, 2013; Vremir et al., 2000, and others).

The oldest references to caves with bones in the south of Ukraine are found in the notes of Turkish traveler of the 17th century (Çelebi, 1961). Research of the so-called "bear caves" on the territory of modern Ukraine began in the middle of the 19th century (von Nordmann, 1858). A little later, the first cave with Pleistocene fauna in western Ukraine was investigated by Łomnicki (1881). But still the cave sites from this region are unfamiliar to most foreign researchers.

3. Geological and Karstological settings

The described area covers the southwestern part of the East European Platform. The soluble rocks and karst make up most of the area.

The territory of modern Ukraine is rich in karst areas. In the south of Eastern Europe, namely on the territory of Ukraine and Moldova, there are five main karst regions and, accordingly, the same number of areas of cave locations with faunistic fossil remains. Two of them are mountain karst regions: the carbonate karst of the Ukrainian Carpathians and the Mountainous Crimea. The other three represent plain karst: the Podillia-Bukovinian Area with a predominance of gypsum karst, the Eastern Podillia-Bessarabian and Black Sea-Azov Areas with carbonate karst (Ridush, 2013). Up to 2008, the total number of known caves in Ukraine was 1690 (Klimchuk et al., 2008). According to our calculations, there are about 100 caves on the territory of Moldova. Despite the significant distribution of karst caves in these territories, the number of caves with finds of fossil bears is quite limited, compared to Central and Western Europe.

The Podillia-Bukovinian Karst-Speleological Area is one of the karst regions in the south of Eastern Europe. It is famous for the huge maze caves in gypsum (Andreychouk and Klimchouk, 2017; Klimchouk, 2012; Klimchouk and Andreychouk, 2017; and others). According to the karst-speleological regionalization of the south of Eastern Europe (Ridush, 2013) (Fig. 1B), it consists of two regions: the West Podolian, to the north of the Middle Dniester, and Pokuttia-Bukovinian (CЄ-VI-Б-1), to the south of the Middle Dniester reach, with the part of the Prut River valley (CЄ-VI-Б-2).

4. Results (cave sites description)

Back in 1881, the bones of cave bear and cave hyena were found in the cave of unknown genesis, near the village of Melna (now Rohatyn district, Ivano-Frankivsk region) (Łomnicki, 1881). Quaternary vertebrate fauna locations near the villages of Stradch and Vynyava were associated with ancient dens of cave hyenas and cave bears. These dens were found in niches filled with redeposited sand (Bachynskyi, 1970; Goretskiy, 1957; Tatarinov, 1958). One sample of the *Crocuta spelaea* from the Horosno-Vynjava site was radiocarbon dated to 40,770 ± 350 yrs BP (MAMS 31302) (cal 42,480 yrs BP) (Westbury et al., 2020).

Karst-speleological regionalization of a territory means considering not only the karst cavities, but also caves of other origin, like suffosion, weathering etc.

Most of the gypsum caves in the region are of hypogenic origin (Klimchouk and Andreychouk, 2017), but after the area underwent uplift,

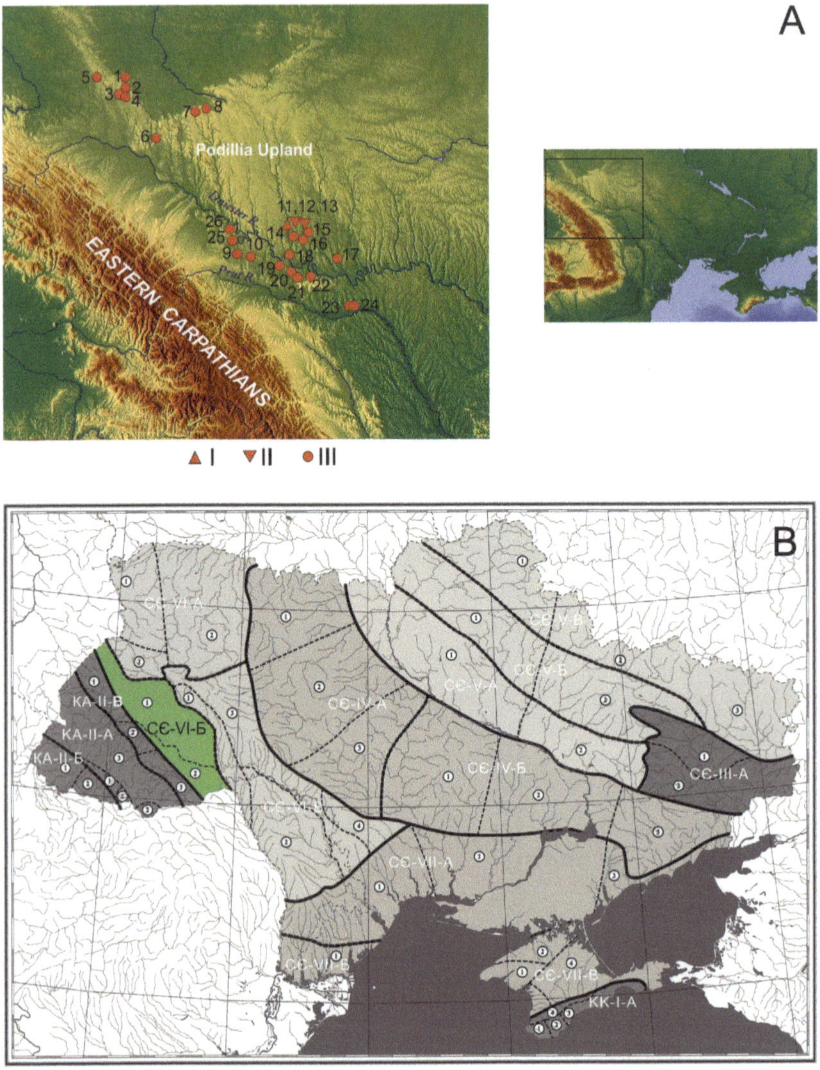

Fig. 1 (A) Location of the main Quaternary paleontological cave sites of the Podillia-Bukovinian Karst-Speleological Area. Legend: The age of site: I—Early Quaternary; II—Middle Quaternary; III—Late Quaternary age. Sites: 1—Chortovi Skeli; 2—Pryima; 3—Shchyrets; 4—Vyniava-Horosno; 5—Stradch; 6—Melna; 7—Divychi Skeli; 8—Mala Ilovitsia; 9—Cholivs'ka Gora (Chortovets'); 10—Ostrivets'; 11—Chortkiv; 12—Horyshnya Vygnanka; 13—Syniakove, Syniakove-1; 14—Ozerna; 15—Borschiv; 16—Kryvchenska (Kryshtaleva); 17—Atlantyda, Ta Dirka; 18—Kasperivtsi; 19—Chun'kiv; 20—Tovtry; 21—Martynivka; 22—Balamutivs'ka; 23—Bukovynka; 24—Malimon-Canyon; 25—Trygolovatka. (B) Location of the Podillia-Bukovinian Karst-Speleological Area (CЄ-VI-Б) (highlighted in green) on the scheme of karst-speleological regionalization of the south of Eastern Europe according to Dublianskaya and Dubliansky (1992) with our changes.

causing the the erosional opening of the cave systems, they quite quickly became filled with allochthonous loose material, consisting mainly of the Quaternary loess and eluvium of the overlying Neogene clays and limestones.

4.1 Atlantyda Cave

The cave is situated on the left bank of the Zbruch River (left tributary of the Dniester River), near Zavallja Village, Kamianets-Podilsky Rayon, Khmelnytska Oblast (48°35′55.43″N, 26°20′42.90″E) (Fig. 1, *17*). The cave, with a total length of up to 8300 m, is of a maze type and has three levels. It is developed in the Miocene gypsum strata.

In the early 1980s, comprehensive research, including a detailed study of secondary deposits, was conducted under the leadership of O. Klymchouk in the Atlantyda Cave. In particular, a morphogenetic analysis of the cave together with cartographic, stratigraphic, granulometric, mineralogical, palynological, and paleomagnetic studies of the cave deposits were carried out (Bahmutov and Lagutin, 1985; Klimchouk, 1984; Klimchouk and Rogozhnikov, 1982; Rogozhnikov, 1984; Rogozhnikov and Lomaev, 1985). It was noted that bone and plant remains were not found in numerous exploration pits, and in 70 samples taken for palynological analysis, spores and pollen, even redeposited ones, were not found (Klimchouk and Rogozhnikov, 1982). However, a little earlier A. Lomayev indicated that the spore-pollen analysis of sediments in the Grotto of the Snow Queen in the cave yielded a "Mindel-Riss—Mindel spectrum" (Lomayev, 1979, p. 73). An attempt was made to conduct a paleomagnetic study of water-mechanical cave deposits (a 4.5-m section in one of the pits). It was established that the formation of deposits took place in the Brunes polarity era (Bahmutov and Lagutin, 1985). The more recent paleomagnetic research of the cave sediments established that the light-gray laminated aleuritic clays in the Globe Gallery, that are characteristic also to other galleries in the cave, were deposited below Matuyama/Brunhes boundary (Bondar et al., 2010).

A complete skeleton of a brown bear (*Ursus arctos* L.) was discovered in 2013 on the lower floor of the cave. The found skeleton lay along the narrow gallery, head to the suggested ancient entrance. It was bedded on the surface of the dense ancient clay, covered with a crust of gypsum crystals, iron, and manganese hydroxide, and dated to pre-Brunes.

The bones were covered with loose fluvial thin-layered loam sediments 25–40 cm in thickness, including also remains of Chiroptera, Rodents,

Lagomorphs, and Mollusks, and impure with charcoal fragments up to 4–5 cm in diameter. The occipital part of the cranium was accidentally partly destroyed by cavers, who discovered the skeleton, and some few bones were partly removed by flowing water from the initial position. The remains belong to a large old male individual with strongly worn teeth. It was dated to the Young Dryas (GS-1).

In other parts of the cave, numerous bones of small Carnivora like *Vulpes*, *Martes* sp., and *Felis silvestris* are bedded on the surface of the cave floor. Also, the tibia fragment of red deer (*Cervus elaphus*) with marks of gnawing by a large Carnivora was found.

4.2 Balamutivs'ka Cave

It is situated near the Balamutivka Village, Chernivtsi Oblast (48°32′40.92″N, 26°4′35.90″E) (Fig. 1, *22*). The cave is developed at the foot of a layer of cracked nodular cryptocrystalline gypsum of the Upper Badenian (Miocene). The lower part of the entrance grotto is cut into the underlying Middle Badenian calcareous sandstones with interlayers of sand underlying the sulfate bed.

The entrance to the Balamutivska cave is in an outcrop of gypsum rock at the slope of the Dniester River valley, which is limited on both sides by erosion basins, and has the appearance of an arch up to 4 m high and up to 6 m wide (Fig. 2). The entrance altitude is 235 m a.s.l. and 70–75 m above the Dniester River level. The cave begins with a large grotto 16 m long and up to 14 m wide. The total length of the grotto, together with several small branches, is 40.5 m, the area is 163 m^2, and the volume is 978 m^3. The average height of the grotto is 6–7 m, with a width of 8–12 m. At the end of the grotto, behind the large-block collapse, a gallery with a width of 1.5–2.0 m and a height of 1.0–1.5 m (sometimes up to 2.0 m) begins. A stream flows almost constantly at the bottom of the gallery, which in the hall of the entrance grotto gets lost among the accumulation of large blocks that have fallen from the walls and ceiling.

In 1951, during explorations of the Dniester Archeological Expedition of the Institute of Social Sciences of the Lviv Branch of the Academy of Sciences of the Ukrainian SSR, O. Chernysh discovered ancient petroglyphs in the Balamutivska Cave (in the entrance grotto). Based on analogies with the monuments of Kamiana Mogyla, eastern Spain and others, he attributed this drawings to the Mesolithic (Chernysh, 1953, 1959). An additional survey of the cave was carried out in 1968 by the same expedition

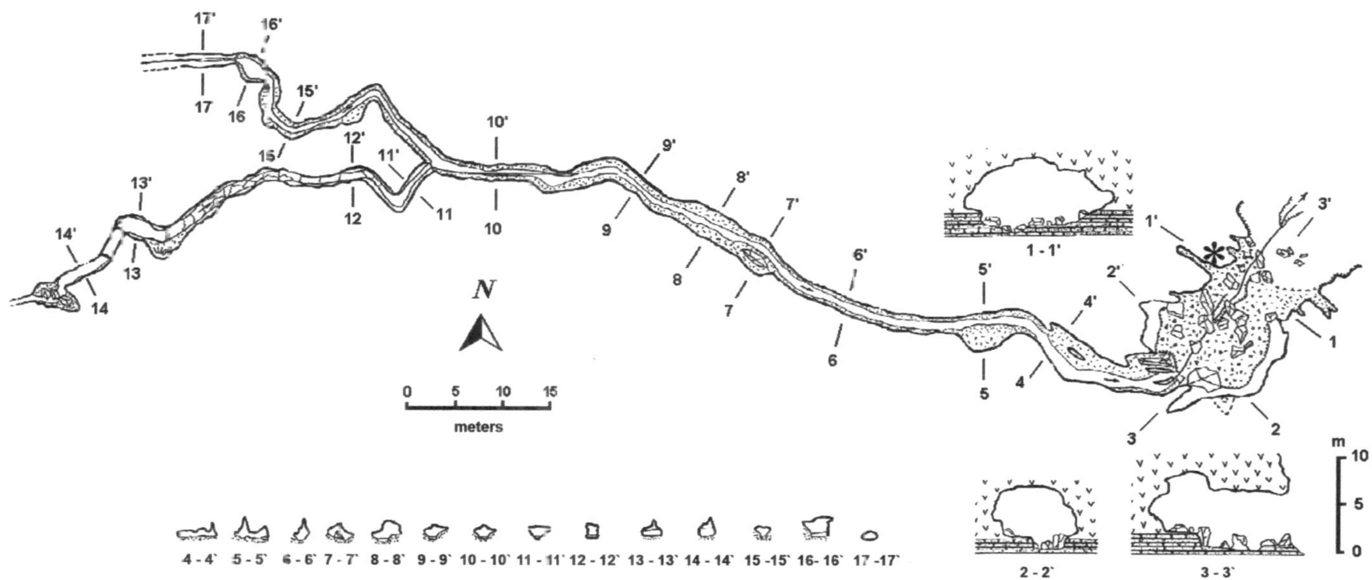

Fig. 2 Plan and cross sections of the Balamutivs'ka Cave (according to Ridush and Kuprich, 2003). *The place with the faunal remains.

(Chernysh, 1975). The drawings that were on the walls of the cave and partly on the ceiling were made on gypsum with black paint—soot mixed with fat and covered with a thin film of secondary gypsum. Areas with images in the form of various drawn lines carved in the gypsum surface were also noted. In total, more than 60 drawings were found in the cave, which were located mainly in groups. The dimensions of these black drawings did not exceed 25 cm. The ancient images consisted of anthropomorphic, animal, and linear geometric drawings. Among the drawings are images of hunting scenes, images of animals, human figures with a bow, etc. In the center is the figure of a shaman. According to O. Chernysh, this cave was a place where the ritual magical ceremonies of the Mesolithic tribes that lived nearby and for which hunting was the main branch of the economy took place (Chernysh, 1975).

In the following years, most of the drawings disappeared due to the collapse of sections of the walls and ceiling caused by frost weathering. When examining the grotto in 1985 O.P. Chernysh only found some fragments. Because of this, some authors doubted the age of the paintings in the Balamutivs'ka Cave, also due to the alleged impermanence of gypsum rocks, high humidity in the grotto, etc., they considered it "unlikely that the cave's destruction went at an accelerated pace only now, and that before the walls of the grotto were preserved the same appearance as in the Mesolithic" (Formozov, 1969, p. 143).

In our opinion, judging by the morphology of the cave and the ground relief around it, in the past (including in the Mesolithic) the cavity of the modern entrance grotto was at some distance (tens of meters) from the land surface. Due to this, the microclimatic conditions (temperature and humidity) in the grotto remained relatively stable for thousands of years, which contributed to the preservation of the drawings. Only over time, because of the slope regression, the grotto was opened and found itself in the zone of daily and seasonal temperature and humidity fluctuations, and as a result the progressive physical weathering was activated in it (Kochergan and Ridush, 2011; Ridush, 2004a).

The few small dead-end galleries adjacent to the main chamber contain slender-thick (up to 1.0 m) deposits of light-yellow loess-like loam and gypsum debris. The screening of the sediments resulted in finding the remains of Chiroptera, *Lepus* sp., *Vulpes* sp., Rodents, Aves, Reptiles, Amphibia, Pisces, Mollusca, and Diptera puparium. The species to be determined. A single little flint flake was also found.

4.3 Borshchiv

The bones of large and small vertebrates were discovered by workers in 1957–1958 and K.A. Tatarinov in 1960 in a large gypsum quarry located on the southern outskirts of the town of Borshchiv (Ternopil Oblast), on the way to the Vysichka village.

Tatarinov (1961) gives the following geological section of outcrop in the quarry:

(1) of burozem soil—0.3 m;
(2) yellowish-brown loess loam—7.0 m;
(3) yellowish-gray slightly cellular limestone—1.0 m;
(4) coarse-crystalline gray gypsum with inclusions of fine-crystalline white gypsum—19.0 m.

Numerous bones (about 10 vertebrae, individual teeth, fragments of limbs, ribs, fragments of the skull and lower jaw) of a woolly rhinoceros (*Coelodonta antiquitatis*) were discovered in a "cleft of gypsum sediments filled with loess loams." The bones were bedded about 8–9 m deep from the soil surface. Likewise, at a depth of 5 m from the surface, "in the gap between the gypsum deposits" were the remains (fragments of tubular bones, ribs, and fragments of teeth) of a mammoth (*Mammuthus primigenius*). A unique find is the jaw of a ram (*Ovis* sp.), which was bedded at the contact of loam with gypsum, 7–8 m deep from the soil surface. The other fauna, *C. elaphus*, *Cricetus*, *Ochotona pusilla*, *Apodemus* sp., *Microtus* sp., *Spalax* sp., are associated with the loam strata (Tatarinov, 1961). Since the remains of the rhinoceros and mammoth are represented by almost complete skeletons, there is reason to believe that they fell into a karst cavity in gypsum, where they died and were buried.

4.4 Bukovynka Cave

Geological and speleological conditions of the site. Bukovynka Cave is located near the. Stalnivtsi village of the Chernivtsi Oblast (48°17′9.00″N, 26°32′4.00″E). It is developed in the upper part of the Middle Badenian gypsum strata (N_1bd_2) with a total thickness of up to 35 m. In the area of the cave, above the gypsum are bedded "Ratin" limestones (up to 1 m) and gray and gray-green clays. Quaternary terrace deposits are represented by pebbles and loams. The modern entrances to the cave, of which there are several, are in an inactive gypsum quarry cut into the left side of the Matka River valley, 1.5 km southwest of the village. Two entrances are located at the foot of the quarry ledge, and another entrance, in the form of a vertical 5-m well, is on the surface of the same ledge. The total length of the cave passages known today is 5460 m, the vertical amplitude is about 15 m. The altitude of the entrances at the foot of the escarpment is about 143 m a.s.l., which is about two meters above the bottom of the adjacent valley of the Prut River tributary, and 25 m above the level of the Prut River.

The cave was discovered and investigated by the Chernivtsi Speleological Club "Troglodite" since 1976 (Ridush and Kuprich, 2003; idush et al., 1998).

Bukovynka Cave is situated in the southeastern part of the Podillia-Bukovynian karst area (Ukraine), in the middle part of the Prut River valley, close to the Romania and Moldova borders (48°17′9.00″N, 26° 32′4.00″E) (Figs. 1, *23*). The cave is developed in the upper part of the Upper Badenian gypsum strata (N_{1bd2}), which are up to 35 m thick. The gypsum deposit in this area consists of two lithofacies. The lower lithofacies are represented by stromatolitic gypsum with intercalations of clastic gypsum, and the upper one is represented by saber and crystalline gypsum (Peryt, 2001). The sulfate bed is underlain by few meters of sand (with 0.5 to 1.0 m sandstone cap) of Lower Badenian age. The top of gypsum layer is covered with up to 1 m of Ratyn Limestone and up to 8 m of the dark-gray clay of the Kosiv Formation (N_{1ks}), with abundant algae (*Lithothamnion*) limestone inclusions.

Quaternary deposits include river-terrace alluvium of the Prut River, composed of so-called "Carpathian Pebbles" of Early Pleistocene age, and gray sand, covered with Middle- and Late Pleistocene pale-yellow loess. Due to the different erosional opening of the massif, locally the Quaternary cover is directly on top of the Ratyn Limestone. The bedding of overlying deposits is strongly distorted by slope failures, which affect both Quaternary deposits and the upper Neogene clays.

Modern cave entrances are artificial. They are in the inactive gypsum quarry, cut in the foot of the left slope of the valley of the Matka River, a left tributary of the Prut River. Two cave entrances can be observed at the foot of the quarry bench, and another entrance is a 5-m vertical shaft on the surface of the same bench (Fig. 3). The total length of the presently known cave passages is 5155 m; the height is about 15 m, the total area is about 7000 m^2, and the volume is 11,250 m^3. The altitude of entrances at the foot of the bench is 135 m a.s.l., 2 m above the adjacent valley bottom, and 25 m above the Prut River (Ridush and Kuprich, 2003).

The cave is of a maze type. It was developed in three levels/floors. The main (middle) floor consists of a series of sub-parallel galleries (2–3 m wide), connected by narrow (up to 1 m) passages. The upper floor is represented by

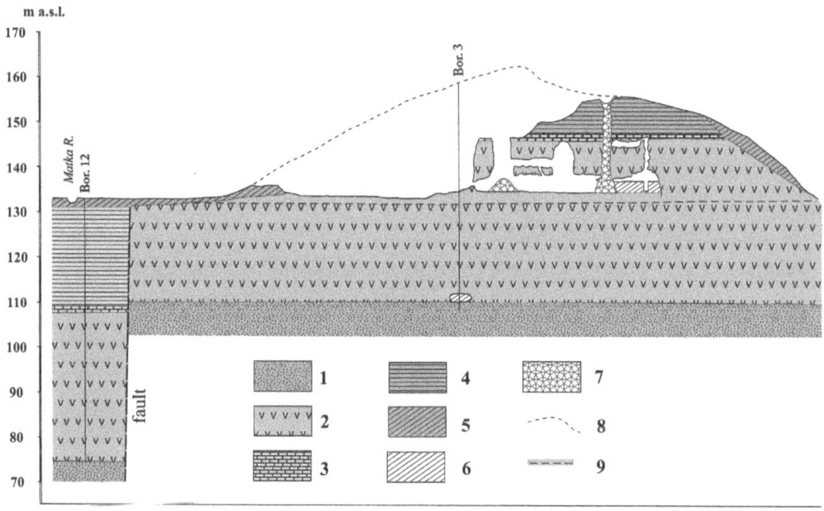

Fig. 3 Geological section of the gypsum karst massif, enclosing the Bukovynka Cave. Legend: 1—Lower Badenian sands; 2—Lower Badenian gypsum; 3—"Ratyn" Limestone; 4—clays of the Kosiv Formation (Miocene); 5—Quaternary terrace loamy deposits; 6—Quaternary loamy deposits in the cave; 7—collapse deposits in the cave; 8—surface line of the massif before gypsum developing; 9—the average groundwater table. Bor.—borehole (according to Bondar and Ridush, 2015).

mostly narrow fissures (0.5–1.0 m) of corrosive origin, narrowing upwards. Commonly, this floor is fragmented and is connected to the main floor by vertical pits/chimneys of 7–10-m in height. The lower floor is represented by mostly narrow (1.0–1.5 m) galleries with rounded vaults. They are partially filled with clay-loamy sediments. This floor is located at the zone of long-term groundwater table fluctuations, and therefore from time to time it is completely flooded. It connects to the main floor through short "windows" in the form of lakes-siphons (with areas to 1.3 m^2). Fluctuations of the water table in cave lakes and, consequently, aquifer varies within 0.2–1.0 m. There are both seasonal and longterm groundwater table fluctuations. The water temperature in the aquifer is 8.5 °C. Water is highly mineralized, with a hydrocarbonate-sulfate content of up to 2.0–2.5 g/L.

The bottom and partly the walls of corridors are covered with thick (up to 4 m) loam-clay deposits. Thin red and black interlayers can be observed in the cave lacustrine clay sequences, which correspond to the oxidizing and reducing hydrochemical conditions in underground paleoponds (Andreychouk, 2007). The upper unit of the Quaternary sequence usually is composed of light loam. It often includes fragments of white, gray, and sometimes reddish calcite flowstone crusts that detached from the ceiling. Today, the interior parts of the cave are thermoconstant with a year-round temperature of 9–10 °C. At the entrance, because of the existence of several inputs, there is an active air exchange with the surface, and temperatures inside the cave depend on the temperature conditions outside. In the winter due to the cooling of the near-entrance galleries, the seasonal ice forms stalactite-like icicles, columns are formed, and clay-loamy sediments are frozen to 0.5 m depth (Ridush and Levytska, 2005).

Geological conditions of the cave location, galleries morphology, and the maze structure of the cavity indicate the hypogenic (artesian) origin of the cave (Klimchouk and Andreychouk, 2017). Most of the maze system was developed before the gypsum strata, and the cave became exposed by erosion.

The topographic relief above the cave is a high and ancient (Early Pleistocene) terrace of the Prut River, 70–80 m above the modern water level in the river. During Middle-Late Pleistocene, the local Matka River, a tributary of the Prut River, cut into this surface and exposed the gypsum strata. Sometimes the stream flowed underground, transiting through the cave chambers. The river stream was flowing through a ponor, which was located at the foot of board of the valley, to the north of the cave. Probably, the discharge of the cave stream on the surface occurred 100 m southwest from the present cave entrance, at the edge of the valley. This part of the cave, from the modern entrance to the place of former underground stream discharge, was destroyed by quarrying of gypsum deposits. The part of the Bukovynka Cave, which consists of the Entrance-, Trapeznyi- and Sukhyi (Dry) chambers, differ from other parts of the cave in morphology and Quaternary succession (Fig. 4). These three chambers were, in the past, a single gallery, which later was divided by collapses (Fig. 4B).

To characterize clastic deposits in the cave, three sections were selected in alignment with the paleostream riverbed. Sections 1 (Fig. 5A) and 2 (Fig. 5B) are situated in the Trapeznyi Chamber and can be observed in natural outcrops on the sides of the chamber. They became exposed due to the subsidence of sediments in the central part of the gallery-like chamber. The stratigraphic and granulometric data of these sections were published (Gerasimenko et al., 2019). The succession in both sections is similar but the description is given for Section 2 as it is the more complete one (Table 1 and Fig. 5B).

Unit T2 (Fig. 5A2), containing bones and coprolites, could be considered as alluvial because of the inclusions of individual gravel size clasts.

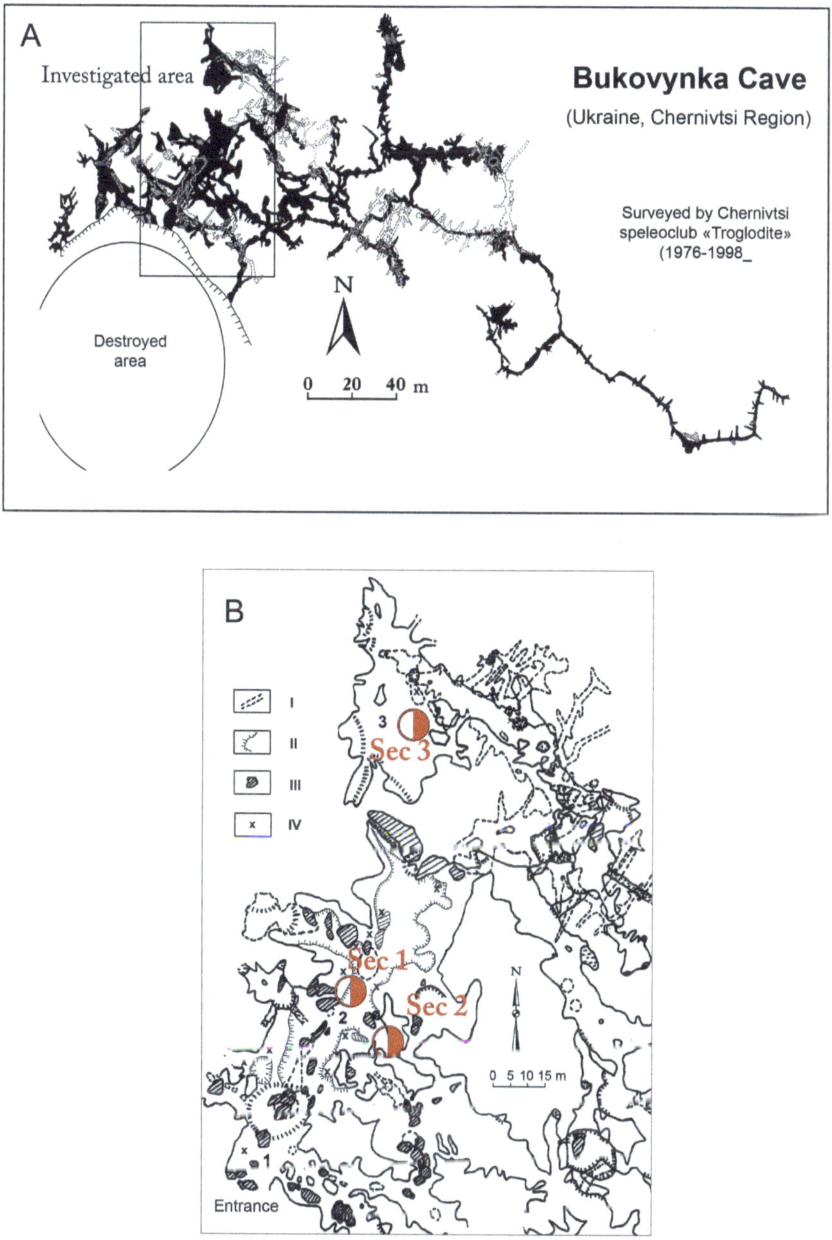

Fig. 4 Map of Bukovynka Cave with location of the sections studied. (A) general plan of the cave. (B) investigated area of the cave maze. I—upper level of galleries, II—inner benches formed by uncemented deposits, III—fallen rock blocks, IV—places of bone finds; 1—Entrance Chamber, 2—Trapeznyi Chamber, 3—Dry Chamber; "Sec 1"—places and numbers of Quaternary deposits section (according to Bondar and Ridush, 2015).

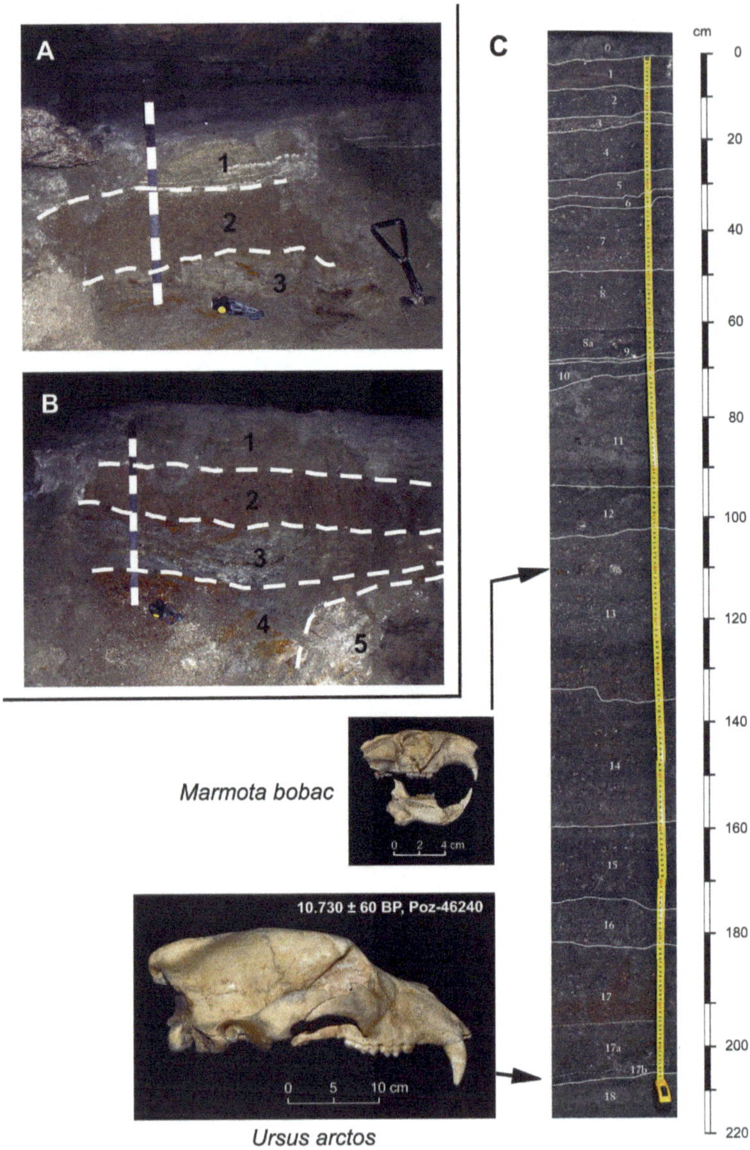

Fig. 5 (A and B) Late Pleistocene deposits of the Sections 1 and 2. 1—Unit T1, Eolian deposits ("cave less"); 2—Unit T2, zoogenic layer ("hyena layer"); 3—Unit T3, fine alluvial deposits of transit stream; 4—Unit T4, gravel alluvial; 5—Unit T5, gypsum bedrock. (C) Late Pleistocene—Holocene deposits in Dry Chamber. 1–18—Units S1–S18 (according to Bondar and Ridush, 2015; Gerasimenko et al., 2019).

Table 1 Section 2 of clastic sediments in Trapeznyi Chamber (Figs. 4 and 5) (according to Bondar and Ridush, 2015).

Unit	Depth, m	Description
T1	0.0–0.3	Light loam, loess-like, laminated, light yellow, with numerous inclusions of secondary gypsum crystals, with rare inclusions of coprolites, small bone fragments, with interbeds enriched with fine-crystalline gypsum (up to 1–2 cm).
T2	0.3–0.6	Light-brown loam, cloddy, with rare inclusions of gruss of limestone and gruss solid laminated (argillite-like) clay, occasionally inclusions of quartz gravel and Lithothamnion limestone debris, as well as coprolites and bone remains from small (up to 0.5e2.0 cm) to individual Mammals teeth and large bones fragments.
T3	0.6–0.9	Gray laminated sandy clay, sometimes as gruss of the same clay. In Section 2 the gravel interlayer (less than 10 cm) with a predominance of Lithothamnion limestone gruss and with inclusions of pebbles and gravel can be observed on top of the unit.
T4	0.96–1.2	Gravel-pebble deposits (predominance of gravel of quartz and sandstone rocks), with an admixture of limestone gravel (*Lithothamnion*) up to 30%, reddish, enriched with FeO2 (redeposited river terrace pebble)
T5	>1.2	Bedrock: macrocrystalline, grayish-brown gypsum

However, stratification was not observed, and all bones and coprolites found in it have no rolling traces, while most of the coprolites are quite fragile and would be destroyed if moved by water-flow. All these remains were found in situ and were not transported and redeposited by water flows. The genesis of these unit is probably almost entirely zoogenic, with only minor eolian input. The lower part of the Sections (units T3-T4) is composed of alluvial cave deposits containing redeposited Quaternary and Neogene sediments.

The ^{14}C age of Unit T2, based on an included cave hyena tooth is 41,300 (+1300/−1100) BP (VERA-2529), cal 12,765–12,621 yrs. BP, which corresponds with the warm Vitachiv Stage (MIS-3) (Gerasimenko, 2010). Zoogenic accumulation took place probably continuously during this phase. The overlying eolian sediments (Unit T1) could correlate to one of cold episodes during MIS-3 (Gerasimenko et al., 2019).

To avoid confusion in units from different localities, the lithofacies units in the Trapeznyi Chamber are designated "T," and units from the Dry

Chamber are marked "S." The characteristic features of sediments of Unit T1 (Fig. 5A1), such as silty composition, stratification, and lack of gravel inclusions, as well as the presence of gypsified intercalations, indicate that the top of Sections 1 and 2 (Unit T1) include airborne sediments (Gerasimenko et al., 2014b).

The sediments of Section 3 were studied by Bondar and Ridush, 2015; Gerasimenko et al., 2014a, 2019). They are represented mainly with grayish-brown loams with limestone grus inclusions (Fig. 5C). A prospecting shaft 2.5 m deep was excavated for this purpose. The screening of loams from Unit S17b showed impurities of travertine grus up to 5%. A similar type of travertine, which was formed underwater, can be observed on the walls and ceiling of the Zolushka Cave (7 km to the east (Andreychouk, 2007; Andreychouk and Klimchouk, 2017). The latter was drained by a quarry only a few decades ago, and preserved many subaquatic features, including this fragile travertine, only 1–3 mm thick.

There are five bone sites in the cave. The first (**Bukovynka-1**) contains the *Trapezny Chamber* and the adjacent galleries. It is the largest chamber in the cave. It is elongated for 40 m from SW to NE. From both ends it's limited by collapses. It can be considered as a wide gallery, that formerly was continued by the Sukhyi Chamber, that now is separated by the large collapse (Fig. 5). The palaeontological site consists of numerous cave hyena dens, located mainly close to the walls of the chamber and galleries, containing numerous coprolites and less numerous bone fragments. In the Unit T2, the remains of several genera and species of large Late Pleistocene mammals were found: cave hyena (*Crocuta spelaea*), woolly rhinoceros (*Coelodonta antiquitatis*), cave bear (*Ursus spelaeus*), brown bear (*U. arctos*), horse (*Equus caballus*), steppe bison (*Bison priscus*), red deer (*C. elaphus*), giant deer (*Megaceros giganteus*), wild pig (*Sus scrofa*), and red fox (*V. vulpes*). Cave hyena remains are represented mainly by numerous fragments of mandibles, maxillae, and separate teeth of at least 10 individuals. In the same unit that contains the bone remains, cave hyena coprolites are quite abundant. Some coprolite material forms continuous layers that apparently formed floors of the cave hyena den. The remains of most species, excluding fox, are the remnants of hyena prey, including old and young individuals of cave bear. Evidence of cave hyena activity as unidentified fragments of crushed bones and coprolites were found in a broad area of cave maze, more than 100 m from all possible ancient entrances (Ridush, 2004b; Vremir et al., 2000).

The second site (**Bukovynka-2**) is in the northern corner of the Sukhyi Chamber. It is represented by only one find—the proximal end of the radius of a bison (*Bison priscus*) of Pleistocene age. The bone was discovered in cone-shaped body made of loose collapsed roof material. A crust of secondary gypsum crystals 1–3 mm thick formed on the surface of the bone and in its internal canal. Judging by the conditions of the find and the regularities of cave-in formation above the gypsum caves (for details on the mechanism of cave-in formation in the gypsum caves of Bukovyna, see (Klimchouk and Andrejchuk, 2002), this bone did not enter the cave from the Quaternary terrace loams, since no traces of these loams were found within the collapse cone. The latter is composed of Neogene clays lying directly above the gypsum. The bone fragment was captured by a collapse from the upper floor of the cave, an unknown part of which is apparently located above the Sukhyi Chamber. Along with the bone, fragments of flowstone calcite formations were brought from the upper floor. In the Bukovynka Cave, this flowstone is mostly formed in the axial cracks on the ceiling of the galleries of the upper floor due to the redeposition of carbonates from the overlying Ratin limestones by condensation waters.

The third paleontological site **Bukovynka-3** (Section 3 according to (Bondar and Ridush, 2015)) is situated in the Dry (Sukhyi) Chamber (Fig. 5C). In the Dry Chamber and adjoining galleries, Holocene faunal remains were found. In the trench (Section 3) at a depth of 1.1 m in Unit S13 a partly preserved cranium and a mandible branch together with some postcranial bones of marmot (*Marmota bobac*) were found. In the same section at 1.8 m depth, at the bottom of Unit S17b, an almost complete cranium (without mandible) of brown bear (*U. arctos*) was excavated (Marciszak et al., 2015) (Fig. 5C). Direct ^{14}C dating provided an age of the Late Pleistocene to Early Holocene transition (10,730 ± 60 BP, Poz-46,240, cal 10,815 BP) (Ridush et al., 2012). At a depth of 2.2–2.5 m the "hyena layer" was found, that is similar to the one in the Trapeznyi Chamber. It is lithologically represented by pale-yellow loam including cave hyena coprolites and bone fragments.

The site of **Bukovynka-4** is on the upper floor of the cave. Here, in the Stolovyi Chamber, a partial skeleton of a red fox (*V. vulpes*) of the Holocene age was found in clay deposits at a depth of 20 cm. Lumps of burnt red clay and shells of terrestrial mollusks are found in the overlying layer. In another place on this floor, a humus layer (3–5 cm) with numerous inclusions of small mammal bone fragments, which belong to foxes' prey (also of Holocene), is observed under a 5–10-cm layer of cave loam.

4.5 Cholivska Gora (Chortivets)

O.M. Adamenko together with D.O. Prunko and P.I. Arsenych examined in the gypsum quarry a karst dolina with bones of large mammals (Table 2). The site is on the north-eastern outskirts of the Chortivets village, Horodenka District, Ivano-Frankivsk Oblast, on the right slope of the valley of the Zagreblia Stream (Cholivska Gora) (Fig. 1A 9), at an altitude of 40 m on the bed of a stream, the valley of which is deeply cut into a plateau composed of Badenian gypsum and limestone (Fig. 6) (Adamenko, 1976).

The fauna from layer 3 was collected by D.O. Prunko and is stored in the Ivano-Frankivsk Museum. Parts of the collection has been taxonomically determined. Among the bones are *Bison priscus*, *Canis lupus*, *Crocuta spelaea*. The cultural layer belongs to the late Paleolithic with Aurignacian features. According to L. Matskevyi its faunistic content and flint artifacts make up a single complex (Matskevyi, 2005).

Table 2 Cross-section of the sediments of the buried cave in the Chortivets Village (according to Adamenko, 1976).

nn	Description	Depth, m
1.	Modern soil of the chernozem type	0.0–0.2
2.	Debris and boulder deposits with loamy filling	0.2–0.4
3.	Debris-clay deposits with a mass accumulation of large mammals' bones, which compose up to 60–70% of the sediment. Bones are badly destroyed, loose, and fall apart when trying to extract them (fauna of Chortivets)	0.4–1.2
4.	Loams are laminated, light brown and gray, dense, with rare fragments of mammal bones. At the bottom of the layer, traces of firing (the fireplace of ancient man) and several scrapers made of patinated flint were found	0.4–0.8
5.	gray to dark gray loams, earthy soil deposits	0.8–0.9
6.	Light brown and gray loams that fill the bottom of the karst dolina and therefore lie with different slopes towards the center of the funnel, parallel to its slopes	0.9–1.6

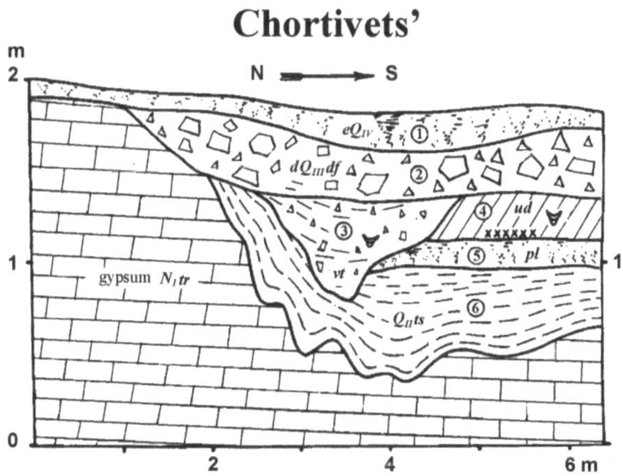

Fig. 6 Section of the filling of the buried karst dolina in the Chortivets village (Cholivska Gora) (according to Adamenko, 1976).

4.6 Chortkiv

In the western region, the oldest Quaternary fauna was recovered in 1953 near the town of Chortkiv, Ternopil Oblast (Fig. 1A, *11*), by participants of the paleontological expedition of the Institute of Zoology of the Academy of Sciences of the Ukrainian SSR. The location is located on a high terrace, on the left bank of the Seret River, in an ancient grotto, which was formed on the site of a karst dolina in calcareous sandstones of the Badenian (Miocene) and is filled with reddish-yellow medium-grained sands. The discovered fauna dates to the Eopleistocene ("early Homycene" according to I.H. Pidoplichko). The lens, where the fauna was found, was about 3 m long, 0.8 m thick and located at a depth of 3 m from the surface. Reddish loams and clays of colluvial origin are near the lens. The composition of the fauna was described by Pidoplichko (1955, 1956b) (Table 3). Recently, based on the fauna of Rodents the site was dated to the second half of the Odesian Fauna (Kryzhanivka Stage, MIS 41–62) (Krokhmal' et al., 2021).

4.7 Chortova Skelia (Lviv VII)

It is located on the eastern outskirts of Lviv, on the land of the "Pid Skeleyu" park which is part of the biosphere-landscape reserve Chortova (Chatova) Skelya (49°19′43.77″N, 24°7′2.74″E). The Marunka River flows to the south of the exposure, and the Poltva River to the north (Fig. 1A, *1*). The site is at an altitude of 65 m above the river level. The rock shelter is exposed to the north.

Table 3 The species composition and number of skeletal remains of mammals from Chortkiv site (according to Pidoplichko, 1956b).

Species	NISP	MNI
Sorex araneus L.	1	1
Sorex macropygmaeus Pall.	1	1
Blarina (ver. Beremendia) ucrainica Pidop.	1	1
Myotis sp.	1	1
Miniopterus sp.	4	3
Ochotona pusilla Pall.	4	3
Citellus cf. *undulates* Pall.	1	1
Citellus cf. *suslica* Guld.	1	1
Glis L.	4	2
Sicista sp.	1	1
Microspalax cf. *leucodon* Nord.	1	1
Cricetullus migratorius Pall.	2	2
Lemmus cf. *lemmus* L.	1	1
Mimomys sp.	23	5
Allophajomys sp.	219	60
Meles meles L.	1	1
Mustelidae	1	1
Hyaena sp.	2	1
Rhinoceros sp.	1	1
Cervus elaphus L.	1	1
Bos primigenius Boj.	4	1

The lowest cultural layer (A), with a thickness of 15–20 cm, is traced at a depth of 240–260–350–360 cm from the rapper. It consists of numerous blocks of light sandstone of Miocene, the spaces inbetween are filled with loess-like loam. About 6000 osteological remains were found in the layer, which were only partly determined by K.A. Tatarinov and V.V. Gumenyuk. The fauna is represented by the mammals *Mammuthus primigenius*, *Rangifer tarandus*, *Lepus timidus*, *C. cricetus*, *Arvicola terrestris*, and birds, *Anser*, *Anas plathyrhyncha*, *Limosa*, *Vanellus*, *Anas guerguedula*, and *Turdus* sp. The Paleolithic industry is assigned by Matskevyi to Aurignacian tradition. This corresponds to the ^{14}C data of 27,200 ± 170 yrs. BP (Ki-5415) (cal 31,596–31,061 yrs. BP) (Matskevyi, 2001, 2005).

The middle layer (B), with a thickness of 15–20 cm, was traced at a depth of 190–210–270–320 cm. It is represented by loess-like loam, pale yellow in color, homogeneous, macroporous, and non-calcareous. There are large blocks that reach more than 50 cm over the long axis in the bedding. The contact to the underlyering bed is unclear and only sometimes discernible. More than 4000 bones were found in the layer. Only a part of the remains was determined (Matskevyi, 2005). According to radiocarbon dating, the layer dates to 13,500 ± 110 yrs BP (Ki-5412) (cal 16,635-15,933 BP).

The Upper Paleolithic layer (C), with a thickness of 20–25 cm, is traced in the third lithological horizon, at depths of 130 (140)–190 (210) cm. The deposits are represented by dark gray, homogeneous loam. There are inclusions of sandstone rock fragments, rubble, and debris in the layer. The lower contact with the underlying deposits is clear in color and uneven. More than 2000 osteological remains belong to the cultural layer. Only 36 bones from 3 individuals of *Mammuthus primigenius* and 52 bones from 6 individuals of *R. tarandus* were identified. Gnawing marks are visible on some bones. A piece of Paleolithic mobile art, an image of a reindeer antler, was found. Based on stratigraphy, osteological remains and industry, as well as radiocarbon dates, the layer is assigned to the Late Paleolithic with elements of the Late Aurignacian tradition, although recorded in Gravettian times. The radiocarbon dates of the animal bones provided an age of 11800 ± 90 BP (Ki-5414) (cal 13,991-13,466 BP), which corresponds to the Dryas 2 (Older Dryas), and possibly the beginning of the Allerød interstadial (Matskevyi, 2005).

Unfortunately, the list of fauna from this site in the text and in table by Matskevyi (2005) does not match, and therefore we will not include the table here. The faunistic material is partly stored in the Krypiakevych Institute of the Ukrainian Studies of the Ukrainian Academy of Science, in Lviv, and can be revised (Table 4).

4.8 Chun'kiv

In 2009, when digging a water supplement well in the Chun'kiv village of Zastavna District, Chernivtsi Oblast (Table 1, 19), the bones of a woolly rhinoceros (*Coelodonta antiquitatis*) were found. The bones were bedded at a

Table 4 The species composition and number of skeletal remains of animals from Divychi Skaly (according to Pidoplichko, 1956a, p. 128).

Species	NISP	MNI
Mammuthus primigenius	4	2
Equus	5	2
Bison priscus	7	2
Cervus elaphus	8	1
Rangifer tarandus	38	3
Marmota bobac	40	4
Ursus spelaeus	5	1
Citellus suskicus	1	1
Dicrostonyx torquatus	1	1
Microtus oeconomus	3	2
Microtus ex. gr. *arvalis*	4	2
Lepus sp.	12	1

depth of 5 m, obviously in the filling of a karst cavity, since the depth of gypsum in the surrounding area does not exceed 1.5–2.0 m. Few metapodials were collected, but the workers told us that it was a complete skeleton, that is still lying there. We are probably dealing with the fall of a heavy animal into the karst sinkhole.

4.9 Divychi Skaly

In 1953, I.H. Podoplichko conducted excavations at the site of Divochi Skeli (Divychi Skaly), situated in the rocky massif of the Kremenetski Mountain, to the NW from the Kremenets town, of Ternopil Oblast (50°7′5.54″N, 25°43′38.86″E) (Tables 1, 7). The bones were bedded in the loess-like loam that filled the karst cavity, at a depth of 1–2 m from the surface. As the researcher believed, the large bones were washed into the fissure from above, and the small ones came from the bird pellets (Pidoplichko, 1956a, p. 128). Since charcoal and ash stains were also found in the loam, it is possible that some of the bones were associated with the Paleolithic cave site.

To the list were added also some bones stored in the Kremenets Museum.

Also, in the Zhytomyr museum some bones of cave hyena and cave bear from Kremenets are stored (Pidoplichko, 1956a, p. 128). We consider it to be quite probable that these bones derived from the same locality.

4.10 Divychi Skaly—Cave Bear Grotto

In 1956, K. Tatarinov excavated a karst niche in the same place of Divychi Skaly, on its northeastern outskirts, which he called Cave Bear Grotto (Fig. 1A, *7*). Under the large block that forms the rock shelter, in loess-like loams up to 1.5 m thick, filling a wide (up to 3–4 m) crack in calcareous sandstones, the remains of large mammals were concentrated (Fig. 7 and Table 5) (Tatarinov, 2000, pp. 70–74).

4.11 Horishnia Vygnanka

The site was in a sand quarry near the Ternopil—Chernivtsi highway, 0.5 km from the Gorishnya Vygnanka village near Chortkiv, Ternopil Oblast (Fig. 1A, *12*). It's represented by the infilling of an ancient grotto, formed in a layer of Lower Badenian white quartz sands, under a cornice of limestones of the same age (up to 1 m). The remains of 30 species of

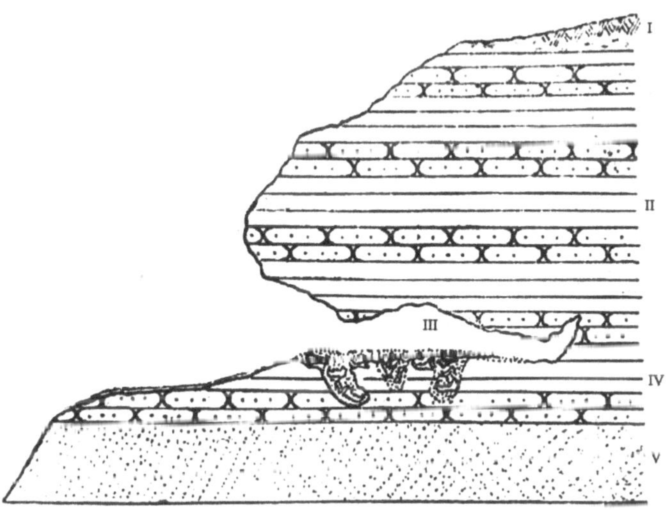

Fig. 7 Schematic section of the Divychi Skeli site, where the vertebrate fauna was found. I—modern soil layer, 0.3 m; II—porous calcareous sandstone (Sarmatian, Miocene), 7.5 m; III—a crack in the sandstone in which fauna was found; IV—bone-bearing layer—fissures in calcareous sandstone filled with terrestrial sediments containing vertebrate fauna; V—Badenian fine-grained quartz sand, visible thickness 1.5 m (according to Tatarinov, 1962).

Table 5 The species composition and number of skeletal remains of animals from Cave Bear Grotto (according to Tatarinov, 2000, p. 72).

Species	NISP	MNI
Dicrostonyx torquatus	1	1
Alopex lagopus	4	1
Ursus spelaeus	168	21
Crocuta spelaea	48	2
Mammuthus primigenius	3	1
Equus	12	1
Coelodonta antiquitatis	4	1
Megaloceros giganteus	2	1
Cervus elaphus	1	1
Rangifer tarandus	52	3
Bison priscus	7	2

mammals and other vertebrates were found here: mouse-eared bat (*Myotis* sp.), Pliocene bear (*Ursus* cf. *wenzensis*(=*minimus*) Stach.), forest cat (*Felis* cf. *silvestris* Schreber.), Pliocene badger (*Arctomeles* (= *Parameles*) *pliocaenicus* Stach.), fox (*Vulpes* sp.), jackal (*Canis* cf. *aureus* L.), eared hedgehog (cf. *Hemiechinus* sp.), small mole (*Talpa minor* Freudenberg), Ukrainian blarina (*Blarina ucrainica* Pidoplitshko), hypolagus hare (*Hypolagus* sp.), small pica (*Ochotona* cf. *pseudopusilla* Gureev et Schevtschenko), marmot (*Marmota bobac podolicus*), small shelf dormouse (*Glis minor* Kowalski), parapodemus mouse (*Parapodemus* cf. *coronensis* Schaub.), large hamster (*C. cricetus* aff. *Major* Woldrich.), Seret hamster (*C. cricetus seretensis* subsp. n.), Mimomys vole (*Mimomys intermedius* Newton.), Allophajomys vole (*Allophajomys pliocaenicus* Kormos.), mole-rat (*Spalax minor*), deer (*Cervus* sp.), bull (*Bos* sp. vel *Leptobos* sp.), Etruscan rhinoceros (*Dicerorhinus etruscus* Falconer) (Tatarinov, 2000; Tatarinov and Bachynskyi, 1968) (Table 6).

The site dates to the Late Pliocene (Villafranchian). Wagner identifies *Ursus wenzensis* Stach. with *Ursus minimus* Devéze et Bouillet, 1827, and notes that these bears survive the Ruscinian–Villanian boundary without changes and occur from MN 15b to MN 16b (Wagner, 2010). According to the fauna of small mammals, L. Rekovets, refers the site to the Zhevakhov phase of the Early Eopleistocene (Rekovetz, 1994, p. 13).

Table 6 The species composition and number of skeletal remains of animals from Horishna Vygnanka (according to Tatarinov, 2000).

Species	NISP	MNI
PISCES		
Cyprinidae gen. et sp.	1	1
AMPHIBIA		
Eopelobates cf. *bajeri* Spinar	6	1
Archipelobates giganteum	1	1
Hyla sp.	16	3
Rana sp.	34	3
Indet	82	–
REPTILIA		
Lacertilia	3	2
Serpentes	1	1
MAMMALIA		
cf. *Hemiechinus* sp.	1	1
Blarina ucrainica Pidop.	1	1
Talpa minor Freud.	2	2
Mytis sp.	1	1
Hypolagus sp.	5	1
Ochotona cf. *pseudopusilla* Yur. et Schev.	7	2
Marmota bobac podolicus	8	1
Glis minor Kowal.	17	7
Spalax minor Kowal.	1	1
Parapodemus cf. *coronensis* Schaub.	1	1
Cricetus cricetus cf. *major* Wold.	31	4
C. cricetus seretensis	12	1
Mimomys intermedius Newt.	7	3
Allopajomys pliocaenicus Korm.	4	1
Canis cf. *aureus* L.	1	1

Continued

Table 6 The species composition and number of skeletal remains of animals from Horishna Vygnanka (according to Tatarinov, 2000).—cont'd

Species	NISP	MNI
Vulpes sp.	14	2
Arctomeles cf. *pliocaenicus* Stach.	16	2
Ursus cf. *wenzensis* Stach.	34	2
Felis cf. *silvesrtis* Schreb.	1	1
Dicerorhinus etruscus Falconer	47	1
Cervus sp.	24	1
Bos sp. vel *Leptobos* sp.	55	1
Indet	212	–

In another work, K. Tatarinov wrote that the accumulation of faunal remains was found in two places: (1) in red-orange alluvial poorly sorted sands with an admixture of red clay; (2) in white medium-grained Lower Tortonian sands (redeposited, of course). The first fossil accumulation was located 25–30 m from the second, which gives reason to assume that these locations are of different ages (Tatarinov, 2000).

4.12 Kasperivtsi

In the 1920s, Yu. Polyansky investigated a Paleolithic site under a ruined rock shelter on the bank of the Seret River, in the Kasperivtsi village (Zalischytsky District, Ternopil Oblast) (Fig. 1A, *18*). The cultural layer was associated with a paleosol, covered by the layer of loess and modern soil. Bones of mammoth, woolly rhinoceros, bison, deer, horse, and wolf were found. Except for the mammoth's thick bones, all other remains were splintered, and some bore the marks of blows and cuts. Along with the bones, quartzite stone implements were found (in our opinion, this meant sandblasted flint, which is often found in Cenomanian sediments). The location dates to Mousterian times. Yu. Polyanskyi attributed the site to the interstadial of Würm I and Würm II. I.K. Ivanova believes that, according to modern ideas, this site belongs to one of the early Würm interstadials (Ivanova, 1969; Polians'kyi, 1929).

The exact location of this site is currently unknown. However, we assume that the mentioned rock shelter or grotto was formed in the massive calcareous sandstones of the Cenomanian (Cretaceous), which are common in the vicinity of Kasperivtsi village. Since these sandstones are karstified in some places, the cavity probably had a karst origin.

4.13 Kryshtaleva (Nyzhniokryvchanska) Cave

The Kryshtaleva (Nyzhniokryvchanska) cave is located near the Nyzhnye Kryvche village (Chortkiv District, Ternopil Oblast), on the narrow interfluve between the Tsyganka River and its tributary the Semeniv Potik River (Fig. 8); both are the tributaries of the Dniester River (48°41′20.30″N, 26° 5′16.00″E) (Fig. 1A, *16*). The cave is developed in the 15-m-thick strata of Miocene gypsum, overlain by 1-m-thick crystallin Ratyn limestone. The total length of the cave is 22.6 km. The altitude of the entrances is 235 m a.s.l and 80 above the river valley bottom (Zimels, 2008).

There are four entrances to the cave. In two of them, Serednia (Middle) and Mala (Small) caves K. Tatarinov found some faunal remains. G. Bachynsky had found some bones also in the so-called Palaeozoological Labyrinth. According to these researchers, the bones overflowed the upper 20-cm layer of gypsum-clay debris covering the bottom of the cave gallery (Tatarinov and Bachynskyi, 1968). All these burials are associated with the marginal part of the Crystal Cave, where the cavities were occasionally open to the surface and therefore intensively filled with debris material. The remains

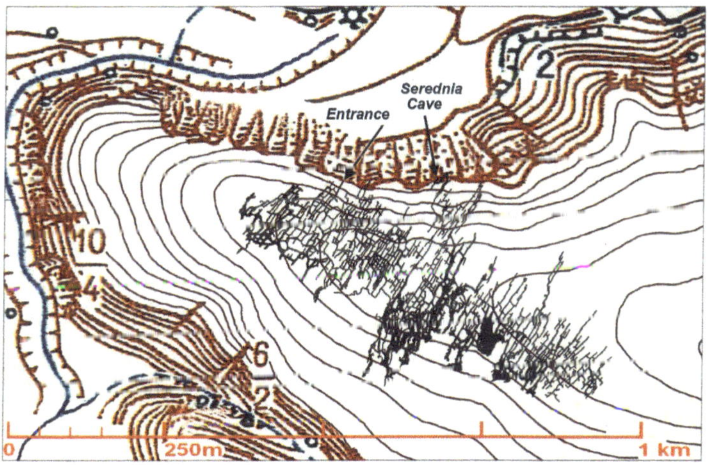

Fig. 8 The location of the Kryshtaleva Cave in relief, according to Zimels (2008).

of small animals predominate here, and among them the remains of pikas and bats. According to K. Tatarinov, the burial was formed mainly from the pellets of diurnal birds of prey. Osteological material was determined by G. Bachynsky, K. Tatarinov, V. Topachevkyi (mammals), M. Voinstvenskyi, I. Marisova (Marysova, 1962), and A. Umanska (birds).

The cave labyrinths near the entrance are almost filled with allochthonous loamy deposits. Remains of the Late Pleistocene (*L. timidus, O. pusilla, Vulpes lagopus, R. tarandus*) and Holocene fauna were found in the cave (Bachynskyi and Tatarinov, 1966) (Table 7), but their stratigraphical position was not known exactly.

Table 7 The species composition and number of skeletal remains of mammals from Nyzhniokryvchanske site (according to Tatarinov, 2000).

Species	NISP	MNI
Talpa europaea L.	2	1
Sorex sp.	2	1
Crocidura sp.	15	3
Myotis cf. *bechsteini* Kühl.	31	2
M. cf. *myotis* Borkhausen	4	1
M. cf. *nattereri* Kühl.	61	9
Myotis sp.	1240	30
Plecotus auritus L.	22	8
Lepus timidus L.	13	3
Ochotona pusilla Pall.	2607	1225
Ochotona sp.	134	100
Marmota cf. *bobak* Müller	15	4
Citellus cf. *citelloides* Kormos	46	15
Glis glis L.	12	4
Dyromys sp.	1	1
Apodemus sp.	2	1
Dicrostonys torquatus Pall.	26	14
Cricetus cf. *cricetus* L.	16	6

Table 7 The species composition and number of skeletal remains of mammals from Nyzhniokryvchanske site (according to Tatarinov, 2000).—cont'd

Species	NISP	MNI
Lagurus lagurus Pall.	3	2
Arvicola terrestris L.	202	66
Microtus oeconomus Pall.	58	17
Microtus cf. *arvalis* Pall.	3	2
Microtus gregalis Pall.	113	47
Clethrionomys glareolus Schreber	7	2
Spalax cf. *podolicus* Pennat.	7	3
Mustela cf. *nivalis* L.	35	6
Mustela nivalis minima	7	4
Mustela erminea L.	19	6
Meles meles sinjakovensis	3	1
Alopex lagopus L.	23	2
Felis sp.	13	1
Rangifer tarandus L.	16	2
Equus caballus L.	1	1

The taxonomical composition was complex. As picas are troglophiles, they probably arranged dens in the cave. Besides them, reliable troglophiles are bats (*Myotis* cf. *bechsteini*, *M.* cf. *myotis*, *M.* cf. *natterei*, and *Plecotus auritus*), badgers (*Meles* cf. *meles*), and probable weasels (*Mustela* cf. *nivalis*, *M. nivalis minima*), ermine (*Mustela erminea*) and several rodents.

On the other hand, the findings of a relatively large number of reindeer bones, represented by fragments of limb bones, phalanges, and the right mandible—a fragment with satisfactorily preserved teeth, are remarkable. In addition, cut marks of a flint tool were found on the jaw. The fossil horse is represented by one well-preserved tooth from the upper jaw, found on a rocky outcrop near the entrance to the Middle Cave (Tatarinov, 1965). Such finds may indicate the habitation of the cave by people.

Unfortunately, the entrance parts of the Serednia and Mala Caves were destroyed in the 1960s by a small gypsum quarry. K. Tatarinov believed that the bones in the Middle Nyzhniokryvchanska cave were washed in by water because until 1959 (prior to the industrial gypsum production), the cave was isolated and therefore inaccessible to predatory mammals or birds (Tatarinov, 1965). We believe that in the late Pleistocene, not one but several entrances to the cave were accessible not only to animals but also to humans. Only later, due to the accumulation of loess layers and activation of slope processes, these entrances were blocked by colluvial deposits.

The new section is studied in the narrow gallery which forms the entrance to the Serednya-2 tunnel of the cave. This is a vertical fissure-like gallery, 0.8–1.0 m wide, 8–9 m high. It is almost completely filled with loose sediments. Only the upper 0.5 m of the profile, through which the gallery connects to the main labyrinth, is unfilled. Up to now, only the upper 4.5 m of sediments were subjected to cleaning. The upper ledge of loose sediments (cave colluvium), about 2.20 m in thickness, has a layered texture and is composed of limestone and gypsum debris and rubble, "crumbs," loam and contains bones of rodents and birds (not studied). The lower ledge, is formed by a 1.0–1.5-m layer of sediments with traces of human activity in the form of numerous charcoal inclusions throughout the layer and three distinct continuous charcoal layers, as well as burnt gypsum (Table 8 and Fig. 9). The profile of all horizons is strongly concave, and the difference in thickness between the central and lateral parts of the layer reaches up to 0.5 m vertically. Several brown bear bones were found in the 12-cm layer separating the anthropogenic sediment pack from the "silent" sandy sediments. The chronostratigraphic subdivision of the section (3 m in thickness) is based on the ^{14}C date of 12,240 $\pm$ 70 yrs BP (Poz-59413) obtained on the *U. arctos* bones (Ridush, 2014), and ^{14}C-dates 11,710 $\pm$ 60 BP (Poz-51431), 11,890 $\pm$ 60 BP (Poz-51430) yielded from charcoal (Nadachowski et al., 2015). The obtained pollen data indicate that the cave infilling was formed during the end of the Middle Pleniglacial, the Late Pleniglacial, the Late Glacial and the Early Holocene. Grain-size studies of the deposits have been also carried out in the section (Avdieienko et al., 2018). According to palynological data (Avdieienko et al., 2018), the bone bearing layer belongs to the Bølling interstadial (12,240 $\pm$ 70 yrs BP (14,6–14,1 cal ka BP)).

4.14 Mala Ilovytsya (Lypa IX, Mount Pustelnya)

In the 1960s, the Lypa IX (Mount Pustelnya) site, located on Mount Pustelnya near Mala Ilovytsya Village (Kremenetsky District, Ternopil

Table 8 Section of the sediments of the lower (Pleistocene) ledge of the excavation in the cave Serednya-2 (Kryshtaleva, Nyzhniokryvchanske Site) (Fig. 9D).

N of layer	Layer description	Depth of the layer base, *m*	Layer thickness, *m*
1	gray loam with inclusions of limestone and gypsum debris ("crush"), transitional to the upper ledge of the section	0.0–0.2	>0.20
2	dark gray loam (gley?), with inclusions of charcoal and debris of burnt gypsum; in the roof there is a carbonaceous layer (1 cm), above and below it—1 cm—interbeds of reddish burnt loam	0.2–0.4	0.1–0.2
3	white and soft burnt gypsum	0.4–0.6	0.15–0.2
4	gray-brown loam, with inclusions and interbeds (up to 1 cm) of charcoal, and debris of soft white gypsum	0.6–0.8	0.1–0.2
5	debris of white soft burnt gypsum with an admixture of loam and with rare carbon inclusions	0.8–0.95	0.1–0.15
6	light brown loam, with the inclusion of charcoal, burnt white gypsum, debris of gypsum and limestone	0.95–1.1	0.1–0.15
7	light yellow sandy loam, with a 1 cm carbonaceous interbed in the roof (layer 7a)	1.1–1.5	0.03–0.06
8	brownish-gray loam, coarse-grained, with inclusions of debris of bedrock clays, gypsum, and limestone	1.15–1.25	0.07–0.1
9	is light brown loam, with inclusions of gypsum eluvium, carbonates, and individual gypsum crystals, with bone remains of *Ursus arctos*, *Rangifer tarandus*, Aves sp. etc. Late Glacial, Bølling interstadial	1.25–1.45	0.15–0.2
10	brownish-yellow loam, with rare inclusions of gypsum debris	1.45–1.6	>0.2

Oblast), was investigated. Mount Pustelnya is a part of the Kremenets Mountains and is located in the border zone between Podillia and Volyn (50°11′49.78"N, 25°59′49.03″E) (Fig. 1A, *8*). The top of the mountain is flat, covered with a layer of loess. On the north-western edge of the

Fig. 9 Section of Upper Pleistocene sediments in the Middle Cave (Nizhniokryvchanske site). (A) The general view of the entrances: on the left—the Serednya Cave, on the right—the newly discovered section in the Serednya-2 Cave; (B) the upper (Holocene) part of the section; (C) the general view of the Serednia-2 entrance; (D) the lower (Pleistocene) part of the section.

mountain is a quarry. A cavity 1.5 m high, filled with sand, silty deposits, and sandstone fragments, was discovered during industrial the limestone production. Animal bone fragments and three flint plates were also found in the filling of this cavity. In 1962 and 1963, M.I. Ostrovsky collected flints (more than 70 specimens) and bone remains of animals there. Most of the material, apart from one point and one lithic core, was picked up at the surface during the pit inspection. According to Grigoriev and Ostrovsky, the site can be attributed to the local Lipska Upper Paleolithic Culture (Ostrovskiy and Grigoriev, 1966), which dates back to the end of the Mologa-Sheksnya interglacial (MIS 3) and the beginning of the Ostashkiv Glaciation (MIS 2) (30–20 ky BP). It is obvious that the tools collected during the inspection of the quarry could have come from the destroyed part of the buried cave.

The taphonomic study of the location was conducted by I.V. Marisova and K.A. Tatarinov. According to K.A. Tatarynov, the fossil bones were in brown loam with inclusions of fragments of oolitic limestone and sandstone in a niche of the southern wall of the quarry at a depth of 7 m from the surface. The thickness of the bone-bearing lens reached 0.8 m with a length of 5–6 m. The species identified are: cave bear (*Ursus spelaeus*)—14 individuals; reindeer (*R. tarandus*)—4 individuals; brown bear (*U. arctos*), and Norway lemming (*Lemmus* cf. *lemmus*)—2 individuals each; wolf (*C. lupus*), arctic fox (*Alopex lagopus*), *cave lion* (*Panthera spelaea*), horse (*E. caballus*)—1 individual each (Tatarinov and Bachynskyi, 1968). The 1966 publication mentions the findings of the cave hyena (*Hyaena spelaea*) and the Arctic lemming (*Dicrostonyx torguatus*), but the 1968 publication does not list these species anymore. K. Tatarinov and G. Bachynsky believed that since the remains of cave bears predominate in the cave, they and some other predators lived in this cave, and also that for a particular time, the cave served as a shelter for Late Paleolithic people who possibly hunted cave bears (Tatarinov and Bachynskyi, 1968).

4.15 Malimon-Canyon Cave

The cave is an artificially opened isolated part of the huge karst system of the Zolushka (Popeliushka) Cave (located on both sides of the border—near the village of Podvirne, Novoseltsky District, Chernivtsi Oblast, Ukraine, and the village of Kryva, Brychany District, Republic of Moldova). It is developed within the gypsum of the middle Badenian (N_{1bd2}). (More details about the Zolushka Cave and the regional karst are provided by Andreychouk (2007) and Andreychouk and Klimchouk, (2017). Gypsums, like other bedrocks, are not exposed in the river valley, as they are covered by a 2–3 m thick colluvium. The entrance to the cave is through an artificial well 15 m deep, located near the Podvirne village of Chernivtsi Oblast (Fig. 1A, *24*), on the left bank of the Patsak River (a left tributary of the Prut), 10 m above the river level. The cave consists of two wide (up to 2–3 m) sub-parallel galleries

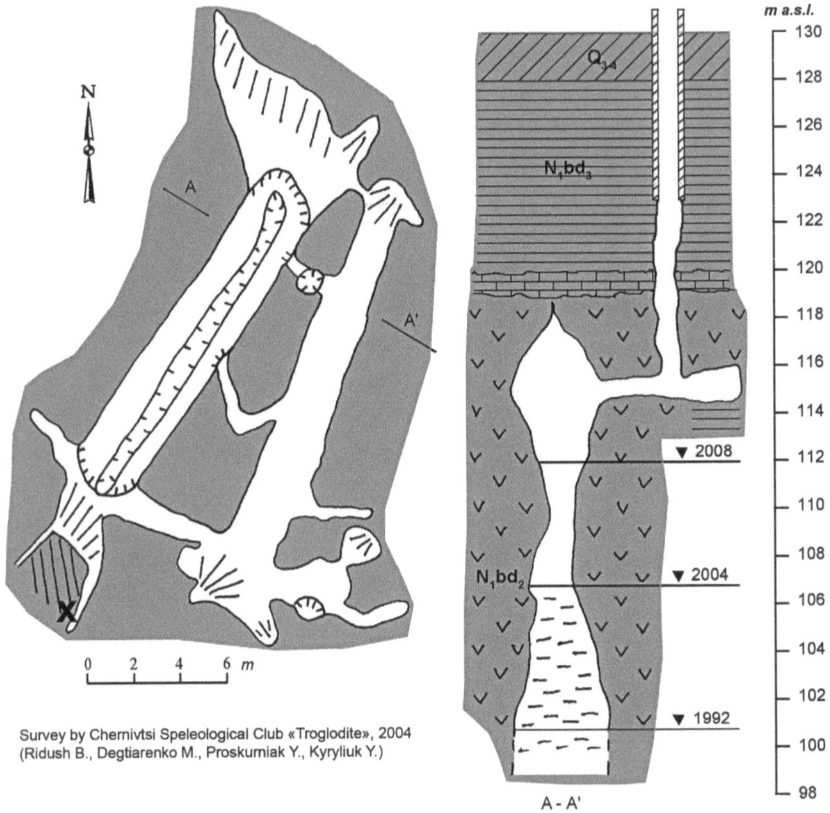

Fig. 10 Plan and cross section of the Malimon-Canyon Cave (according to Ridush and Kuprich, 2003). X—the place of Pleistocene bones bedding.

running south-west and north-east, connected by four narrow passages approximately perpendicular to the main galleries (Fig. 10) (Ridush and Kuprich, 2003). In fact, the cave is a part of the more extensive labyrinth, but it is bounded on all sides by landslides. The north-western gallery through a cracked well (feeder) connects with the cavities of the lower tier, which is located 15 m below, and is now beneath the groundwater table. (In general, the groundwater table in the massif has been lowered to 28 m due to water-pumping from a gypsum quarry located nearby on the territory of Moldova). The upper level of the cave is located in the upper part of the gypsum stratum, but almost 5–6 m below the current water table of the Patsak River. Some of our observations in the sediments inside the Zolushka Cave provide evidence for the active descending of this tectonic block at the end of Late Pleistocene.

During speleological excavations, a woolly rhinoceros (*Coelodonta antiquitatis*) humerus was discovered in one of the galleries of the upper level of the cave. A thin (4–5 cm) layer of brownish loam was found under the landslide deposits, which can be correlated with the humus layer in the Bukovynka Cave (see above). It is covered with moist, light loam, close to the eolian "cave forest" from the Dining Hall in the Bukovynka Cave. The geological analysis of the roof of the cavity in this place showed that the collapse had, so to speak, an internal nature and affected only the rocks containing the cave. Therefore, it remains to assume that the bone got into the cave by the large predators (*Crocuta splelaea*), and the cave used to have a horizontal entrance. The radiocarbon data of the bone is 38.000 ± 800 BP (Poz-47727). Therefore, the site can be dated to the Late Pleistocene, particularly the Vytachiv Stage (MIS-3), like the cave hyena den in the Bukovynka cave, located just 6.5 km away.

4.16 Martynivka Cave

It is a karst cave developed in the upper part of the Miocene gypsum strata. It is situated between Pogorylivka and Chornyi Potik villages of the Chernivtsi region, in the Middle Dniester area (48°32′45.49″N, 25° 59′32.19″) (Fig. 1A, *21*). The two cave entrances have human height and are in the gypsum cliff, 30 m above the valley bottom, on the right bank of the Chornyi Potik River, a right tributary of the Dniester River. This area's 10 m thick gypsum strata is represented by white and light-gray microcrystalline nodular gypsum. Initially, it had a hypogenic origin but was later strongly modified by the weathering processes. The total length of all galleries is 80 m, but the main gallery is just 25 m long, 2–5 m wide, and 2–3 m high (Ridush and Kuprich, 2003). The gallery ends with the accumulation of rock blocks formed by the ceiling collapse.

Since 1960s the archeological layers of the 7–8th, 13th, and 17th centuries are known (Ridush, 2000). The up to 1.0 m thick cultural layer of the 17th century overlaps the rock-fall. The natural sediments in the entrance part of the cave were removed in the 13th century when the cave was inhabited. The cultural layer of this epoch developed directly on the rock floor. In the 17th century, most of the 13th-century cultural layer was also removed. The small fragments of undisturbed sediments were preserved only below the large blocks in the inner part of the gallery. The excavation of the gypsum blocks produced numerous faunal remains of small mammals wich were enclosed in the loose sediments between the blocks and the cave wall as well as between the blocks themselvs. The bone-bearing sediments are represented by 0.3–0.6 m thick alevritic loess-like yellow and light brown sediment and fine cryogenic gypsum debris, with inclusions of coarse debris. There are three taxonomical orders represented: Insectivora, Chiroptera and Rodentia. Pikas (*O. pusilla*) and hamsters (*C. cricetus*) dominate; fewer water voles (*Arvicola amphibius*) and voles (*Alexandromys oeconomus* = *Microtus oeconomus*). Forest forms are sporadic: dormice (*Glis glis*), squirrels (*Sciurus vulgaris*), and red forest voles (*Clethrionomys glareolus*). Very few voles (*Microtus arvalis*). The presence of lemmings (*Dicrostonyx henseli*), steppe lemming (*Lagurus lagurus*) and narrow-headed vole (*Lasiopodomys gregalis*) is unique. The species assemblage indicates the ecologically mixed nature of the fauna with the dominance of the species of mesophilic habitats and the predominance of steppe species over cold-loving species (lemmings and narrow-headed voles) (Ridush et al., 2021).

Nowadays, the site is situated in the semi-aphotic part of the cave. Considering the regular slope regression that is characteristic for the area, 20–30 kys BP, the site was further from the entrances, in the aphotic zone. The bone accumulation could be related to the activity of troglophilic small carnivores like foxes or mustelids, which are inhabiting the cave or visiting it. Animals such as picas and dormice are also troglophilic species and could die during winter hibernation. The nesting of birds of prey in this place is very unlikely because of the low ceiling of the gallery.

4.17 Melna

This is the first paleontological cave site in Western Ukraine. It was revealed in a deep valley, 7 km east of the Melna village (Rohatyn District, Ivano-Frankivsk Oblast) (Fig. 1A, *6*). Here, beneath the layers of Miocene sandy limestones, in the upper part of the fine-grained Miocene sands, caves were formed by suffosion. The latter are filled almost to the ceiling with yellowish sandy loam. In 1879, M. Lomnytskyi discovered here a significant number of remains of a cave bear (*Ursus spelaues*), a cave hyena (*Crocuta (Hyena) spelaea*), their coprolites, as well as the remains of other unidentified mammals (Łomnicki, 1881) (Fig. 11). The materials were transferred to the Dzieduszycki Museum in Lviv (now it is the State Museum of Natural History), but Tatarinov notes that the fate of this material is unknown, since it is not in the museum's collections (Tatarinov, 1958).

In addition, Podoplichko notes for the Melna village three other finds: (1) in Quaternary deposits (?)—*C. lupus, Ursus spelaeus*; (2) skull of *Crocuta*

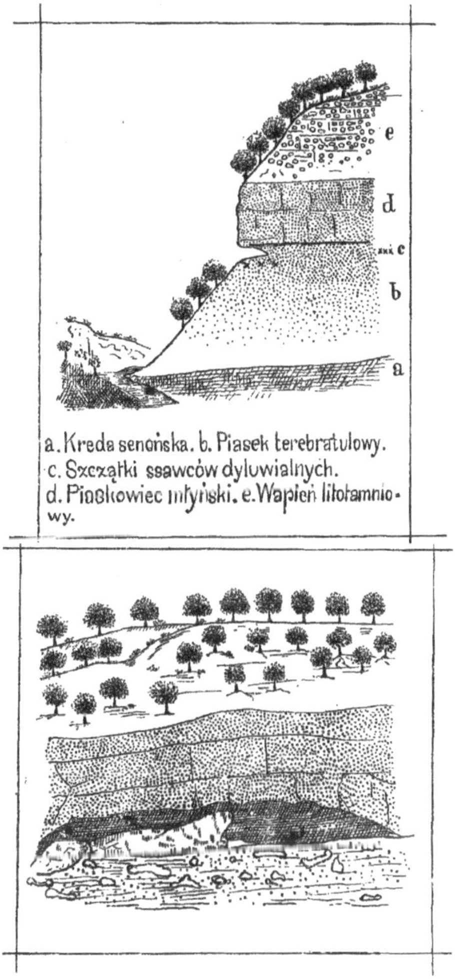

Fig. 11 The geological sections of the Melna Cave paleontological site (according to Łomnicki, 1881).

(*Hyaena*) *spelaea*; (3) skulls, teeth, limb bones, sacrum, vertebrae, and other bones from five individuals of the cave bear *Ursus spelaeus* (Pidoplichko, 1956a, p. 120). Again, it is unknown whether we are talking about Łomnicki's or other finds.

4.18 Ostrivets'

The Late Pleistocene fauna was described by Tatarinov (1959) near the Ostrivets' village of the Stanislav Oblast (now Horodenka District,

Ivano-Frankivsk Oblast) (Fig. 1A, *10*): *Bison priscus, E. caballus, Coelodonta antiquitatis, Lepus europaeus, Hyaena* (*Crocuta*) *spelaea*, and Aves. The bones were discovered in May 1958 during the production of sandstone for building stone. The sandstone quarry is located on the Masyok Mountain. A typical karst landscape can be observed in the vicinity of this village. Fossil animal bones were found in two shallow karst dolinas, which were uncovered because of the clearing the sandstone slab. The geological section of the quarry is as follows: soil—0.5 m; weathered *Lithothamnia* limestone (Miocene)—0.3 m; sandstone (lithified)—3–5 m. The karst dolina was filled with soil and fragments of weathered rock. The bones of fossil animals lay on the contact of the soil and weathered fraction with the sandstone slab (Tatarinov, 1959).

This region is an area of gypsum karst, where Miocene gypsum is underlain by limestones and sandstones of the same age. Judging the description by Tatarinov, the site was associated with a destroyed and buried karst cave in gypsum, in which, during the Late Pleistocene, there was a cave hyena's den.

4.19 Ozerna (Popova Yama) Cave

The giant labyrinth, with a total length of 144 km, is located near the village of Strilkivtsi, Borschiv District, Ternopil Oblast, at the confluence of the Seret and Nichlava rivers (48°45′51.47″N, 25°57′56.99″E) (Fig. 1A, *14*). Like other giant labyrinths of the region, it is formed within Miocene gypsum (Klimchouk, 2012; Klimchouk and Andreychouk, 2017). The modern entrance is located at the bottom of a large karst dolina that cuts into the labyrinth in its farthest northwestern part (Fig. 12A).

In 2009, in the southeast of the cave system, speleologists from Ternopil discovered a new part of a labyrinth with numerous bones of a cave bear (*Ursus spelaeus* or *U.ingressus*) (Fig. 12B–D). According to speleologists, the bones lie both on the surface of the floor and inside the clay aggregate. The taphonomy and faunal composition of the location require a special study, but certain conclusions can be drawn from the already available data.

Due to the surface bedding of the examined bones, the collagen content was insufficient for the radiocarbon analysis. A significant number of remains of cave bears indicate that this part of the cave served as a winter den for bears during certain stages of the Late Pleistocene (not later than MIS-2). Since the site today is too far from the modern entrance, it is logical to conclude that in the Late Pleistocene, another entrance or a few entrances to the labyrinth existed. In this area, the roof of the gypsum layer is bedded at the level of

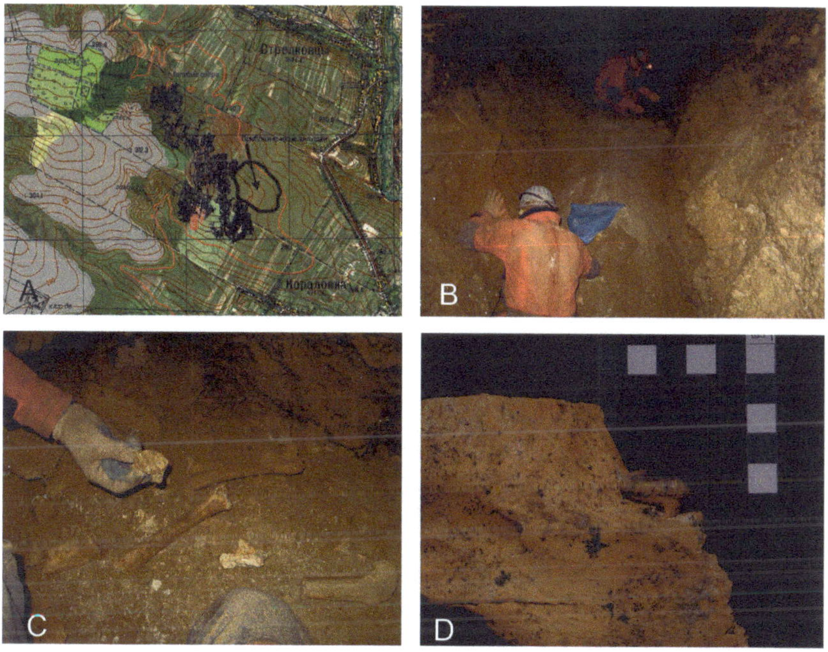

Fig. 12 The site with *Ursus spelaeus* L. in the Ozerna Cave. (A) Location of the cave; the arrow points to the location of the new area of the cave; 260 m isohypse, which approximately matches with the level of the roof of the gypsum layer, is highlighted in red. (B) The nature of the galleries where the bone remains were found; (C) bones of *Uloborus spelaeus* lying on the bottom of the gallery; (D) a fragment of the cave bear mandible with a molar (M2) (based on the materials by V. Apostoliuk and V. Andrash).

the basement of the high river terrace of the Dniester valley, dated to the Early Pleistocene. However, the continuation of the gypsum roof over the nearest adjacent drain, the Nichlava River, is only 30 m. Therefore, the erosive dissection of the labyrinth took place only in the Late Pleistocene. Also, complete drainage of the cave system has not occurred yet, as the cave experience multiannual fluctuations of the groundwater table. Therefore, the closing of the entrances exposed by denudation could have taken place during the Last Glacial Maximum (LGM, MIS 2, Bug Stage), when an intensive loess accumulation was characteristic of this area.

4.20 Pryima I (Nimecha)

Up to now, Pryima I is the oldest archeological site in the caves of the Western Ukraine. It was investigated by L. Matskevyi in the upper grotto of the multi-tiered cavity. The site is located 1 km southwest of the Pryima village,

Mykolaiv District, Lviv Oblast, in the Nimecha tract (Fig. 1A, *2*). The rock with the grottoes is located on the valley shoulder, on the left (northern) slope of the deep valley of the drying Nameless stream, the left tributary of the Dniester. The site is located 2.3 km east of the Dniester, and 1.2 km east of the Mykolaiv—Rozdol highway, in the area between the villages of Rozvadiv and Veryn. This rock has the local name Diryava (the other name is Biny Cave). The cave is of the karst-sufossion type, and developed in the residual composed of Neogene sandstones. The relative altitude of its fifth floor (grotto) above the thalweg of the stream is 45 m, the absolute altitude is 372 m a.s.l. In addition to the grotto (which, judging by the section given by L. Matskevyi, is now almost destroyed), the cave complex has four more levels—the upper (I) and lower (II) caves, as well as two rock shelters—the upper (I) and lower (II) (Fig. 13).

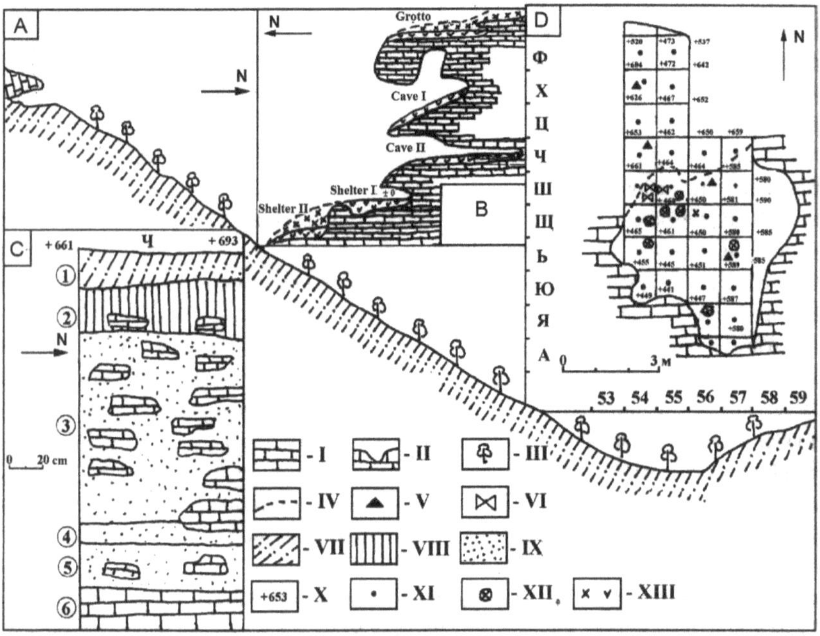

Fig. 13 Settlement in the multi-tiered cave-rock ensemble of Priyma I (Niemecha): (A) section of sediments from south to north between squares 53 and 54; (B) scheme of the ensemble; (C) stratigraphic column of the western wall of square Ч-54; (D) grotto plan; I—bedrock; II—cavities in bedrock; III—forest massif; IV—the overhang limit of the rock cavity (drip line?); V—point; VI—bear tooth; VII—modern soil; VIII—loess-like loams; IX—sand; X—mark from the zero point (reference); XI—excavated squares; XII—*Homo* sp. bones; XIII—colluvium and cultural layer (according to Matskevyi, 2005).

The cave sediments included several cultural layers, including Middle Paleolithic, Late Paleolithic, Mesolithic, Eneolithic, Early Bronze, Early Iron, and Middle Ages layers. Here we discuss only layers related to Pleistocene. The fauna on the site contains up to two thousand specimens, which were determined by K. Tatarinov. The bone collection is stored in the Krypiakevych Institute of Ukrainian Studies of the Ukrainian Academy of Science. Unfortunately, the detailed list of determined bones was lost.

The **Mousterian Cultural Layer** has a thickness of 10–20 cm and is located in the base of the cave deposits, at 160–185 cm deep in front of the cave and 120–145 cm deep in the grotto. The layer comprises fine-grained clayey dark gray to black sand and yellow loam. The latter is of medium phosphorus and low humus. Among the two thousand specimens of osteological remains, the bones of the cave bear (*Ursus spelaeus*) prevail. Also, the bones of the European hare (*L. europaeus*) and other animals are present. A significant number of teeth of the cave bear was located at the entrance to the grotto, and the main number of human remains was embedded nearby. The complex is dated to 45,600 ± 450 yrs. BP (Ki-4583) (cal 49,079–46,775 BP), and may possibly relate to the Moershoft episode (OIS 3) (Matskevyi, 2005).

The numerous human bones, including fragments of the occipital bone of the skull and an "upper part of the skull, accompanied by a flit point, were claimed to be found in this layer. Another five Neanderthal bones, including the radius, were found closely packed near the north-western part of the grotto, on the rock floor near the three points, the teeth of the cave bear, and the remains of a fireplace, in situ within the cultural layer (Matskevyi, 2005). Unfortunately, no special publications about these human remains are available, except for some interviews in the press. So, we strongly doubt and question the taxonomical determination of these remains. Also, we do not know where these bones are stored now, after L. Matskevyi passed away in 2019.

The **Upper Paleolithic Layer** is 15–20-30 cm thick, at a depth of 280–350 cm. The materials embedded in light gray and brownish loamy sand with elements of hygromorphic paleosol. The layer contains the most numerous bones (see Table 9). Seven bones are suggested to be human (*Homo* sp.). The Paleolithic culture was identified to be Gravettian. However, several ^{14}C dates, for the layer (13,900 ± 130 (Ki-4145), 13,600 + 120 (Ki-4146), 14,100 ± 105 (Ki-4147), 14,200 ± 90 (Ki-4163), 13,800 ± 105 (Ki-4178)), indicate the end of the Pleistocene, the Younger Dryas (Matskevyi, 2005). Archeological and faunistic material from the LP Layer suggest some earlier

Table 9 The species composition and number of skeletal remains of mammals from Pryima I, composed (according to Matskevyi, 1998, 2005).

	Mousterian		Late Paleolithic		Late Paleolithic–Mesolithic	
Species	**NISP**	**MNI**	**NISP**	**MNI**	**NISP**	**MNI**
Mammuthus primigenius			9	7		
Coelodonta antiquitatis			42	6		
Alces alces			377	6		
Cervus elaphus			112	5	17	1
Rangifer tarandus			2650	26	21	2
Cervus sp.			14	1		
Capreolus capreolus			83	16	16	1
Bison priscus			90	5		
Bison sp.			287	2		
Equus equus (ferus)			1197	23		
Sus scrofa			61	4	11	1
Ursus arctos			462	12		
Ursus spelaeus	56	1	3063	22		
Crocuta spelaea			531	8		
Panthera spelae			73	5		
Vulpes vulpes			67	4	2	1
Vulpes corsac			1	1		
Alopex lagopus			139	6		
Meles meles			198	6		
Lutra sp.			165	14		
Putorius putorius			23	1		
Martes martes			8	1		
Lepus timidus			33	5		
Lepus europaeus	12	1	70	4	6	1
Ochotona pusilla			11	3		
Marmota bobac			41	5		

Table 9 The species composition and number of skeletal remains of mammals from Pryima I, composed (according to Matskevyi, 1998, 2005).—cont'd

	Mousterian		Late Paleolithic		Late Paleolithic–Mesolithic	
Species	**NISP**	**MNI**	**NISP**	**MNI**	**NISP**	**MNI**
Dycrostonyx torguatus			39	13		
Microtus subterraneus			1	1		
Microtus arvalis					4	2
Microtus gregalis			1	1		
Arvicola terrestris			38	5		
Cricetus sp.			14	1		
Cricetus cricetus			36	4		
Sciurus sp.			51	1		
Erinaceus europaeus			8	1		
Talpa europaea			1	1		
Homo sp.			7	2		

time. The presence of *Ursus speleus* and *Crocuta spelaea* indicates the time not later than the beginning of the LGM. Such species as *S. scrofa* and *Capreolus capreolus* usually indicate interglacials, although the wild pig could be present also in interstadials. That is why we believe that the fauna from this layer accumulated during the several paleoclimatic stages.

4.21 Shchyrets (Tserkovna Gora)

In the Shchyrets village (Nemyriv District, Lviv Oblast) in 1950, employees of the Lviv Museum of Natural History of the Academy of Sciences of the Ukrainian SSR investigated the bone site with Quaternary mammals in a karst sinkhole (ruined cave?) during work in a gypsum quarry located on Tserkovna Gora (Church Mount) (Fig. 1, *3*). The bones were embedded in yellowish-gray loam at a depth of 4–4.5 m from the land surface. The bony horizon of about 0.5 m in thickness located on the gypsum and was covered by loess about 1 m thick. During the excavations, the remains of the following animals were found (Table 10). Among 387 bone fragments, there were ten mammoth molars. The teeth of a woolly rhinoceros, teeth

Table 10 The species composition and number of skeletal remains of mammals from Shchyrets site after (Pasternak and Tatarinov, 1952; Pidoplichko, 1956a, 1956b; Tatarinov, 2000).

Species	NISP	MNI
Mammuthus primigenius	60	5
Equus equus	1	1
Ursus spelaeus	1	1
Equus equus	?	?
Bison priscus	?	?
Coelodonta antiquitatis	?	?

and bones of a horse, a cave bear, and a bison were also found (Pasternak and Tatarinov, 1952; Pidoplichko, 1956a; Tatarinov, 2000, pp. 68–70).

4.22 Shchyrets-1

In 1949, 2 km south of Shchyrets (Fig. 1A, *3*), the remains of *Equus, Alces, Bison priscus, Ursus spelaeus* were found in loams filling a karst dolina in gypsum (Pidoplichko, 1956a).

4.23 Stalactite (Lokitka) Cave

The cave is situated near the Lokitka village, Tlumach District of the Ivano-Frankivsk Oblast (Fig. 1A, *26*). In former times the cave has numerous stalactites because of the 2-m thick limestone strata overlaying the gypsum. It was first described by Łomnicki (1896). The cave has an epigenic origin and was developed in Miocene gypsum by the small underground stream (Dublianskyi and Smolnikov, 1969). I. Pidoplichko collected in the cave some faunistic remains: Rodents—*O. pusilla, A. amphibius, M. oeconomus, M. arvalis,* small Carnivora—*Mustela nivalis,* Chiroptera—*Vespertilio,* large and small birds (*Aves*) (Pidoplichko, 1956a). The most numerous remains of pika (30 bones of 15 individuals) indicate a Pleistocene age of the assemblage. The collection is kept in the Lviv Natural History Museum of the National Academy of Sciences of Ukraine. At present the entrance to the cave is obliterated.

4.24 Stradch

It was discovered in 1949 by Goretskiy (Goretskiy, 1957). From 1953 to 1955, under his leadership, excavations of this location were carried out, in which also Bachynsky took part. Remains of cave bear (*Ursus spelaeus*), cave hyena (*Crocuta spelaea*), wolf (*C. lupus*), horse (*E. caballus*), woolly rhinoceros (*Coelodonta antiquitatis*), red deer (*C. elaphus*), giant deer (*Megaloceros* sp.), *reindeer* (*R. tarandus*), mammoth (*Mammuthus primigenius*, early form) and gopher (*Citellus* sp.) were found. The remains of the cave hyena and the taphonomic features of the burial were studied by G. Bachynskyi (1967).

The site is in the vicinity of the Stradch village of the Lviv Oblast (15 km from Lviv) (Fig. 1A, *5*), in the precipice of the steep left bank of the Vereshchytsia River, 25 m above its level. The fossil bones were found in a small niche filled with medium-grained white quartz sand under the cornice formed by a layer of sandstone. Both the sand layer and the overlying sandstone layer are sediments of the Badenian Stage (Miocene). The main accumulation of bones had the shape of a lens with a width of up to 1.5 m and a thickness of about 0.5 m which continued below a slight slope, deepunder the sandstone cornice. In this lens, isolated limb bones, jaw fragments, teeth, and even whole skulls (4 hyena skulls), together with clusters of coprolites of various sizes and fragments of sandstone and limestone, were embedded in the main mass of sandy material. Isolated fossil bones were also found in the sand on either side of the main cluster beneath the cornice. Most bones were light yellow and white in color. The cavities of tubular bones are filled with calcified sand, and the surface of many bones is covered with a thin limestone crust. Examination under a microscope showed that the marginal areas of the bone tissue of some remains are partially mineralized with calcite. The site was formed mainly during the Late Pleistocene (Riss-Würm) (Bachynskyi, 1967, pp. 14–15).

4.25 Syniakove

In the period between 1936 and 1939, during the paleontological research of Siniakove Cave (Fig. 14), animal remains of the Pleistocene fauna were found (now stored in the Lviv Natural History Museum). Podoplichko identified the taxonomical composition of this fauna (Table 11). The exact location of the cave is unknown. It is very possible that it was in the same place, where the Syniakovo-1 Cave is located (see below) (Fig. 1A, *13*).

4.26 Synyakove-1

The site was discovered in a sand quarry near the Synyakovo Village, which is near the town of Chortkiv, Ternopil Oblast, 500 m east of the Synyakove bus station (49°0′14.58″N, 25°48′41.09″E) (Fig. 1, *13*). Now it is within

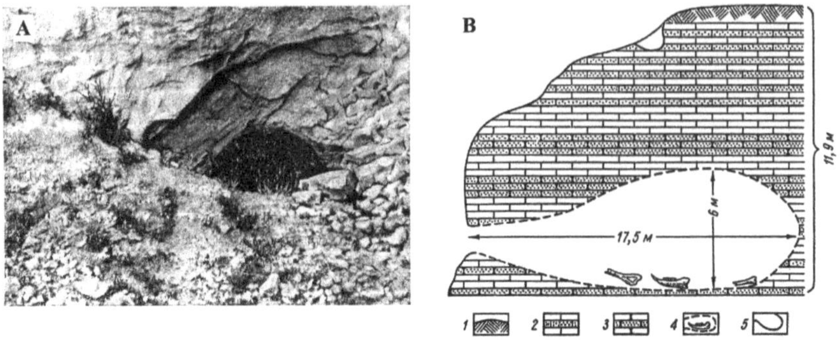

Fig. 14 (A) The entrance to the cave (photo by K. Tatarinov); (B) Schematic geological section of deposits in the cave near the Syniakove Village: (1) soil horizon; (2) limestones; (3) carbonate sandstones with cracks filled with loams with animal remains; (4) a cave where the remains of Pleistocene animals were found; (5) a niche that contained the remains of fossil fauna (according to David et al., 1990).

Table 11 The species composition and number of skeletal remains of mammals from Syniakove site (Pidoplichko, 1956a, p. 130).

Species	NISP	MNI
Coelodonta antiquitatis	16	1
Bison priscus	1	1
Cervus elaphus	93	5
Capreolus capreolus	2	1
Ursus arctos	34	3
Meles meles	13	3
Putorius sp.	2	1
Crocuta spelaea	3	1
Canis lupus	8	2
Desmana moschata ternopolitana	6	1
Talpa europaea	2	2
Vespertilionidae	1	1
Lepus europaeus	1	1
Ochotona pusilla	4	2
Castor fiber	2	1
Cricetus cricetus	25	5
Cricetulus migratoris	2	1
Arvicola amphibious	1	1
Microtus arvalis	3	1

Chortkiv limits. It is located at an altitude of 295 m above sea level, at the level of a Pliocene terrace of the Dniester. The material was collected in 1962 by A.B. Bogutsky and in 1963 K.A. Tatarinov.

The geological section of exposed sediments is as follows: a) thin soil layer—0.1–0.2 m; (b) highly weathered (to the state of fragments) Ratyn limestone—1 m; c) Lower Tortonian (middle Badenian?) gray with a yellowish tinge quartz layered sands of different densities—more than 10 m. The Badenian sandstones have not only vertical (6–8 m deep and up to 0.5 m wide), but also horizontal cracks-niches, reaching 0.5 m in height. The cracks are filled with white quartz sand and clastic material (sandstone, Ratin limestone). The accumulation of bones of large animals and their coprolites is connected with the filling of an ancient grotto (12 m long, 10 m wide and 1.5 m high), which was formed in Middle Badenian white quartz medium-grained skewed sands, under the cornice of calcareous sandstones (Fig. 15). The remains of animals (separated limb bones, phalanges, individual teeth and fragments of skull bones) were scattered in a layer of yellow unlaminated sand that completely filled the ancient grotto (David et al., 1990; Tatarinov, 2000, pp. 65–68) (Table 12).

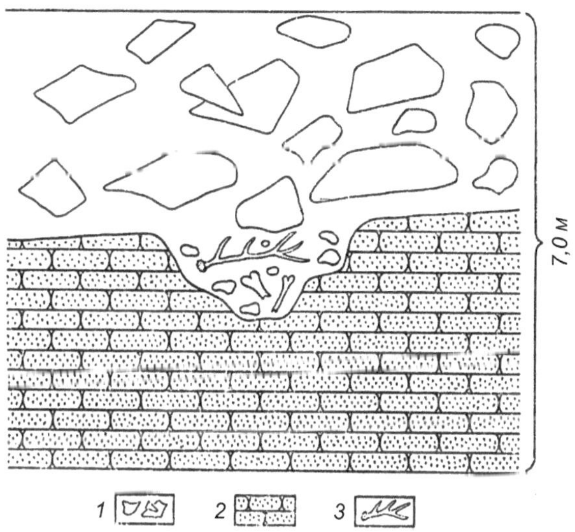

Fig. 15 Schematic section of sediments in the Synyakovo-1 quarry: 1—heavily destroyed limestone; 2—medium- and fine grained quartz sandstones; 3—karst dolina filled with sand, limestone fragments and remains of Pleistocene animals (David et al., 1990).

Table 12 The species composition and number of skeletal remains of mammals from Syniakove-1 site (according to David et al., 1990; Tatarinov, 2000).

Species	**NISP**	**MNI**
Insectivora		
Desmana moschata ternopolitana Pidop.	1	1
Chiroptera		
Myotis cf. *danutae Kowal.*	8	6
Plecotus sp.	7	4
Pipistrellus sp.	4	1
Lagomorpha		
Lepus sp.	4	4
Ochotona pseudopusilla Gur. *et* Schev.	7	2
Rodentia		
Marmota sp.	1	1
Citellus sp.	3	1
Castor fiber L.	8	3
Dyromys sp.	5	1
Rattus sp.	1	1
Apodemus sp.	1	1
Parapodemus cf. *coroneris* Schaub.	3	3
Cricetus cricetus major L.	6	1
Allocricetus sp.	6	3
Cricetulus sp.	3	3
Arvicola cf. *hintoni* Ahar.	4	1
Microtus (Pitymys) arvaloides Hint.	5	2
Microtus ex gr. *arvalis* Pall.	5	5
Spalax sp.	1	1
Carnivora		
Canis sp.	12	1

Table 12 The species composition and number of skeletal remains of mammals from Syniakove-1 site (according to David et al., 1990; Tatarinov, 2000).—cont'd

Species	NISP	MNI
Vulpes vulpes L.	12	1
Cuon sp.	11	3
Mustela erminea L.	2	2
Mustela nivalis minima Tatar.	2	2
Putoris sp.	1	1
Lutreola lutreola L.	1	1
Martes foina homizeniensis Tatar.	2	1
Gulo sp.	1	1
Meles meles sinjakovensis Tatar.	34	10
Lutra lutra L.	1	1
Ursus spelaeus rossicus Bor. (U. savini Andrews)	61	7
Crocuta cf. *spelaea* Gold.	6	2
Felis sp.	2	1
Panthera spelaea Gold.	11	3
Equus cf. *mosbachensis* Reich.	5	3
Dicerorhinus aff. Mercki Jager	10	2
Megaloceros sp.	4	2
Capreolus sp.	8	3
Cervus cf. *elaphus* L.	31	5
Bos sp. vel *Bison* sp.*)*	3	1

Tatarinov believed that various predators lived in the ancient grotto at different times and died there themselves. Their remains were mixed with the bone material of their prey. The bones at the bottom of the grotto were covered with sand that gradually fell from the walls and ceiling of the grotto and were buried under slabs that fell from the ceiling. Burials in vertical cracks were formed mainly from the pellets of birds of prey, in particular owls, which settled in the cracks before they were filled with sand (Tatarinov, 2000, pp. 65–68).

Bachynskyi and Tatarinov believed that the fauna corresponds to the Tiraspol faunal complex and the burial was formed over a long period (Bachynskyi, 1965; Tatarinov, 1966a; Tatarinov and Bachynskyi, 1968). Later Tatarinov recognized that this fauna is close to the "middle anthropogenic" (Tatarinov, 2000), i.e. middle Pleistocene according to current concepts. David considers this fauna to be similar to that of Tiraspol, with which he considers the common species to be a cave lion, a horse close to Mosbach, a Kirchberg rhinoceros, a red deer, etc., noting the absence of Late Pliocene forms in it (David, 1977). In fact, this fauna is mixed with rather heterogeneous elements, some of which are characteristic of the Singilian Fauna of the Middle Pleistocene. So, *U. spelaues rossicus* Bor. today is confidently identified with *Ursus savini* Andrews, which appears at the very beginning of the Middle Pleistocene (Baryshnikov and Foronova, 2001). It is possible that the given list contains mixed assemblages from the filling of various cavities and fissures at different times, and fauna from different times was included in the list. Recently, based on the fauna of Rodents, the site was related to the last stage of the Tiraspolian Fauna (Tiligul-Lubny Stages, MIS 14—MIS 15) (Krokhmal' et al., 2021).

Unfortunately, now the site does not exist anymore since the sandstone bank of the terrace is strongly destroyed by the local development of sand (Fig. 16).

4.27 Tadirka Cave

This recently discovered small cave is close to the Atlantyda Cave (48°35′52.14″N, 26°20′42.67″E) (Fig. 1A, *17*). Evidently, it is one of the ancient entrances to the Atlantyda Cave system, just separated from it by sediment infill. It was partly destroyed by the industrial gypsum production in the 1960s. The remains of at least two individuals of brown bear (*U. arctos*) and rodents were found. In the narrow gallery where bones were embedded, the walls were partly burnt, and charcoal fragment were found in the sediments. As the cave is too narrow for continuous human habitation, it is very probable that it was temporarily inhabited by humans during the hunting seasons and that open fire was being used. The faunal material is currently being studied.

Fig. 16 The current state of the Synyakovo-1 site. (A and B) Sand quarry with entrances to arbitrary tunnels; (C) scratches from the claws of a bear (?) on the wall of a ruined cave; (D) vertical cracks widened by sufossion, cutting both sandstones and the upper part of the sands, with patches of reddish paleosol.

4.28 Trygolovatka Cave

The cave is located in the rocky Trigolovatka Mount, near the Khotymyr village (Tlumatsky District, Ivano-Frankivsk Oblast) (Fig. 1A, *25*). A. Kirkor mentions it for the first time. In the big cave there was a lake with a stream deriving from it. The entrance to the cave, a few years before A. Kirkor's exploration, was destroyed due to the industrial production of gypsum stone from this mountain. During gypsum stone production from the cave itself, human skulls and bones were found at a depth of 3m. The bones found during the gypsum exploration, some already fossilized, were transferred by A. Kirkor to the museum in Krakow. The researcher suggested that a Stone Age burial could have existed in this cave (Kirkor, 1876, p. 8).

As our research has shown, in fact, this massif is composed of coarse-crystalline gypsum. The mentioned stream is a karst spring with clean transparent water and a flow rate of ≈10 L/s, flowing out from under the gypsum blocks. In the alluvium of the stream, a few meters far from its exit from the underground, we discovered several fragments of strongly fossilized bones of large mammals.

The significant fossilization of the bones extracted from the cave in the 19th century, together with the fragments we found, may indicate their rather ancient, at least Late Pleistocene, age. It is possible that these human remains are still stored in the collections of one the Krakow museums.

4.29 Tovtry Cave

The cave is situated at the northern outskirt of the Tovtry village in Chernivtsi Oblast (48°34′6.86″N, 25°52′3.13″E) (Fig. 1, *20*). It is 26 m in length and near 8 m in height (Ridush and Kuprich, 2003). This small cave forms part of an ancient hypogenic labyrinth developed at the base of the 30-m thick Miocene gypsum strata. Initially, this part of the cave was filled with loose sediments consisting of light-brown loam with numerous inclusions of debris-like large gypsum crystals. In the 1960s, the cavity was transected by a small gypsum quarry, and the cave filling became exposed. The sediments from the depth of 1–4.1 m were examined for their pollen content, which resulted in their dating to the Last Interglacial, Early Glacial stadials, and interstadials (Avdieienko, 2018, 2019). Only at a depth of 6.3–7.5 m rare bone fragments and bones belonging to bison (*Bison priscus*) and horse (*Equus ferus*) were recovered from the cave infill. Some bones exhibit traces of gnawing by carnivores.

4.30 Vyniava—Horosno

The site is situated in a sand quarry, between the Vyniava and Horosno villages, about 25 km to SW of Lviv (49°39′44.77"N, 23°57′52.59″E). It includes several buried niches and small grottoes inhabited during different episodes of Pleistocene, that are discovered from time to time because of sand quarrying (Fig. 17). The first finds of *Crocuta spelaea* from this site were published by K. Tatarinov (Tatarinov, 1958) (Table 13). Later the list of species was added (Tatarinov, 2000, pp. 76–78) (Table 14). Some bones from this quarry got into private collections. One of the cave hyena teeth was dated to 40,770 ± 350 BP (MAMS 31302) (Westbury et al., 2020) (Table 15), that place the site in the timescale of the Late Pleistocene, MIS 3. The recent investigation of the small mammals from a cave in this quarry revealed the next list of species (a number means the MNI): *Sorex*

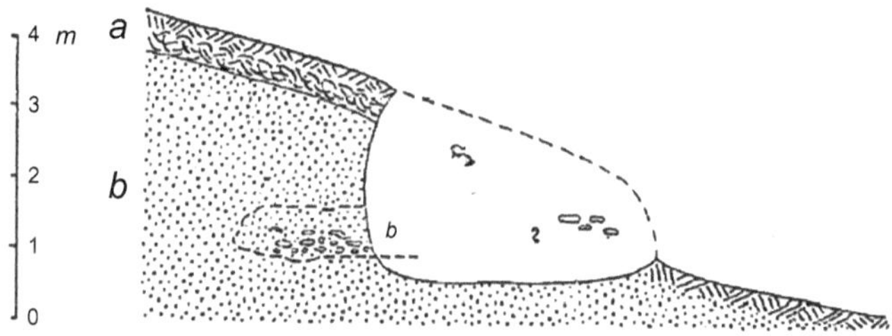

Fig. 17 The section of the Vyniava site (according to Tatarinov, 1958).

Table 13 Geological section of the Vyniava site (according to Tatarinov, 1958).

Unit	Thickness, m	Description
1	0.6	colluvial deposits containing sand, *Lithothamnion* limestone fine debris, and humified loam
2	1.2	grayish-white quartz sand of lower Badenian (Miocene)
3	0.5–0.7	greenish glauconite sand (Miocene?) with the bone remains

Table 14 The species composition and number of skeletal remains of animals from Vyniava (according to Tatarinov, 2000).

Species	NISP	MNI
AVES		
Lyrurus tetrix	1	1
MAMMALIA		
Spalax cf. *polonicus*	3	1
Apodemus sp.	17	2
Microtus sp.	34	2
Canis lupus	5	2
Ursus spelaeus	4	1
Crocuta crocuta spelaea	62	17
Mammuthus primigenius	7	2
Equus equus	87	6
Coelodonta antiquitatis	51	5
Cervus elaphus	7	3
Bison priscus	2	1

sp.—1; *Crocidura* sp.—1; *Ochotona* sp.—2; *Sicista* sp.—2; *Spermophilus odessanus*—10; *Lemmus* sp.—1; *Arvicola chosaricus*—9; *M. oeconomus*—2; *M. gregalis*—11; *M. arvalis*—1. This fauna was dated to the Middle Pleistocene (first half of MIS 6) (Popova et al., 2019).

From the taphonomical point of view it becomes apparent that the faunistic remains of large mammals are associated with the cave hyenas' dens and are represented by the bones of hyenas and their prey. Small mammals relate to nests of birds of prey. The cavities inhabited by the predators can be of suffusion and weathering origin, as well as zoogenic. They were developed in low cemented sandstone under a cover of strongly cemented calcareous sandstones.

5. Discussion and conclusions

The region is specific for its karst and caves. But the very long gypsum labyrinths contain not too many bone sites. At the same time numerous caves in sandstones of suffosional and even weathering origin contain quite old fauna, like Chortkiv, Synyakove, Horishnia Vygnanka, Melna and others. Paleolithic cave sites, up to now, are known only from the suffosional cavities and rock shelters of Pryima and Chortovi Skeli. The evidence of human presence in form of charcoal in sediments and burnt walls or stones, were noted in the Kryshtaleva and Tadirka caves. The investigated sections of cave sediments are usually quite short and do not cover long periods. Only Kryshtaleva and Tovtry have a potential for successions covering more than 50 kyr. There is a lack of radiocarbon dates in the region, especially for the direct dating of the fauna (Table 15), as well as other physical dating.

The character of the bone accumulations can usually be related to the activity of large and small carnivores. The sediments contain the bones and remains of their prey. Accumulations of small vertebrates (Rodents, Reptiles, Amphibia) usually are related also to the activity of the birds of prey.

The typical inhabitants of caves, so-called troglophiles, are cave hyenas, cave and brown bears, foxes, badgers, mustelids, and red foxes. Among Rodents, the most troglophilic are picas (*Ochotona*) (Ridush, 2004c).

Table 15 Results of AMS radiocarbon dating of Quaternary paleontological cave sites in Podillia-Bukovinian Karst-Speleological Area (Ukraine).

Site	Bone	Species	Lab code	%C	%N	14C	cal BP	Source
Horosno-Vynjava	tooth	*Crocuta spelaea*	MAMS 31302			40,770 ± 350	44,430–43,125	Westbury et al. (2020)
Bukovynka Cave, Trapeznyi Chamber	tooth	*Crocuta spelaea*	VERA-2529			41,300 + 1300/− 1100	46,629–42,660	Ridush (2009)
Bukovynka, Sukhyi Chamber	tooth	*Ursus arctos*	Poz-46240			10,730 ± 60	12,765–12,621	Ridush (2012)
Malimon-Canyon Cave	humerus	*Coelodonta antiquitatis*	Poz-47727	5.5%C,	1.2%N	38,000 ± 800	43,145–41,276	Nadachowski et al. (2015)
Kryshtaleva (Kryvchenska) Cave	fibula	*U. arctos*	Poz-59413	6.4%C	2.4%N	12,240 ± 70	14,812–13,894	Nadachowski et al. (2015)
Chortova Skelia, Layer A		*Unknown*	Ki-5415			27,200 ± 170	31,596–31,061	Matskevyi (2005)
Chortova Skelia, Layer B		*Unknown*	Ki-5412			13,500 ± 110	16,635–15,933	Matskevyi (2005)
Chortova Skelia, Layer C		*Unknown*	Ki-5414			11,800 ± 90	13,991–13,466	Matskevyi (2005)
Pryima I, Middle Paleolithic		*Unknown*	Ki-4583			45,600 ± 450	49,079–46,775	Matskevyi (1998)
Pryima I		*Unknown*	Ki-4145			13,900 ± 130	17,302–16,485	Matskevyi (2005)
Pryima I		*Unknown*	Ki-4146			13,600 + 120	16,868–16,059	Matskevyi (2005)
Pryima I		*Unknown*	Ki-4147			14,100 ± 105	17,419–16,891	Matskevyi (2005)
Pryima I		*Unknown*	Ki-4163			14,200 + 90	17,479–17,026	Matskevyi (2005)
Pryima I		*Unknown*	Ki-4178			13,800 ± 105	17,038–16,407	Matskevyi (2005)

All the14C ages are calibrated with 95.4% probability using the IntCal20 atmospheric curve (Reimer et al., 2020) in the OxCal 4.4 software (Brock et al., 2013). Abbreviations: cal BP—calibrated before present.

References

Adamenko, O.M., 1976. Chertovets. Obnazhenie 507. In: Matskevyi, L. (Ed.), Zvit pro roboty u Lvivskiy, Zakarpatskiy ta Ivano-Frankivskiy oblastiakh u 1976 r. Naukovyi Arkhiv IA NANU, Kyiv-Lviv, pp. 6–7.

Andreychouk, V., 2007. Peshchera Zolushka. Uniwersyte Śląski, Wydział Nauk o Ziemi, Украинский Институт спелеологии и карстологии, Sosnowiec-Simferopol.

Andreychouk, V., Klimchouk, A., 2017. Zoloushka cave (Ukraine–Moldova)—a prime example of hypogene artesian Speleogenesis in gypsum. In: Klimchouk, A., et al. (Eds.), Hypogene Karst Regions and Caves of the World. Springer International Publishing AG, pp. 387–406, https://doi.org/10.1007/978-3-319-53348-3_24.

Avdieienko, Y., 2018. Late Pleistocene Vegetational and climate changes recorded in the clastic deposits of the Tovtry cave (Western Ukraine). In: XXV Konferencja Stratygrafia Plejstocenu Polski. Plejstocen Gór Świętorzyskich, 3-7 września 20018 r. Huta Szklana, Poland, pp. 151–153.

Avdieienko, Y., 2019. Revealing of palaeogeographic changes during the Pryluky and Tyasmyn stages in the middle Dniester area for reconstructions of the Paleolithic man environments. Phys Geogr Geomorphol 95 (3), 62–68. https://doi.org/10.17721/phgg.2019.3.06.

Avdieienko, Y., Gerasimenko, N., Ridush, B., 2018. Paleontological and lithological study of the Kryshtaleva cave (Ukraine). In: Hajna, N.Z., Mihevc, A., Năpăruş-Aljančič, M. (Eds.), Quaternary Stratigraphy in Karst and Cave Sediments. Karst Research Institute, Scientific Research Centre of the Slovenian Academy of Sciences and Arts, Postojna, Slovenia, pp. 11–13.

Bachynskyi, G.O., 1965. Vertebrates taphonomical peculiarities of Syniakove and Tarkhankut cave sites of fossil terrestrial. Dopovidi Akademii Nauk Ukraiskoi RSR 5, 658–661.

Bachynskyi, G.O., 1967. In: Pidoplichko, I.G. (Ed.), Tafonomia antropogenovykh i neogenovykh mistseznahodzhen nazemnykh khrebetnykh Ukrainy. Naukova Dumka, Kyiv.

Bachynskyi, G.A., 1970. Tafonomichna kharakterystyka mistseznakhodzhen vykopnych kherebetnykh v karstovykh pecherakh Ukrainy [Taphonomic characteristics of the findings of the fossil vertebrates in the caves of Ukraine]. Fizychna Geografia ta Geomorfologia 4, 153–160.

Bachynskyi, G.O., Tatarinov, K.A., 1966. Taphonomical peculiarities of Nizhne Kryvche spelean site of fossil vertebrates. Dopovidi Akademii Nauk Ukrainskoi RSR 10, 1348–1351.

Bachynskyi, G.A., Dublianskyi, V.N., Shtengelov, E.S., 1964. Krivchenska Cristall cave in gypsum odf Podolia. Peshchery 4 (5), 49–55.

Bahmutov, V.G., Lagutin, K.P., 1985. Experience of Paleomagnetic study of cave sediments. Fizychna Geografia ta Geomorfologia 32, 46–50.

Baryshnikov, G., Foronova, I., 2001. Pleistocene small cave bear (Ursus rossicus) from the South Siberia, Russia | un pequeño Oso de las Cavernas (Ursus rossicus) del Sur de Siberia, Rusia. Cadernos do Laboratorio Xeoloxico de Laxe 26, 373–398.

Bondar, K., Ridush, B., 2015. Rockmagnetic and palaeomagnetic studies of unconsolidated sediments of Bukovynka cave (Chernivtsi region, Ukraine). Quat. Int. 357, 125–135. https://doi.org/10.1016/j.quaint.2014.04.025.

Bondar, K., Virshylo, I., Gordienko, T., 2010. 'Magnetostratigraphy and anisotropy of the magnetic susceptibility of the hydro-mechanical sediments of the karst cave of Atlantyda. Visnyk Kyivskogo Universytetu. Geologia 49, 24–28.

Brock, F., Ramsey, C., Lee, S., 2013. Recent and planned developments of the program OxCal. Radiocarbon 55, 720–730.

Çelebi, E., 1961. In: Tveritinova, A.S. (Ed.), Kniga puteshestvia. Vyp.1. Zemli Moldavii i Ukrainy. [Travel Book. Iss. 1. Lands of Moldova and Ukraine]. Izdatelstvo Vostochnoi Literatury, Moscow.

Chernysh, A.P., 1953. A cave with ancient drawings. Bulleten Komissii po izucheniyu chetvertichnogo perioda 18, 99–103.

Chernysh, O., 1959. A new monument of primitive art. Materialy i doslidzhennia z arkheologii Prykarpattyia i Volyni 2, 40–53.

Chernysh, O., 1975. Starodavnie Naselennia Podnistrovia Za Doby Mezolitu [The Ancient Population of Podnistrovia during the Mesolithic Period]. Naukova Dumka, Kyiv.

Croitor, R., et al., 2014. Giant deer megaloceros giganteus blumenbach, 1799 (cervidae, mammalia) from palaeolithic of Eastern Europe. Quat. Int. 326–327, 91–104. https://doi.org/10.1016/j.quaint.2013.10.068.

David, A.I., 1977. Comparison of the theriofauna of the early Anthropogene of Moldavia with the corresponding fauna of Eurasia. In: David, A.I., et al. (Eds.), Faunisticheskije Kompleksy i Flora Kainozoya Prichernomoria. Chişinău, Ştiinţă, pp. 3–23.

David, A.I., Tatarinov, K.A., Svistun, V.I., 1990. In: Negadayev-Nikonov, K.N. (Ed.), Khishnyje, Hobotnyje I Kopytnyje Rannego Pleistotsena Yugo-zapada SSSR [Carnivores, Proboscidea, and Ungulates of the Early Pleistocene of the South-West of the USSR]. Ştiinţă, Chişinău.

Dublianskaya, G., Dubliansky, V., 1992. Kartografirovanie, rayonirovanie i inzhenerno-geologicheskaya otsenka zakarstovannykh territoriy. Izdatelstvo RAN, Novosibirsk, pp. 1–144.

Dublianskyi, V.N., Smolnikov, B.M., 1969. Karstologo-geofizicheskiye issledovania karstovykh polstej Pridnestrovskoi Podolii i Pokutia. Naukova Dumka, Kyiv.

Formozov, A.A., 1969. Ocherki Po Pervobytnomu Iskusstvu: Naskalnyje Izobrazhenia I Kamennyje Izvajania Epohi Kamnia I Bronzy na Teritorii SSSR [Essays on Primitive Art: Petrogliphs and Stone Sculptures of the Stone and Bronze Age in the Territory of the USSR]. Nauka, Moscow.

Gerasimenko, N.P., 2010. Koreliatsia korotkoperiodychnykh etapiv pleistocenu za paleolandshaftnymy danymy [correlation of short stages of Pleistocene after palaeolandscape data]. In: Matviishyna, Z.M., et al. (Eds.), Prostorovo-Chasova Koreliatsia Paleogeographichnykh Umov Chetvertynnogo Periodu na Terytorii Ukrainy [Spatio-Temporal Correlation of Palaeogeographical Conditions of Quaternary on the Territory of Ukraine]. Naukova Dumka, Kyiv, pp. 104–129.

Gerasimenko, N.P., Korzun, Y.L., Ridush, B., 2014a. The late glacial and Holocene environmental changes in the middle Prut area (palaeontological and lithological data from the Bukovynka cave, the Sukhy chamber). Fizychna Geografia ta Geomorfologia 2 (74), 68–74.

Gerasimenko, N., Ridush, B., Korzun, J., 2014b. Pollen and lithological data from the Bukovynka cave deposits as recorders of the late Pleistocene and Holocene climatic change in the eastern foothills of the Carpathian Mountains (Ukraine). In: Late Pleistocene and Holocene Climatic Variability in the Carpathian-Balkan Region. 2nd International Workshop, Cluj Napoca, Romania, 6–9 November 2014. Abstracts volume, pp. 54–58.

Gerasimenko, N., Ridush, B., Avdeyenko, Y., 2019. Late Pleistocene and Holocene environmental changes recorded in deposits of the Bukovynka cave (the east-Carpathian foreland, Ukraine). Quat. Int. 504, 96–107. https://doi.org/10.1016/j.quaint.2018.03.028.

Goretskiy, V.A., 1957. About the find of Pleistocene vertebrates near Stradch Village of Lviv Oblast. Bulleten Komissii po izucheniyu chetvertichnogo perioda 21, 125–131.

Ivanov, B.N., 1956. On the typology of the karst relief of the plains on the example of the Podillia-Bukovinian karst area. Voprosy Karsta na Yugie Evropeiskoy Chasti SSSR Yalta, 131–156.

Ivanov, B.N., 1961. Recent tectonic movements and development of karst phenomena in Ukraine. Chetvertichnyi period 13–15, 240–249.

Ivanov, B.N., 1965. Problems of paleogeography of karst in Ukraine. In: Materialy po chetvertichnomu periodu Ukrainy: K VII kongressu INQA. Naukova Dumka, Kyiv, pp. 298–311.

Ivanova, I.K., 1969. Geomorfologia i paleogeografia Pridnestrovia v paleolitie. In: Grichuk, V.P., Ivanova, I.K., Grigoriev, G.P. (Eds.), Priroda i razvitie pervobytnogo obshchestva na teritorii Evropeiskoi chasti SSSR. Nauka, Moscow, pp. 111–119.

Kirkor, A.H., 1876. Pokucie pod względem archeologicznym. Kraków. Available at https://polona.pl/item/pokucie-pod-wzgledem-archeologicznym-badania-a-h-kirkora, Nzk5NDE2NTc/10/#info:metadata.

Klimchouk, A.B., 1984. Experience in a detailed study of hydromechanical sediments of a large cave system. Peshchery 19, 70–88.

Klimchouk, A., 2012. Ukraine giant gypsum caves. In: Encyclopedia of Caves. Elsevier Inc, pp. 827–833, https://doi.org/10.1016/B978-0-12-383832-2.00119-5.

Klimchouk, A., Andrejchuk, V., 2002. Karst breakdown mechanisms from observations in the gypsum caves of the Western Ukraine: implications for subsidence hazard assessment. Int. J. Speleol. 31 (1/4), 55–88. https://doi.org/10.1007/s00254-005-1279-1.

Klimchouk, A., Andreychouk, V., 2017. Gypsum karst in the southwest outskirts of the eastern European platform (Western Ukraine): a type region of artesian transverse Speleogenesis. In: Klimchouk, A., et al. (Eds.), Hypogene Karst Regions and Caves of the World. Springer International Publishing AG, pp. 1–910, https://doi.org/10.1007/978-3-319-53348-3_23.

Klimchouk, A.B., Rogozhnikov, V.Y., 1982. Sopriazhonnyj Analiz Istorii Formirovania Peshchernoi Sistemy (na Primierie Peshchery Atlantida) [Conjugate Analysis of the History of the Formation of the Cave System (on the Example of the Atlantyda Cave)]. Institute of Geological Science AN UkrSSR, Kyiv.

Klimchouk, A.B., Andreychouk, V.N., Turchinov, I.I., 2009. The Structural Prerequisites of Speleogenesis in Gypsum in the Western Ukraine. Ukrainian Institute of Speleology and Karsotology, Sosnowiec-Simferopol.

Klimchuk, A.B., et al., 2008. Kadastr Peshcher Ukrainy: Metodicheskiye Materialy I Perechen [Inventory of Caves in Ukraine: Methodological Materials and List]. Ukrainian Institute of Speleology and Karsotology–Ukranian Speleological Association, Simferopol.

Kochergan, Y., Ridush, B., 2011. Cryogenic weathering in karst caves of the middle Dnister area. *Naukovyi visnyk Chernivetskoho universytetu* Geografiia. 587–588, 30–35.

Korzhyk, V.P., 2007. Karst i Pechery Bukovyny. Zelena Bukovyna, Chernivtsi.

Krokhmal', O., Rekovets, L., Kovalchuk, O., 2021. An updated biochronology of Ukrainian small mammal faunas of the past 1.8 million years based on voles (Rodentia, Arvicolidae): a review. Boreas 50 (3), 619–630. https://doi.org/10.1111/bor.12511.

Lomayev, A.A., 1979. Geologia Karsta Volyno-Podolii. Naukova Dumka, Kyiv.

Łomnicki, M., 1881. Zapiski do dyluwialnej fauny ssakow w Galicyi Wschodniej. Kosmos: Czasopismo Towarzystwa przyrodników imienia Kopernika 6 (12), 560–562.

Łomnicki, A.M., 1896. Pieczary Stalaktytowe w Łokutkach. Kosmos: Czasopismo Towarzystwa przyrodników imienia Kopernika 21, 373–378.

Marciszak, A., et al., 2015. Ursus arctos L., 1758 from Bukovynka cave (W Ukraine) in an overview on the fossil brown bears size variability based on cranial material. Quat. Int. 357, 136–148. https://doi.org/10.1016/j.quaint.2014.09.052.

Marysova, I.V., 1962. Pleistocene birds of the Kryvchenska cave. Naukovi Zapysky Kremenetskogo Pedagogichnogo Instytutu 7, 63–75.

Matskevyi, L., 1998. Pytannia naidavnishogo zaselennia Mykolaivshchyny [the question of the earliest settlement of Mykolaiv District]. In: Mykolaivshchyna. Lviv, Instytut Ukrainoznavstva im. I. Krypiakevycha, pp. 8–38.

Matskevyi, L.G., 2001. Paleolithic and Mesolithic on the territory of modern Lviv. Vita Antiqua 3–4, 108–124.

Matskevyi, L., 2005. Pidsumky doslidzhen paleolitu i mezolitu v pecherakh zakhodu Ukrainy [Results of investigations of Paleolithic and Mesolithic in caves of west of Ukraine]. Materialy i doslidzhennia z arkheologii Prykarpattyia i Volyni 9, 90–134.

Nadachowski, A., et al., 2015. Eksploatacja Zasobów Fauny Przez Paleolityczne Spoleczności Łowiecko-Zbierackie na Przykładzie Strefy Strefy Pery-i Metakarpatskiej. In: Łanczont, M., Madeyska, T. (Eds.), Paleolityczna Ekumena Strefy Pery—I Metakarpackiej [Palaeolithic Ekumene of Pery- and Metacarpathian Zone]. Lunlin, Wyd. UMCS, pp. 837–909.

Ostrovskiy, M.I., Grigoriev, G.P., 1966. Lipskaya palaeolithic culture. Sovetskaya Arkheologiya 4, 3–13.

Pasternak, S.I., Tatarinov, K.A., 1952. New finds of quaternary mammals in the west Podillia. Zbirnyk prats Zoologichnogo muzeyu Akademii Nauk Ukrainskoyi RSR 25, 89–93.

Peryt, T.M., 2001. Gypsum facies transitions in basin-marginal evaporites: middle miocene (Badenian) of West Ukraine. Sedimentology 48 (5), 1103–1119. https://doi.org/10.1046/j.1365-3091.2001.00410.x.

Pidoplichko, I.G., 1955. New data on the fauna of vertebrates from quaternary deposits of Ternopil oblast. Dokl. Akad. Nauk SSSR 100 (5), 989–991.

Pidoplichko, I.G., 1956a. Materialy Do Vyvchennia Mynulykh Faun URSR [Materials to the Study of Former Faunas of UkrSSR]. Vydavnyctvo AN URSR, Kyiv.

Pidoplichko, I.G., 1956b. To the study of quaternary vertebrates of the Ternopil oblast. Naukovi Zapysky Pryrodoznavchogo Muzeyu Lvivskogo Filialu AN URSR 5, 45–52.

Polians'kyi, Y., 1929. Podilski Etiudy. Zbirnyk Pryrodopysnolikarskoi Sektsii Nauk. Tovaryst. Shevchenka 20, 43–53.

Popova, L., et al., 2019. Middle Pleistocene small mammal Fauna of the cave locality at Vinyavy (L'Viv region). *Visnyk of Taras Shevchenko National University of Kyiv* Geology 2 (85), 23–27. https://doi.org/10.17721/1728-2713.85.03.

Reimer, P., Austin, W., Bard, E., et al., 2020. The IntCal20 northern hemisphere radiocarbon age calibration curve (0-55 cal kBP). Radiocarbon 62 (4), 725–757.

Rekovetz, L.I., 1994. Melkiye Mlekopitayushchije Antropogena Yuga Vostochnoi Europy [Small Mammals of the Anthropogenic of the South of Eastern Europe]. Naukova Dumka, Kyiv.

Ridush, B., 1998. Taphonomic sites in the Bukovynka cave. Svet 1 (18), 30–31.

Ridush, B., 2000. To the question about cult caves of Slavs in the Dnister area. Arheologichni Studii. 1 Prut, Chernivtsi, pp. 184–193.

Ridush, B., 2004a. Paleokarstologichni rekonstruktsii ta naskelne mystetstvo Seredniogo Podnistrovia (na prykladi Balamutivskoi pechery) [Paleokarstological reconstructions and rock-art in the middle Dniester area (on example of Balamutivska cave)]. In: Balukh, V. (Ed.), Materialy V kongresu Mizhnarodnoi asotsiatsii ukrainistiv: Sotsialno-gumanitarni nauky. Chernivtsi, Ruta, pp. 369–372.

Ridush, B., 2004b. Quaternary deposits in Bukovinka cave and their paleogeographical importance. *Naukovyi visnyk Chernivetskoho universytetu* Geografiia. 199, 105–115.

Ridush, B., 2004c. Vertebrates as component of fossil troglofaunas. In: Zagorodniuk, I. (Ed.), Fauna Pecher Urkrainy [Cave Fauna of Ukraine]. Kyiv, pp. 102–116.

Ridush, B., 2009. "Bear caves" in Ukraine. Slovenský Kras Acta Carsologica Slovaca 47, 67–84.

Ridush, B., 2012. Cave sites of Crocuta spelaea in Ukraine. In: Constantin, S. (Ed.), The 18th International Cave Bear Symposium (ICBS) and the "Fossils in Karst" International Workshop, September 20–22, 2012. Băile Herculane, Romania, p. 60.

Ridush, B., 2013. Paleogeographic Reconstruction of the Natural Environment of the Late Cenozoic at the South of Eastern Europe after Research of Cave Sediments. Manuscript.

Thesis for the degree of Doctor of Science in Geography, Institute of Geography of Ukrainian Academy of Science, Kyiv.
Ridush, B., 2014. "Bear caves" of South-Eastern Europe. Speleol Karstol 12, 26–41.
Ridush, B., Kuprich, P., 2003. Pechery Chernivetskoi Oblasti [Caves of Chernivtsi Region]. Prut, Chernivtsi.
Ridush, B., Levytska, O., 2005. The microclimate of the Bukovynka karst cave. *Naukovyi visnyk Chernivetskoho universytetu* Geografiia. 246, 44–53.
Ridush, B., Bobylev, A., Kuprich, P., 1998. Bukovynka Cave. Svet 1 (18), 26–29.
Ridush, O., Stefaniak, K., Marciszak, A., 2012. Complete skull of the brown bear Ursus arctos Linnaeus, 1758 from Bukovynka cave (Ukraine) and distribution of the species in the latest Pleistocene and Pleistocene/Holocene boundary in south-East Europe. In: The 18th International Cave Bear Symposium, Program—Abstracts—Field Trip Guide. The International Workshop "Fossil Remains in Karst and their Role in Reconstructing Quaternary Paleoclimate and Paleoenvironment". Baile Herculane, Romania, p. 61.
Ridush, B., Marciszak, A., Stefaniak, K., 2013. Complete skeleton of Pleistocene Brown bear (Ursus arctos) from Atlantyda cave (Western Ukraine). In: Rabeder, G., Kavcik, N. (Eds.), 19th International Cave Bear Symposium, Semriach (Styria, Austria). Abstracts & Excursions, Semriach, Styria, Austria, p. 17.
Ridush, B., et al., 2021. Martynivka: A New Cave Site with Late Pleistocene Small Mammals' Fauna In Th E Middle Dniester Area (Ukraine). In: Ratajczak-Skrzatek, U., Kovalchuk, O., Stefaniak, K. (Eds.), Quaternary Stratigraphy—Palaeoenvironment and Humans in Europe. INQUA SEQS 2021. University of Wrocław, Wroclaw, Poland, pp. 98–99.
Rogozhnikov, V.Y., 1984. Water-hemogenic sediments in karst caves-labyrinths of Podolskoye Pridnestrovie. Peshchery 19, 46–55.
Rogozhnikov, V.Y., Lomaev, A.A., 1985. Experience of a comprehensive study of the Atlantyda cave in the light of environmental problems. Fizychna Geografia ta Geomorfologia 32, 33–42.
Tatarinov, K.A., 1958. Cave hyenas from the quaternary sediments of Opillia. Dopovidi Akademii Nauk Ukraiskoi RSR 7, 797–800.
Tatarinov, K.A., 1959. Quaternary mammalian fauna of pre-Carpathians. Nauchnyje Zapiski Uzhgorodskogo Universitteta 40, 59–66.
Tatarinov, K.A., 1961. Remains of fossil animals from the outskirts of Borshchiv. Naukovi Zapysky Kremenetskogo Pedagogichnogo Instytutu 6, 13–21.
Tatarinov, K.A., 1962. Pleistocene and Holocene mammals of the Kremenets Mountains. Naukovi pratsi Ukrayinskogo naukovo-doslidnogo hydrometeorolohichnogo instytutu 7, 34–51.
Tatarinov, K.A., 1965. Plesitocene mammals from the Nyzhnekrivchanska cave (Podolia). Paleontologicheskiy Sbornik Lvovskogo Univesiteta im.I.Franko 2, 30–37.
Tatarinov, K.A., 1966a. Cave fauna of the middle Dniester area. In: Voprosy Izuchenia Karsta Russkoi Ravniny. Materialy Soveshchania Geograficheskoi Sektsii MOIP (Moscow, 25–26.05.1966). Moscow University, Moscow, pp. 110–114.
Tatarinov, K.A., 1966b. Pleistocene vertebrates of Podolia and pre-Carpathians. Bulleten Komissii po izucheniyu chetvertichnogo perioda 32, 51–62.
Tatarinov, K.A., 2000. Pozdniekainozoiskiye Pozvonochnyje Zapada Ukrainy [Later Cenozoic vertebrates of Western Ukraine]. Nadstyrja, Lutsk.
Tatarinov, K.A., Bachynskyi, G.A., 1968. Peshchernyje zakhoronenia pliotsenovykh i antropogenovykh pozvonochnykh v zapadnykh obllastiakh Ukrainy [Cave burials of Pliocene and Anthropogene vertebrates in the western regions of Ukraine]. Bull. Moscow Soc. Naturalist 73 (5), 114–121.

von Nordmann, A., 1858. Palaeontologie Suedrusslands. Ursus spelaeus (odessanus). H.C. Friis, Helsingfors.

Vremir, M., Ridush, B., Codrea, V., 2000. The late Pleistocene vertebrate Taphocenosis of Bukovinka cave (Western Ukraine): preliminary results. In: Karst Studies and Problems: 2000 and beyond: Proceedings of the Joint Meeting of Friends of Karst. Theoretical and Applied Karstology and International Geological Correlation Programm, Cluj–Napoca, pp. 158–161.

Wagner, J., 2010. Pliocene to early middle Pleistocene ursine bears in Europe: a taxonomic overview. J Natl Museum (Prague) 179 (20), 197–215. Natural History Series.

Westbury, M.V., et al., 2020. Hyena paleogenomes reveal a complex evolutionary history of cross-continental gene flow between spotted and cave hyena. Sci. Adv. 6 (11), 1–11. https://doi.org/10.1126/sciadv.aay0456.

Zimels, Y.L., 2008. Kryshtaleva Cave. Aston, Ternopil.

CHAPTER SIX

An overview of the geology of paleoanthropological and archaeological sites of South Africa

Tebogo V. Makhubela[a,*] and Silindokuhle S. Mavuso[b]
[a]Department of Geology, University of Johannesburg, Johannesburg, South Africa
[b]Department of Geology, Rhodes University, Eastern Cape, South Africa
*Corresponding author: e-mail address: tvmakhubela@uj.ac.za

Contents

Stratigraphy & Timescales, Volume 7
ISSN 2468-5178
https://doi.org/10.1016/bs.sats.2022.10.001

Abstract

Paleoanthropological and archaeological remains in South Africa are found in various sites formed in diverse environments of caves, rock shelters and open-air settings. These sites contain deposits dating back mainly to the Quaternary, but also extending back into the late Pliocene. The Cradle of Humankind (Cradle) is the main locality of the *Fossil Hominid Sites of South Africa* UNESCO World Heritage Site. It has 15 sites that have preserved fossils belonging to several hominin species that lived during the late Pliocene and Pleistocene times. Sites in the Cradle were developed in continuously evolving cave systems due to lithological differences and tectonic processes that promote and control the dissolution of the Late Archean-Early Proterozoic dolostone host bedrock. Thus, the deposits in the caves have complex and intermittent stratigraphy, difficult to resolve even with advanced dating techniques. Outside the Cradle, there are innumerable sites with deposits that mostly date later than the Early Pleistocene. Thus, these sites have preserved paleoanthropological and archaeological remains associated with archaic humans and *Homo sapiens*. Here we provide an overview of the stratigraphy and timescales of six sites in the Cradle and twenty sites outside the Cradle to highlight the differences between the sites, their importance and need for further research at those sites.

1. Introduction

South Africa is a significant location for paleoanthropological and archaeological research as it hosts multiple sites that comprise Pliocene and Pleistocene geological deposits (Figs. 1–3). These sites have contributed to the understanding of the hominid evolution and development through time through the discovery and study of preserved fossils and artifacts and other indications of culture. Most of the research conducted in these sites is centered on the discovered paleoanthropological (hominid fossils) and archaeological (material culture). In most cases, the geology is not comprehensive and largely focuses on describing and dating the deposits closely associated with the finds. Much of the geological work in South Africa is concentrated in economically significant Precambrian deposits as well as the geographically expansive Paleozoic and Mesozoic, leaving a big gap in work done in Tertiary and Quaternary locations. Additionally, there is

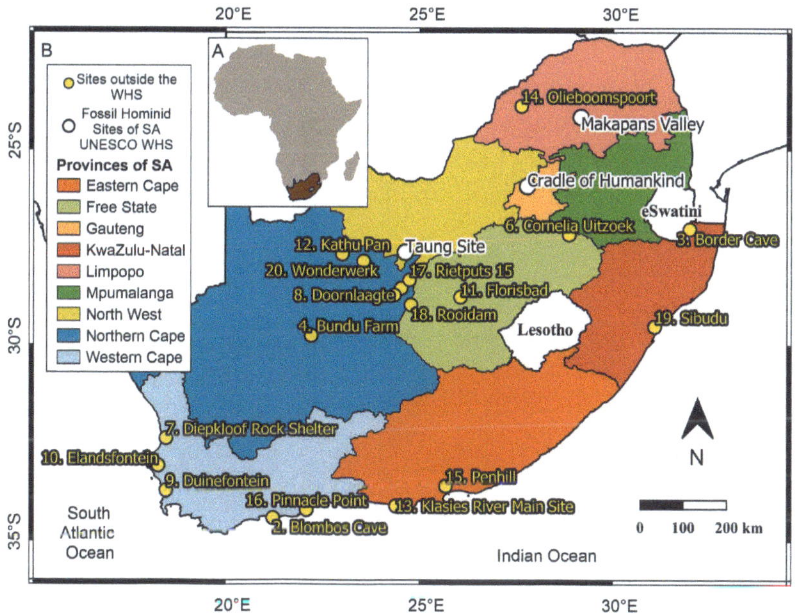

Fig. 1 (A) Map of Africa showing the position of South Africa in the continent. (B) Map of South Africa showing the locations of the sites reviewed in their provinces. The Cradle of Humankind, Makapans Valley and Taung Site together make up the Fossil Hominid Sites of South Africa UNESCO World Heritage Site.

the difficulty of radiometric dating of Plio-Pleistocene deposits in South Africa due to the absence of volcanic tuff layers; this is in contrast to paleoanthropological and archaeological sites of eastern Africa. South African sites occur in diverse lithologies and environments of caves, rock shelters and open-air sites, and most are undated or their ages are poorly constrained because the established radiometric dating methods (e.g., Deino, 2013; Wintle, 2013) are inapplicable. The uranium-lead (U-Pb) and U-series dating methods have been used extensively to date carbonate deposits (e.g., speleothems or tufa), but these methods have analytical challenges and limitations that hinder age determination (e.g., Pickering et al., 2011a; Robbins et al., 2021; Makhubela and Kramers, 2022). Clastic sediments can be dated using radiocarbon, optically stimulated luminescence (OSL), and cosmogenic nuclide dating methods, but they have their limitations. Examples of these limitations are the limit of 50 000 years for radiocarbon (e.g., Wood, 2015), determining the natural radiation accurately for OSL (e.g., Duller, 2012; Dirks et al., 2017), and determining pre-burial inheritance and interpretation of results for cosmogenic nuclides (e.g., Partridge et al., 2003; Kramers and Dirks, 2017; Shaar et al., 2021). In this chapter, we provide an overview of the geology focusing on the nature, stratigraphy and geochronology of deposits in six sites the Cradle of Humankind (Cradle, Figs. 1–3). In addition, we present an overview of the locations, geological deposits and ages of twenty paleoanthropological and archaeological sites around South Africa outside the delimited area of the Cradle (Figs. 1 and 2).

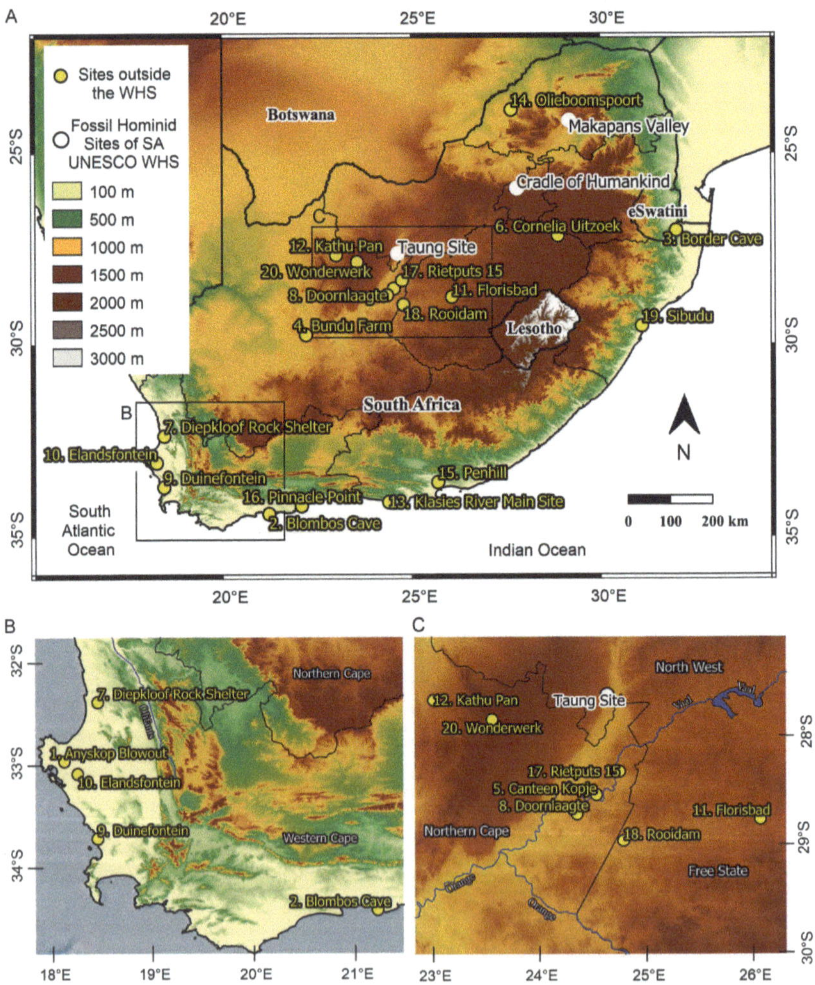

Fig. 2 Digital elevation model maps (SRTM 30 m resolution) showing the topography of South Africa. (A) Sites reviewed are located in different landscapes and elevations. (B) Numerous sites are located at the low relief coastal areas of the Western Cape Province. (C) Central South Africa also has numerous sites located closer to each other.

1.1 Cradle of Humankind

The Cradle is part of the *Fossil Hominid Sites of South Africa*, a United Nations Educational, Scientific and Cultural Organization (UNESCO) World Heritage Site (WHS). It was initially inscribed in 1999 as the Fossil Hominid Sites of Sterkfontein, Swartkrans, Kromdraai, and Environs. The Cradle is a 47 000 hectare delimited area, situated mainly in the NW corner of the Gauteng Province and overlapping into the adjoining North West Province (Fig. 3). In 2005, the *Fossil Hominid Sites of South Africa* UNESCO WHS was expanded to include the Taung Skull Fossil Site situated about 350 km SW of the Cradle in the North West Province and the Makapan Valley situated about 240 km NE of the Cradle in the Limpopo Province (Fig. 1). The Taung Skull Fossil Site is the type locality for both the genus and species of *Australopithecus africanus* following the 1924 discovery of the Taung Child skull (Dart, 1925) in the Buxton-Norlim Limeworks (Hopley et al., 2013; Kuhn et al., 2016). The Makapan Valley consists of numerous caves and sites such as the Makapansgat Limeworks, which produced some of the oldest *Australopithecus africanus* fossils older than 3 Ma, and Cave of Hearths, which preserves deeply stratified evidence of Earlier, Middle and Later Stone Ages and the Iron Ages (McNabb and Sinclair, 1996; Reed et al., 2022).

The Cradle is underlain mainly by rocks of the Neoarchaean to Paleoproterozoic supracrustal Transvaal Supergroup on the NW margin of the Johannesburg Dome Archean basement rocks (Fig. 3A). The Cradle area is an erosional, karst landscape of rolling or undulating low relief topography formed by stromatolite-rich dolomite of the Malmani Subgroup and accentuated by intercalated chert ridges and quartzite of the overlying Pretoria Group (Figs. 3 and 4; Dirks and Berger, 2013). The karst landscape of the Cradle is different from the classic models due to the scarcity of surface features such as sinkholes, polje and surficial fluvial networks (Martini et al., 2003; Martini, 2006). However, the area contains abundant caves and paleo-caves formed by hyperphreatic processes that have facilitated an excess of 300 cave systems characterized by complex mazes of passages and fissures controlled by N to NNE and E to ESE trending joints and fractures (Martini and Kavalieris, 1976; Martini et al., 2003; Dirks and Berger, 2013). To date, 15 of these cave systems are regarded as major sites that have yielded paleoanthropological fossils (Fig. 5) as well as artifacts classified as Oldowan (e.g., Kuman et al., 2021) and Acheulean (e.g., Kuman and Clarke, 2000). Here we only provide an overview of some of six sites (Fig. 3). Many of these cave systems occur along the contacts of the Malmani Subgroup Formations

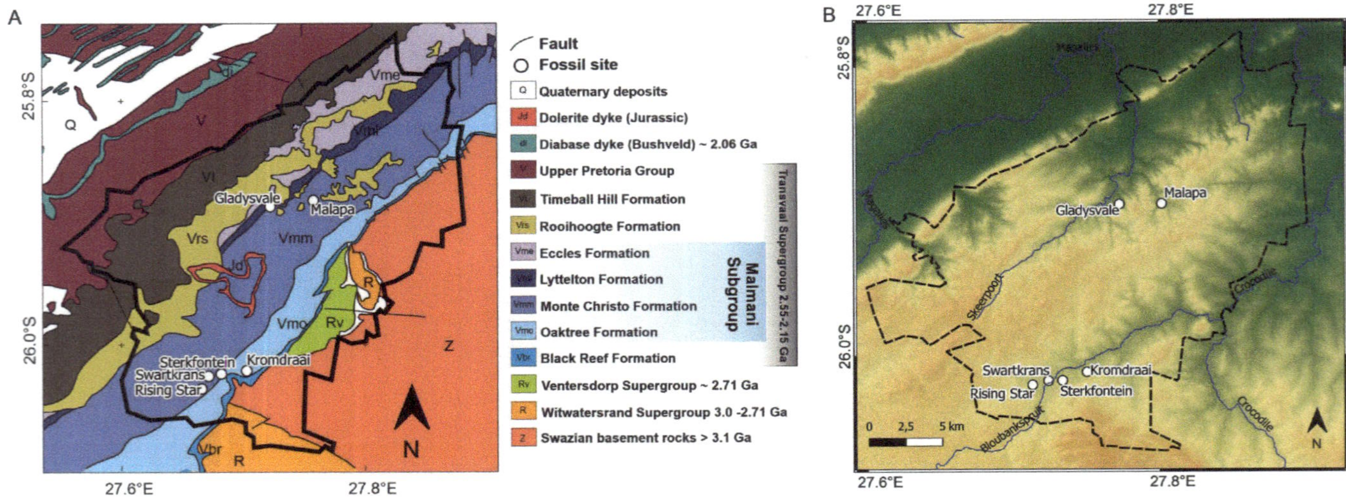

Fig. 3 Maps of the Cradle of Humankind. (A) Geological map showing the bedrock geology of the six sites reviewed in the Cradle. (B) Digital elevation model map (SRTM 30 m resolution) showing the topography of the Cradle and sites reviewed.

Fig. 4 Photograph showing the undulating low relief topography of the Cradle of Humankind's karst landscape. Chert and quartzite ridges locally stand out in the landscape as they are more resistant to erosion than the dolomite bedrock of the area. *Photo credit: Dirk van Rooyen.*

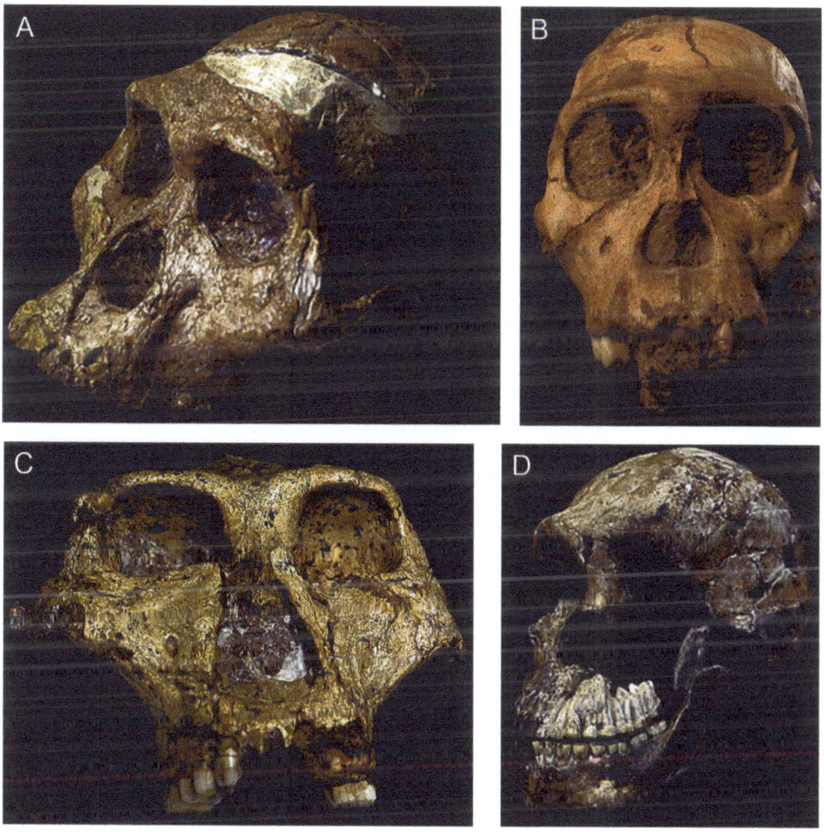

Fig. 5 Paleoanthropological remains from the Cradle of Humankind. (A) Mrs Ples *Australopithecus africanus* fossils (STS 5) from Sterkfontein. (B) *Australopithecus sediba* MH1 skull from Malapa (Berger et al., 2010). (C) Paranthropus robustus SK 48 skull from Swartkrans (Broom and Robinson, 1949). (D) *Homo naledi* LES1 skull from Rising Star (Hawks et al., 2017). *Photo credits: José Braga (A, C: used under CC-BY-4.0), Lee Berger and the University of the Witwatersrand (B, D: used under CC-BY-4.0).*

Fig. 6 (A) Uitkomst rock shelter in the Cradle of Humankind. (B) Extensive 12 m thick tufa deposits in the Cradle of Humankind. *Photo credits: Dirk van Rooyen (A), Tebogo Makhubela (B).*

(Fig. 3A; Dirks and Berger, 2013). Almost all the sites in the Cradle are caves and cave remnants, but there are also less studied rock shelters such as Uitkomst (Fig. 6A) and open-air sites such as the Lefika la Noka tufa deposits (Fig. 6B). The Lefika la Noka tufa deposits are located where the spring-fed Grootvlei River and the Skeerpoort River confluence N of the Malapa and Gladysvale fossil sites, and cover an area of around 170 m by 85 m with a thickness of ca. 12.5 m. The tufa is still actively forming and the deposits seem to span the past 200–300 thousand years. Therefore, it is anticipated that research from this tufa site will be important for further understanding of the paleo-environment of the Cradle during a time when *Homo naledi* was living in the area and *Homo sapiens* was evolving on the African continent.

2. Sites in the Cradle of Humankind

2.1 Sterkfontein

The Sterkfontein cave system is a site with the longest, almost continuous span of paleoanthropological explorations and excavations in the world, since 1895 when lime prospectors started blasting open some parts of the cave system (Thackeray, 2020). Geologist David Draper started scientific work at Sterkfontein in that same year when he described the surface outcrops of the fossil-bearing breccia and flowstone formations that were different from the cave-forming dolomite bedrock (Martini et al., 2003). This work reported the presence of fossils of a primate skull, carnivores, equids, rodents and birds in the fossiliferous breccia rock, but for 40 years until 1935, lime mining activities that destroyed many fossil-bearing breccia deposits and their formation context dominated at the Sterkfontein site

(Thackeray, 2020). Systematic scientific work of paleontological and paleoanthropological context in the underground deposits only started from 1936 with Robert Broom who discovered multiple australopithecine specimens (Broom, 1936, 1938a), prior to the major discovery of a nearly complete cranium of *Australopithecus africanus* (Sts 5, Fig. 5A) dubbed Mrs Ples (Broom, 1947). Over the past 8 decades, a plethora of fossil discoveries of diverse hominids and other faunal assemblages were made at Sterkfontein. During this time, systematic and continuous geological work of the deposits at both surface and underground were also taking place, starting with the mapping and description of the breccia deposits by Cooke (1938). However, the Sterkfontein cave fill deposits are so complex and disordered, which has led to multiple, contentious stratigraphic interpretations published over the past 8 decades. Among all the stratigraphic interpretations, the most appreciated and commonly used as a reference is the lithostratigraphic interpretation of the Sterkfontein deposits titled 'Sterkfontein Formation' proposed by Partridge (1978) and revised by Partridge and Watt (1991) using borehole cores from five boreholes. Stratford (2020) provides a review of the major stratigraphic interpretations proposed for Sterkfontein cave since Cooke (1938) until 2014. The review of Stratford (2020) focuses on the Sterkfontein Formation and examines how recent work has challenged and or improved it towards establishing a stratigraphic interpretation that will provide a comprehensive framework for excavations in the Sterkfontein deposits.

The Sterkfontein Formation uses a member system to classify the deposits into six members: Member 1 (M1) to Member 6 (M6) from oldest to youngest (Partridge, 1978; Partridge and Watt, 1991). The five Sterkfontein borehole cores were re-evaluated by Pickering and Kramers (2010) who suggested a revised stratigraphy based on sedimentary facies classification of five facies (A-E) and U-Pb dating of speleothems. The speleothems were classified as Facies E. M1 was deposited directly on weathered dolomite and lacks paleontological fossils even though it can be as thick as 20 m in some places (Martini et al., 2003). It is thought to have been deposited during the time when the cave was been enlarged by erosion but still lacking a permanent opening to the surface (Partridge, 1978; Martini et al., 2003). Pickering and Kramers (2010) described M1 as "Facies D: cave earth" because it is a mixture of angular dolomite and chert gravel from the host bedrock mixed with externally-derived clastic sediments and bone fragments, containing abundant manganese and iron oxides that give it a dark brown color (Partridge and Watt, 1991). M2 deposits have been described from inside the Silberberg Grotto and from three borehole cores as reddish-brown, horizontally bedded loam sediment containing abundant bone and numerous speleothem fragments (Partridge, 1978; Partridge and Watt, 1991). The nearly complete *Australopithecus* StW 573 skeleton nicknamed Little Foot was found in M2 inside the Silberberg Grotto (Clarke, 1998; Clarke and Kuman, 2019). Later studies identified two facies in M2 deposits. Pickering and Kramers (2010) identified Facies A as a proximal calcified, coarse-grained breccia surrounding StW 573 fossils and Facies C as a distal sediment similar in description to Partridge (1978), but other studies (Clarke, 2006; Bruxelles et al., 2014, 2019) identified two facies that were both proximal sections of an extensive talus deposit. M3 is a 5–11 m thick deposit observed inside the Silberberg Grotto, on surface, and in four of the borehole cores (Martini et al., 2003). It starts with a basal calcite flowstone that can be up to 1.2 m thick and is overlain by ca. 8 m reddish brown muddy sandstone containing angular gravel fragments and bone fragments (Partridge, 1978; Partridge and Watt, 1991). M4 is a reddish-brown, largely calcified massive breccia deposit up to 9 m thick in four beds that differ in terms of loam matrix and abundance of rock debris that range from pebbles to boulders (Partridge, 1978; Partridge and Watt, 1991; Martini et al., 2003). The M4 deposit, now mainly exposed on surface due to cave roof collapse and subsequent erosion, is regarded as the main fossil-bearing deposit because it contains a large collection of *Australopithecus Africanus* (Stratford, 2020). The relationship between M3 and M4 is unclear, which has led to Pickering and Kramers (2010) contending that M3 is a distal endmember of M4 instead of being a distinct deposit as suggested by Partridge (1978). M5 unconformably overlies M4 in the western section of the surface deposits with thickness up to 5 m thick. M5 is similar to M4 in terms of mode of deposition and fabric due to its reddish-brown breccia that happens to contain more sandy loam as well as occasional bones and stone tools (Partridge and Watt, 1991; Martini et al., 2003). The similarities between these two members and the deposition of M5 around boulders of M4 formed by solution cavities has posed challenges with distinguishing between the two members at this western section and further study and dating are still needed (Stratford, 2020; Granger et al., 2022). M6 is also confined to the western section and similar to M5, but it is darker, less calcified, and much smaller with a thickness up to 1.5 m (Martini et al., 2003).

Dating the complex deposits of the Sterkfontein cave has been a long-term trial. M2 in the Silberberg Grotto was initially radiometrically dated between 4.72 ± 1.02 Ma and 3.78 ± 0.31 Ma using cosmogenic aluminium-26 and beryllium-10 ($^{26}Al/^{10}Be$) isotopes in quartz from samples obtained below, above and adjacent the StW 573 skeleton (Partridge et al., 2003). Subsequent to the dating of the sediments, contrasting ages were suggested for M2 around the StW 573 skeleton based on uranium-lead (U-Pb) dating of speleothems. Walker et al. (2006) determined ages of 2.25 ± 0.09 Ma and 2.17 ± 0.17 Ma, respectively, for the speleothems below and above the sediments hosting StW 573. Pickering and Kramers (2010) determined an age of 2.35 ± 0.10 Ma for the lower speleothem and then concluded that the ages of two bracketing speleothems constrain the M2 deposits enclosing the StW 573 skeleton to a time range of 2.46–2.16 Ma. Using the stratigraphy in the borehole cores and the speleothem U-Pb ages, however, Pickering and Kramers (2010) proposed that M2 was deposited rapidly in ca. 200 thousand years between 2.8 and 2.6 Ma. This was because the flowstone that forms the basal layer of M2 in borehole 1 yielded an age of 2.8 ± 0.28 Ma whereas two 1–2 m thick flowstones that cap M2 and separate it from M4 in boreholes 1 and 4 yielded ages of 2.83 ± 0.34 Ma and 2.65 ± 0.30 Ma, respectively (Pickering and Kramers, 2010). Renewed stratigraphic analysis of the deposits around StW 573 by Bruxelles et al. (2014) argued that the speleothems dated were formed by later infilling of cavities in the sediments and, thus, the speleothem dates were out of stratigraphic sequence. Thereafter, Granger et al. (2015) published an $^{26}Al/^{10}Be$ isochron burial age of 3.67 ± 0.16 Ma for the depositional age of the StW 573-bearing M2 breccia. This age was deemed to be reliable against sediment reworking given the principles of the isochron burial method (Balco and Rovey, 2008), but was challenged on the basis of an alternative interpretation of the $^{26}Al/^{10}Be$ data indicating that one of the samples used in the isochron could only have a maximum age of 2.8 Ma (Kramers and Dirks, 2017). M4 deposits were initially dated between 2.6 Ma, based on the speleothems in the boreholes, and 2.1 Ma, based on speleothems from the surface exposures (Pickering and Kramers, 2010). However, recent research by Granger et al. (2022) produced an $^{26}Al/^{10}Be$ isochron burial age of 3.41 ± 0.11 Ma and two simple burial ages of 3.49 ± 0.19 Ma and 3.61 ± 0.09 Ma from M4 sediments in the surface exposures, Fossil Cavern and Jacovec Cavern. The $^{26}Al/^{10}Be$ burial ages are consistent with previous observations of stratigraphy and petrography (Bruxelles et al., 2014, 2019) that described the speleothems dated in the Sterkfontein deposits as intrusive in nature and only deposited much later than the accumulation of the breccia deposits.

2.2 Swartkrans

Swartkrans cave is located ca. 0.8 km west of Sterkfontein cave and the Bloubank River separates the two sites (Fig. 3). Like all fossil sites in the Cradle, Swartkrans was also affected by the commercial lime mining during 1930 to 1935 and again during 1950–1951, leading to the disorientation of the clastic deposits and incomplete stratigraphy (Brain, 1970). Scientific work at Swartkrans started almost half a century later than at Sterkfontein with the work of Robert broom together with John Robinson (Broom and Robinson, 1949, 1950, 1952). This early work from Swartkrans presented spectacular descriptions of specimens Sk 48 cranium and the Sk 23 mandible assigned to *Paranthropus robustus*, a type of a "robust" australopithecine (Fig. 5C; Brain, 1988; de Ruiter, 2003). Then there was a period of inactivity at Swartkrans until C.K. Brain started meticulous excavations that went on for 21 years between 1965 and 1986. These excavations included the removal of the miner's rubble, which were necessary for stratigraphic work that revealed the complex nature of the deposits formed by several phases of accumulation and erosion (Sutton et al., 2009). Following that, the work led to the adoption of the Member system used at Sterkfontein by Partridge (1978), and five Members 1–5 of the Swartkrans Formation were defined for the site (Brain, 1993, 1994). Prior to that, the deposits were initially classified as two distinct masses of breccia: pink breccia that occupied the Outer Cave, and the stratified brown breccia that partially filled the Inner Cave (Brain, 1958). Swartkrans is well known and important, mostly due to Brain's work, for early behavior of hominids including their use of bone tools (Brain, 1981, 1993), their controlled use of fire at ca. 1 Ma (Brain and Sillent, 1988), as well as the second largest collection of Oldowan and Early Acheulean stone tools in Southern Africa (Sutton et al., 2009). Swartkrans has also yielded the largest collection of *Paranthropus robustus* fossils (Brain, 1988). This collection of *Paranthropus robustus* together with the Oldowan stone tools collection at Swartkrans have now been dated as the oldest occurrence in Southern Africa (Kuman et al., 2021).

Member 1 is the oldest unit represented by the steeply dipping, orange-colored and poorly calcified Lower Bank (LB) on the floor of the cave and the pinkish, more fossil-rich breccia of Hanging Remnant (HR), an unsupported mass adhering to the north wall of the cave (Brain, 1988, 1993). LB and HR were part of an infilling debris cone that was calcified to some degree but its middle section was eroded to the lower parts of the cave system (Brain, 1988). The early fossils described by Broom and Robinson were recovered from the HR, while the developed Oldowan artifacts and fossils of *Paranthropus* and early *Homo* (Kuman et al., 2021). Member 1 deposits have been dated to 2.25–1.71 Ma using U-Pb dating of speleothems that bracket the HR (Pickering et al., 2011a), 2.19–1.8 Ma using $^{26}Al/^{10}Be$ simple burial dating of LB sediments (Gibbon et al., 2014), and the LB more precisely to 2.22 ± 0.09 Ma using $^{26}Al/^{10}Be$ isochron burial dating (Kuman et al., 2021). Member 2 unconformably overlie Member 1 and occurs as heavily calcified, fauna-rich brown breccia in the Outer Cave and stratified brown breccia in the Inner Cave (Brain, 1993). It was dated to 1.36 ± 0.29 Ma using U-Pb dating of bovid tooth (Balter et al., 2008). Member 3 is partially preserved on the west wall of the cave as a 6 m infill that unconformably erode into Members 1 and 2 (Brain, 1993), and was dated to 0.96 ± 0.09 Ma using $^{26}Al/^{10}Be$ simple burial dating. Member 4 is a colluvial deposit that occurs in the northeast corner of the cave and contains abundant MSA stone artifacts (Sutton et al., 2009). Member 4 deposits have been the focus of the recent Swartkrans Paleoanthropological Research Project (SPRP), which produced a date of 110.33 ± 1.98 ka using U/Th dating for a flowstone that underlies the deposits (Sutton et al., 2009). Member 5 occurs as a channel infill alongside the northwest wall of the cave with a radiocarbon age of ca. 11 000 years (Brain, 1993; Sutton et al., 2009).

2.3 Kromdraai

The Kromdraai cave is located ca. 2 km E of the Sterkfontein cave (Fig. 3). It consists of two fossiliferous localities: Kromdraai A (KA) that has yielded archaeological remains (Oldowan and Acheulean) and faunal fossils without hominins, and Kromdraai B (KB) that has yielded faunal fossils including hominins (Skinner et al., 2013; Braga et al., 2017). The type specimen of *Paranthropus robustus* was recovered from KB in 1938 as a partial skeleton (Broom, 1938a, 1938b, 1942, 1943). In addition, craniodental and postcranial fossils of over 30 hominin individuals of *Paranthropus robustus* and early *Homo* have been recovered from KB (Braga and Thackeray, 2003; Braga et al., 2017). Today the Ka and KB deposits are exposed at surface due to cave unroofing and erosion. KA and KB have long been treated as deposits from two distinct caves that had limited connection due to the 30 m long dolomite bedrock separating them. New excavations at Kromdraai continue to reveal the extensive KB deposits beneath the soil cover. Currently, KB has three localities identified as KB East, KB West, and Kromdraai Extension

Site (Braga et al., 2017). The stratigraphy of KB has been revised several times since the early works of Brain (1958), Partridge (1982) and Vrba (1981). The currently used stratigraphy was revised by Bruxelles et al. (2016). Brain (1958) developed a basic stratigraphy that described the deposits as either stony breccia or pink breccia. Partridge (1982) and Vrba (1981) adopted the member system used at Sterkfontein and described the deposits as elaborate five members. The stratigraphy of Bruxelles et al. (2016) describes the KB deposits as seven members. Unfortunately, fossil specimens excavated before 1982 by Broom (1938a, 1938b, 1942, 1943, 1947) and Brain (Brain, 1958, 1970, 1981, 1988, 1993, 1994) cannot be assigned to the member system currently used even when the depth they were recovered at is known (Skinner et al., 2013).

A condensed description of the Bruxelles et al. (2016) stratigraphy of KB: Member 1 has been described as gray dark stony breccia. It is composed of sediments that clearly show that during its deposition the cave had limited, indirect opening to the surface. Member 1 sediments are black-brown silty sands with dolomite fragments in the matrix mainly composed of iron and manganese, attesting to their decayed nature as most of the soluble calcium and magnesium have dissolved away. The chert fragments are still angular and without the distinctive altered appearance seen in sediments from the surface. The KB Member 1 is similar to the Sterkfontein Member 1. Member 2 is the typical breccia composed of abundant gravel clasts in a loamy matrix that has been extensively cemented by calcification. The gravel clasts are mostly angular to subrounded dolomite and chert. In places where Member 2 has been decalcified, it occurs as brown silty-sand with gravel and scattered blocks of calcified breccia. Member 3 is a coarse-grained sandstone uncoformably overlying Member 2. It has a pinkish color and contains accessory angular clasts of chert and dolomite. It contains abundant concretions of noticeably crystallized and clean (white) calcite from the cementation, which gives this unit a distinctive appearance. There are calcite veinlets, towards the top of the deposits, which connect to a flowstone layer that overlies Member 3 as if it were a capping flowstone but the contact in some places is obscure, making the flowstone seem like a later infilling. Member 4 has been described using three facies that are treated as submembers. Therefore, Member 4 is a sequence of breccia layers that have differing amounts of gravel types (clasts of chert, dolomite, quartz and calcite). Submember 4.1 is made up of alternating layers of breccia and sandstone. Submember 4.2 is a matrix-supported pinkish sandy breccia that contains clasts of very coarse gravel and fossil bone. Submember 4.3 is similar to 4.2 but the sandy matrix and gravel are less coarse. Member 5 is an orangish-brown, massive sandstone containing appreciable amounts of rounded granulates of chert and ferruginous grains, probably Fe-Mn-oxide coated grains. The ferruginous grains together with the absence of dolomite clasts give Member 5 a distinctive appearance because they are not present in Members 3 and 4. Member 6 is a dark, orangish-brown breccia that contains only gravel of rounded chert. It also contains abundant ferruginous grains in cubic form, interpreted as pyrite grains that have been transformed into goethite or hematite. Member 7 is a finely laminated sandstone with abundant calcite veinlets and a high concentration of large fossil bone. This member has patches or layers that contain gravel of only chert and ferruginous clasts, possibly boulders of Member 6 included into it. The KB deposits have not been radiometrically dated yet, and the age assignment are currently based on comparison to other sites in the Cradle such as Sterkfontein and Swartkrans given the presence of similar fossils and deposits in these sites.

2.4 Malapa

The Malapa cave is located ca. 15 km to the NE of Sterkfontein Cave (Fig. 3) in the Cradle Nature Reserve. It lies near the base of an originally >30 m deep, strata-bound and fracture-controlled cave system now exposed at surface after erosion by valley incision (Fig. 7A–C; Dirks et al., 2010). Two skeletons of *Australopithecus sediba* were discovered at Malapa (Berger et al., 2010; Dirks et al., 2010). The deposits of the Malapa cave were initially described as five distinct sedimentary facies A to E (from bottom to top) with a flowstone covering facies A and B (Dirks et al., 2010), and later revised to include facies F and a second flowstone layer that underlie it (Fig. 7D; Pickering et al., 2011b). Facies A and B occur below central flowstone 1 sheet, which dips moderately to the N-NE suggesting a N-directed water flow during deposition of the clastic sediments (Dirks et al., 2010). Facies A is dark colored, coarse-grained and moderately well sorted, with abundant rounded grains of chert, quartz, dolomite, bone fragments and peloids. Facies B consists of alternating peloidal grainstone and clastic sandstone. A milky quartz vein surrounded by an alteration halo of Fe-Mn-oxides and clay intruded facies A and B but does not appear to intrude the flowstone. Facies C is a thick erosion remnant that drapes over the flowstone and is identical to the facies B peloidal grainstone, but it contains Fe-Mn-oxide-coated subrounded fragments of quartz, dolomite and shale, and some bone fragments towards the top of the layer. Facies D

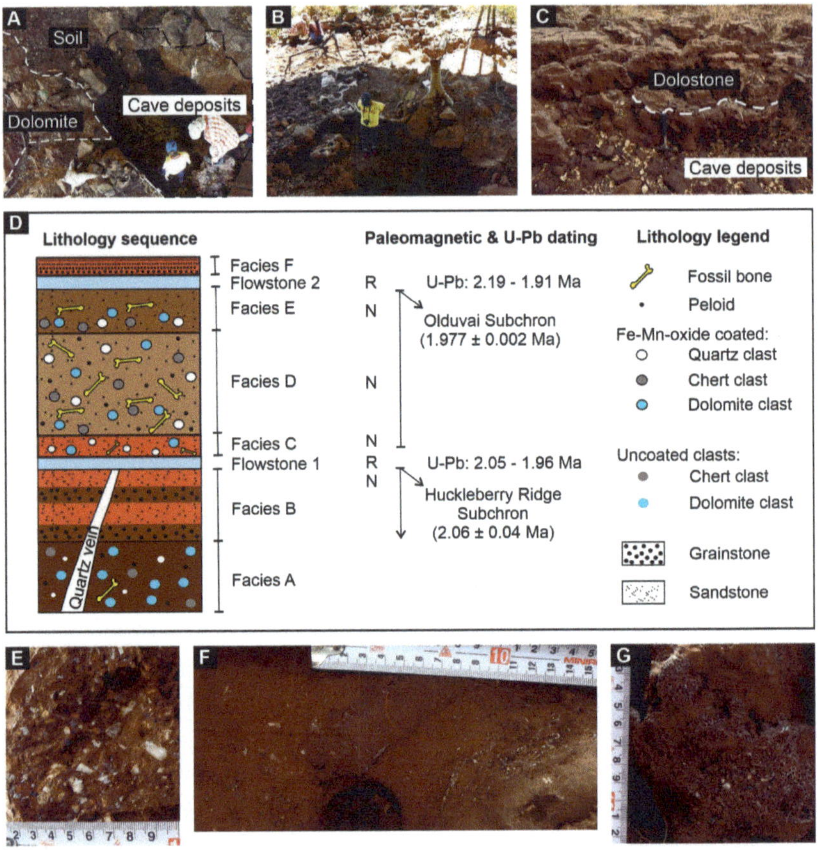

Fig. 7 Photos and stratigraphy of Malapa site. (A–C) Malapa cave deposits are now exposed on surface and covered by soil. The deposits on surface cover an area bigger than 5x5 m (B) and two pits have been excavated where cave deposits are wide and deep (A). In most cases the cave deposits are thin remnants on the dolomite bedrock (C). (D) Lithological sequence of the Malapa deposits and the dating results. (E) The fossil-bearing facies D is coarse-grained and poorly sorted sandstone. (F) Facies E is finer grained with higher degree of sorting compared to facies D. (G) Facies F is a massive peloidal grain stone with horizontal layering and graded bedding. *Photo credits: Tebogo Makhubela (A, B), Paloma de la Peña (C, E–G).*

unconformably overlies facies C and is the unit where well-preserved *Australopithecus sediba* and other macro- and micro-mammal fossils were recovered. It is a 1.5 m thick massive and poorly sorted, light-colored, coarse-grained sandstone cemented by sparite (Fig. 7E). The framework grains are mainly 0.5 to 2.5 mm sized quartz, chert, dolomite and peloids, and less commonly Fe-Mn-oxide-coated grains (Fig. 7E). Facies E consists of calcareous sandstone, similar to facies D, but darker colored and finer-grained, with higher degree of sorting and rounding (Fig. 7F). Fossil bone is also abundant in facies E, but less well preserved than bone content in facies D. This could have been due to damage by prolonged exposure to standing or flowing water prior to final burial and cementation (Brain, 1981). Facies F is a massive peloidal grainstone with horizontal layering and graded bedding (Fig. 7G; Pickering et al., 2011b). The Malapa cave deposits are some of the best-dated deposits in the Cradle based on the combination of U-Pb dating of flowstones with paleomagnetic dating. Flowstone 1 was dated to 2.024 ± 0.062 Ma and 2.026 ± 0.021 Ma, providing maximum age range of 2.05 to 1.96 Ma for the hominin-bearing facies D (Dirks et al., 2010), whereas two subsamples of flowstone 2 were dated to 2.048 ± 0.140 Ma, providing a range of 2.19 to 1.91 Ma for the facies D capping flowstone (Pickering et al., 2011b). Paleomagnetic analysis showed that flowstone 1 mainly preserve a reverse polarity, but has a normal polarity excursion near its base, which was interpreted as the Huckleberry Ridge Subchron at 2.06 ± 0.04 Ma (Fig. 7D; Dirks et al., 2010). Sediments of facies C, D, and E above flowstone 1 recorded normal polarity were interpreted as the beginning of the 1.95- to 1.78 Ma Olduvai Subchron (Dirks et al., 2010). Flowstone 2 recorded a reversed polarity that when combined with the U-Pb ages (>1.91 Ma) indicates that this layer must have formed before the beginning of the Olduvai normal-polarity subchron at 1.977 ± 0.002 Ma (Fig. 7D; Pickering et al., 2011b; Channell et al., 2002).

2.5 Gladysvale

Gladysvale Cave is located in the Cradle Nature Reserve ca. 5 km W of Malapa (Fig. 3). Scientific studies at Gladysvale commenced in 1936 by Robert Broom (Berger and Tobias, 1994), but geological studies started with the first description of the internal cave deposits by Peabody (1948). The Gladysvale cave system is comprised of an upper (main) chamber that connects to at least two lower chambers (Berger and Tobias, 1994; Lacruz et al., 2002; Pickering et al., 2007). On the surface outside the upper chamber there was another chamber that was deroofed and eroded as evidenced by *in situ* cave deposits named Gladysvale external deposit (Lacruz, 2002; Lacruz et al., 2002). Like other sites in the Cradle, Gladysvale was also affected by lime mining between 1902 and 1928 (Berger and Tobias, 1994), but the mining activity only affected the closed chambers and diminished speleothem formation, which would still be extensively active today

(Pickering et al., 2007). Gladysvale is not known for producing many paleoanthropological and archaeological remains. It has, however, has produced a significant faunal assemblage in association with well-stratified cave fill sequence of clastic sediments and flowstones (see below), valuable for understanding the paleoclimatic and paleoenvironmental controls on the nature and timing of sedimentation in other sites of the Cradle (Lacruz et al., 2003; Pickering et al., 2007).

The Gved deposits are fossil-rich (most hominin fossils at Gladysvale were recovered from this deposit) and present in calcified and decalcified form (Lacruz et al., 2002). They are horizontally well-stratified with units (labeled A at the bottom to K at the top) that are underlain and overlain by flowstone layers (Lacruz, 2002). As a result, the Gved deposits have a stratigraphy that is rare in sites of the Cradle and other South African Plio-Pleistocene sites given the organized structure of flowstone bounded units (FBUs). Nine sedimentologically distinct units were identified and described as FBUs by Lacruz et al. (2002). The nine FBUs were grouped into three sedimentation cycles called flowstone bound intervals (FBIs) based on the thicker and cleaner flowstone layers that separated groups of FBUs (Lacruz et al., 2002). Units A and B were grouped together into FBU 1 that is slightly upward coarsening with a thickness of ca. 70 cm. Unit C is FBU 2, an upward fining unit ca. 50 cm thick. FBU 1 and 2 were grouped together as FBI 1. Unit D is ca. 58 cm thick and upward coarsening whereas Unit E is ca. 55 cm thick and upward fining, and these two units are FBU 3 and 4, respectively. Unit F is thin (ca. 16 cm) and occurs as FBU 5, which caps FBI 2 composed of FBU 3-5. Unit G is FBU 6, which is ca. 67 cm thick and starts as an upward coarsening unit but because upward fining towards the top. FBU 7 is made up of two subunits of Unit H, collectively ca. 42 cm thick and upward fining. Unit I is FBU 8 with thickness ca. 37 cm and upward fining. FBU 9 is Unit J, which is ca. 17 cm and contains two discontinuous flowstone layers. FBU 6-9 make up FBI 3. The upward-coarsening and upward-fining packages of the Gved sequence were interpreted as representing proximal and distal sections of a debris cone and a fan, respectively, depending on the amount of sediment washed into the chamber and accumulating over the area (Lacruz et al., 2002). The Gved deposits were dated to 870-578 ka using a combination of paleomagnetic dating of Units B, C, E and F, and electron spin resonance (ESR) dating on bovid tooth enamel of three teeth from Units C, E and I (Lacruz et al., 2002, 2003).

The deposits underground Gladysvale cave, named Gladysvale internal deposits (Gvid), have also been studied by Pickering et al. (2007) in the upper chamber and the stratigraphy has been arranged from three areas of the deposits. Gvid is a fan accumulated up to ca. 4 m at the proximal section through at least two north-facing entry points and thins to <2 m at the distal section. Gvid and Gved are not correlative, and Gvid has seventeen identified FBUs. The clastic sediments have been described using a facies classification from A–D, and the speleothems associated with the clastic sediments are classed as facies E (Pickering et al., 2007). The Gvid deposits are described here based on Pickering et al. (2007). Facies A is a clast-supported breccia composed of large, subrounded clasts over 15 cm and can contain minor matrix of decalcified reddish-brown, medium-grained sand. The composition of the clasts is mainly dolomite, but other types can also be present with little to no bone content. Facies B is different from facies A as it is clast poor and predominantly composed of mostly decalcified brownish to pale reddish-brown fine to medium-grained sand. When facies B contains clasts, they are randomly dispersed throughout the layer, 5–10 cm and subangular to subrounded when the composition is chert, dolomite and that of speleothem. Facies C is well-calcified, poorly to well-layered, pale reddish-brown to brown medium-grained sandstone. It can contain minor amounts of 1–5 cm clasts, but it contains abundant, well-preserved fossil bone. Facies D is grayish-brown to brownish-gray, finely laminated mudstone with a microstratigraphy. It is well-calcified, and clast-poor but when clasts occur they are <1 cm. Minor amount of fossil bone also occurs in well-preserved form.

In the middle section of the Gvid deposit an area called Hinge section has deposits ca. 3.7 m thick, and there facies A is represented by FBUs 11-13 while facies B is represented by FBUs 14-16. In the Western Face 1 (WF1) section, ca. 6 m north of the Hinge section, the deposits are ca. 1.95 m thick but correlate the FBUs 11-16 on the Hinge section in addition to FBU 17. FBUs 12-13 and 15-17 are represented by facies B at WF1 whereas FBUs 11 and 14 are represented by facies D. In the Peabody Chamber where deposits are separated from the WF1 and Hinge sections, the deposits are 3.95 m thick consisting of FBUs 1-10 and 12-17. Here the Peabody Chamber deposits have FBUs 1, 5 and 17 represented by facies B, FBUs 2-4, 6 and 12-16 represented by facies C, FBUs 7-10 represented by facies D. The Gvid deposits were dated between ca. 570 ka and ca. 7 ka using U/Th dating on speleothems (Pickering et al., 2007). The oldest age (571.4 ± 24.9 ka) was obtained from the base of a stalagmite on WF1.

The oldest age in the stratified sediment sequence was 393.03±7.58 ka obtained from the middle of a stalagmite at the base of FBU 14 in WF1. A flowstone ca. 10 cm away from the stalagmite at the base of FBU 14 yielded 255.3±2.51 ka. The capping flowstones yielded 42.82±6.77 ka above FBU 16 at the Hinge section, 14.39±0.06 ka above FBU 17 at WF1 and 16.5±0.15 ka above FBU 17 at Peabody Chamber. Away from the three (Hinge, WF1 and Peabody) areas where the stratigraphy was studied, a capping flowstone above FBU 17 was dated to 7.45±0.35 ka.

2.6 Rising Star

The Rising Star site is well known for the discovery of *Homo naledi* in 2013. The cave system of this site covers an underground area of 300 m by 190 m, located 1.1 km and 2.2 km west of Swartkrans Cave and Sterkfontein Cave, respectively (Fig. 3). This cave system formed in a relatively chert-poor stromatolitic dolomite with a thickness of 15—20 m in the lower part of the Monte Christo Formation (Fig. 3A; Dirks et al., 2015). About 4500 m of passages has so far been mapped in the Rising Star cave system and they connect several fossil-bearing chambers and antechambers (Elliott et al., 2021). Fossils of *Homo naledi* have been found in two main chambers: Dinaledi Chamber (UW101, Dirks et al., 2015) and the Lesedi Chamber (UW102, Hawks et al., 2017), and an area around the Dinaledi Chamber called the Dinaledi Subsystem. The Dinaledi Subsystem is a series of narrow fissure passages denoted UW108 to UW111 due to the presence of *Homo naledi* fossils. These passages connect the Dinaledi Subsystem to the Hill Antechamber (UW107) and Chaos Chamber (Elliott et al., 2021). The highest concentration of the *Homo naledi* fossils, exceeding 2000 specimens representing at least 15 individuals, came from the top 30 cm of the floor sediments in the Dinaledi Chamber, which makes it the central locality of the Rising Star cave where most research has been centered. The Dinaledi Chamber is about 30 m below the surface and about 80 m from the main entrance used to access the cave system, and it is only accessible via the Hill Antechamber. The Hill Antechamber connects with the adjoining Dragon's Back Chamber through a 12 m fracture referred to as the Chute. Lesedi Chamber is the locality with the second most accumulation of *Homo naledi* fossils with over 150 specimens of at least three individuals collected (Hawks et al., 2017). However, the Lesedi Chamber is separated from the Dinaledi Chamber by a straight-line distance of about 50 m and a traversable distance of about 150 m. About 50 specimens of *Homo naledi* fossils have been collected from the floor surface and above chert ledges in passages and chambers of the Dinaledi Subsystem (Elliott et al., 2021).

The sediments in the Rising Star cave originate from a debris cone deposited inside the Postbox Chamber through a vertical shaft that used to connect to the surface (Robbins et al., 2021). Thus, the sediments in most parts of the cave system, particularly the distal areas such as the Dinaledi Chamber and Subsystem, are largely unconsolidated to poorly consolidated mud-rich deposits (Fig. 8A). This makes the Rising Star cave sediments different from the generally lithified and calcified, coarser breccia deposits found in other hominid sites in the Cradle (Dirks et al., 2015, 2017; Wiersma et al., 2020; Robbins et al., 2021). The sediments in the

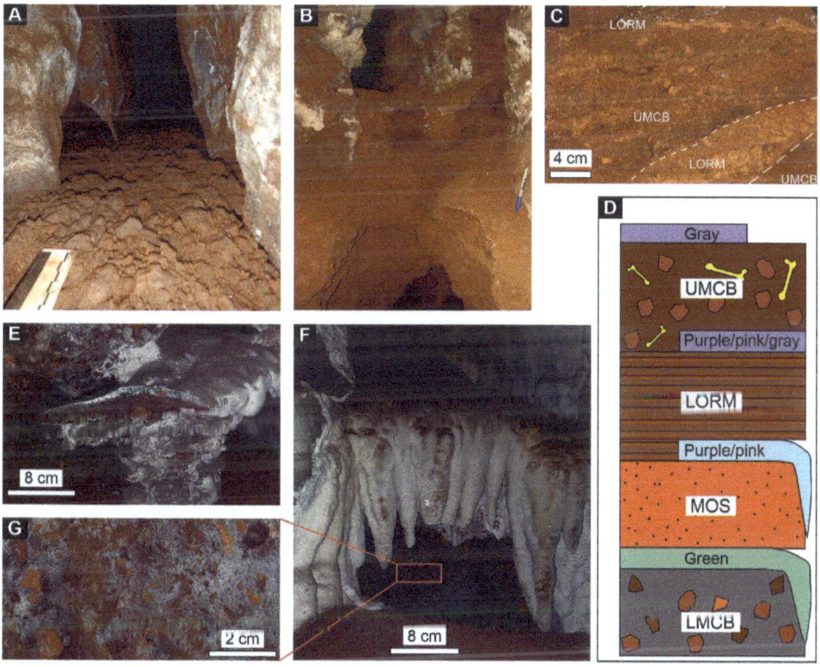

Fig. 8 Photos and stratigraphy of Rising Star cave. (A) Sediments in the Dinaledi Chamber are largely unconsolidated to poorly consolidated mud-rich deposits. (B) Unit of laminated orange red mudstone (LORM). (C) LORM undergoes a process of auto-brecciation to form unlithified mud clast breccia (UMCB; Wiersma et al., 2020). (D) The stratigraphic sequence in the Dinaledi Chamber based on Dirks et al. (2017) and Robbins et al. (2021). (E F) Inside the Dinaledi Chamber, the older flowstones (color code: green, purple and pink) occur as flowstone drapes over erosion remnants of lithified mud clast breccia (LMCB) and massive orange sandstone (MOS) hanging on the walls. (G) zoomed in section of the LMCB, which is calcified and darker than UMCB. *Photo credits: Tebogo Makhubela.*

Dinaledi Chamber and Subsystem are muddy sediments with variable fine to coarse sand content. This is due to the Dragon's Back dolomite block, a 5 m high obstruction that restricted the coarse-grained sediments (>2 mm) from entering the Dinaledi Chamber and Subsystem to deposit breccia sediments seen in other parts of the Rising Star cave (Dirks et al., 2015, 2017). The floor sediments in the Lesedi Chamber, however, contain a layer of gravel not seen anywhere in the Dinaledi Chamber and Subsystem, affirming the separation between the two parts of the cave. With only mud-laden water entering the Dinaledi Chamber and Subsystem, laminated orange-red mudstones (LORM) were deposited but later transformed to unlithified mud clast breccia (UMCB) through auto-brecciation that involved autochthonous to para-autochthonous physical and diagenetic break-up (Fig. 8B and C; Wiersma et al. (2020)). LORM and UMCB were initially described as two distinct depositional facies 1 and facies 2, respectively, by Dirks et al. (2015).

Four lithostratigraphic units have been established in the Dinaledi Chamber and Subsystem based on the floor sediments across the Dinaledi Subsystem, the erosion remnants that occur at a variety of stratigraphic positions and the distribution of laterally continuous flowstone intercalations (Fig. 8D; Dirks et al., 2017; Robbins et al., 2021). The flowstones were dated using the U-Th disequilibrium method and a color-coded classification was developed to represent seven flowstone groups from old to young as green, blue, red, yellow, purple, pink and gray flowstones (Dirks et al., 2017; Robbins et al., 2021). Four groups (green >400 ka, purple 290-180 ka, pink 170-140 ka and gray <125 ka) of flowstones were identified in the Dinaledi Chamber and Subsystem (Dirks et al., 2017; Robbins et al., 2021). The older green, purple and pink flowstones are drapes preserved as hanging erosion remnants with rounded edges (Fig. 8E and F) and were collectively denoted Flowstone 1a-e by Dirks et al. (2015). The gray flowstones with ages 125-88 ka were regarded as representing the most significant phase of flowstone development in the Dinaledi Chamber (Dirks et al., 2015). They form cascades emanating from fractures in cave walls and flow down the cave walls, spreading out across the cave floor to form thin flowstone cover on floor sediments containing *Homo naledi* fossils. The gray flowstones <32 ka are still active and forming presently. They are primarily restricted to the chamber ceiling, forming delicate needle- and rod-like forms that include well-formed crystals of aragonite. The lithostratigraphic units in the Dinaledi Subsystem are from old to young: lithified mud clast breccia (LMCB, Fig. 8G) erosion remnant covered by green flowstone, massive orange sand (MOS) is covered by purple or pink flowstone,

LORM and UMCB, both covered by purple, pink and gray flowstone units (Robbins et al., 2021). UMCB is the only lithostratigraphic unit that contains *Homo naledi* fossils.

3. Sites outside the Cradle of Humankind

3.1 Anyskop Blowout

Anyskop Blowout is an open-air, archaeological site located around Langebaan Lagoon about 10 km from the Atlantic Ocean overlooking the coastal plain (Fig. 2B; Kandel and Conard, 2012). A prominent hill that rises gently above the coastal plain is called Anyskop and offers a vantage point to view the surrounding landscapes of the Saldanha Bay and Langebaan Lagoon region (Kandel et al., 2006). This probably explains why the archaeological remains of the site convincingly record occupation by archaic and modern *Homo sapiens* during all the South African archaeological records from Early Stone Age (ESA) through Middle Stone Age (MSA) to Later Stone Age (LSA) and Howiesons Poort (Kandel and Conard, 2012; Dietl et al., 2005). The hilltop of Anyskop Blowout comprises mostly well-consolidated sediments that being from the Miocene to the Early Pleistocene with rich faunal assemblages (Hendey, 1982; Roberts et al., 2011). These sediments are capped by a concreted calcrete deposit which is overlain by unconsolidated, 3–4 m thick reddish-brown aeolian sandy unit (Kandel et al., 2006). Attempts to date the calcrete using U-series were unsuccessful, but a minimum age of 250–200 ka has been suggested based on the presence of Acheulean artifacts on the flat surface of the calcrete (Dietl et al., 2005). The aeolian sands were dated using OSL and at a point of 0.5 m above the calcrete a depositional age of 182 ± 9 ka was obtained (Roberts et al., 2011). The Blowout refers to an ca. 3.5 ha area of the deposit that has been eroded by wind and removed the aeolian sand, thus, exposing the calcrete deposit (Kandel and Conard, 2012). This implies some level of deflation in the context of the finds reported at the site. Although Kandel and Conard (2012) have suggested that the deflation has not had significant impact on the context of the site. Attempts to determine the timing of the blowout event using OSL dating yielded an age of 0.3 ± 0.6 ka (Roberts et al., 2011).

3.2 Blombos Cave

Blombos Cave is located 25 km west of Still Bay Village at about 34.5 m above sea level and 100 m from the Indian Ocean (Figs. 1 and 2; d'Errico et al., 2005). The cave was formed in an ancient wave-carved calcarenite cliff made up of sediments of the Bredasdorp Group (Henshilwood et al., 2018; d'Errico et al., 2005). Blombos is well known for its records of the MSA, particularly for understanding the development of symbolic behavior in *Homo sapiens* (Jacobs et al., 2020). The site has yielded a large assemblage of faunal finds (humans, macromammals and tortoises, fish and shellfish) and cultural artifact (ochre, lithics, bone tools and abstract art) (Henshilwood et al., 2001, 2011; d'Errico et al., 2005). The site has well-stratified deposits of MSA and LSA separated by a culturally-sterile aeolian dune sand layer (Henshilwood et al., 2018). Generally, all layers are sand units interbedded with shells, organic matter and ash (Jacobs et al., 2020). The MSA sequence has layers grouped into four phases M3 (101-94 ka), M2 lower (85-82 ka), M2 upper (80-74 ka) and M1 (76-70 ka), and the ages of each phase were constrained using a combination of OSL, U/Th and Thermoluminescence (TL) dating (Henshilwood et al., 2011; Jacobs et al., 2020). The sterile sand layer indicating a hiatus was dated to 70 ka and the LSA deposits were dated to 2 ka (Henshilwood et al., 2018).

3.3 Border Cave

Border Cave is a 50 m wide and 35 m long semi-circular rock shelter located in the Lebombo Mountains in the KwaZulu-Natal Province near the eSwatini Border (Figs. 1 and 2; Backwell et al., 2018). The cave formed due to differential weathering within the rhyolitic lava units of the Jozini Formation, which is part of the Lebombo Group. Border Cave is an important MSA site that has yielded significant human remains and impressive archaeological evidence of human cultural behavior and technological innovation of the past 250 ka (Villa et al., 2012; d'Errico et al., 2012; Backwell et al., 2018). The lithological sequence observed at Border Cave is alternating units known as the Brown Sand (BS) and White Ash (WA) (Backwell et al., 2018). BS and WA occur together or individually across 6 Members (Beaumont et al. 1978). Both geological and human processes and materials have contributed to the BS and WA deposits. For example, roof spall and in situ precipitation in the case of the former, and ash, bone, artifacts and layered grass beds in the case of the latter (Backwell et al., 2018). In addition, the host rock lithology is a prominent source of material in the depositional sequence as indicated by the sand-sized particles of the host rock rhyolite, which dominate the BS member. Even the gravel clasts in the deposit are predominantly composed of the angular rhyolite fragments (Backwell et al., 2018).

3.4 Bundu Farm

Bundu Farm is an open-air archaeological pan site located in the about 60 km south of the town of Marydale in the Northern Cape Province (Figs. 1 and 2; Hutson, 2018). It has produced both faunal remain and lithics of the early MSA and transitional Fauresmith (Kiberd, 2006). The pan deposits have a complex sedimentary sequence of seven main sedimentary horizons called, from oldest to youngest, group 7 (g7) to group 1 (g1) (Kiberd, 2006). G7 is a 30 cm thick white powdery calcrete deposited directly onto the basement shales. G6 is a white-gray calcified clay horizon that is also 30 cm thick and has a sharp contact to g7. It contains sporadic artifacts and faunal fossils. G5 is a horizon that is subdivided into the upper 5a and lower 5b due to decomposition that is happening in the upper parts. G5a is a friable pale reddish-brown matrix-supported pebbly breccia with abundant faunal fossils and worked lithics. G5b is a calcified reddish-brown sandstone with minor pebbles, faunal fossils and artifacts within it. G4 is a 30 cm thick calcrete horizon subdivide into two layers g4a and g4b. g4a is a laminated calcrete whereas g4b is massive and grades into breccia and contains numerous artifacts and fossils. The calcrete in g4 transforms into silcrete in some areas. G3 is a 20 cm thick horizon of calcified sediments, and it is subdivided into four subgroups 3a to 3d. 3d is the bottom layer occurring as a sandy-silty layer that has a moderate amount of worked lithics. It is followed by 3c, which is a calcified pebbly breccia that contains abundant lithic artifacts. 3b is made up of columnar calcrete and contains minor amount of worked and unworked lithics. The top layer of 3a is composed of breccia with sub-angular to sub-rounded pebbles and minor cobbles, and contains abundant worked lithics. G2 is a 40 cm thick sand-silt horizon that forms the present surface in some parts of the pan. It is mostly homogenous with discrete bands that have abundant worked lithics and minor fragments of ostrich eggshell. G1 is a reddish sand horizon that can be up to 15 cm thick. It contains some minor LSA lithics in some places. Efforts to date these deposits used ESR+U-series on tooth and bone and the ages obtained ranged between 144 ka and 371 ka for both the early U-uptake and linear U-uptake models (Kiberd, 2006). A mean age of 245 ka was suggested for all the deposits at Bundu Farm based on all the ages obtained (Hutson, 2018).

3.5 Canteen Kopje

Canteen Kopje, also known as Klipdrift, is located 500 m away from the N bank of the Vaal River at a height of 16 m above the river, ca. 1.3 km SE of

Barkly West, Northern Cape Province (Fig. 2C; Chazan et al., 2013). This site was the first alluvial diamond mine in Africa starting 1869 and mining carried on until 1948, and during this time the thick alluvial deposits of the ancient Vaal River were severely disturbed, and a lot of the archaeological remains were lost (Chazan et al., 2013; Li et al., 2017). The deposits of this site, which are temporally extensive and can be up to 7 m thick, are comprised of the basal Vaal River gravels and the overlying Hutton Sands, denoted stratum 2 and stratum 1, respectively, by Beaumont in earlier excavations (Fig. 9; Chazan et al., 2013; Lotter et al., 2016). Li et al. (2017) recognized five stratigraphic units from the excavations of ca. 7 m thick deposits at Pit 6. The four units from the bottom are part of the Vaal River gravels (stratum 2) composed of coarse alluvium deposits. The lowermost unit is ca. 1.4 m, contains large boulders and preserved an early Acheulean assemblage. The second unit from the bottom is ca. 1 m thick with large cobbles and it was followed by an ca. 0.6 m thick pebbly coarse alluvium. The uppermost coarse alluvial deposit is ca. 1.8 m thick and preserves Victoria West core assemblage. A 0.2 to 0.3 m thick unit referred to as

Fig. 9 Photos of Canteen Kopje site. The deposits at this site are comprised of a basal Vaal River gravel ca. 4–5 m thick and the top layer of ca. 2–3 m thick Hutton Sands. (A) Pit showing contact between the gravel and the overlying sands. (B) Wall exposure of the gravel sequence without the sands. (C) Pit showing the red aeolian sands beneath soil. *Photo credits: Matt Lotter.*

the mixed contact zone separates the four units of stratum 2 from the uppermost ca. 2.1 m thick red aeolian Hutton Sands, which preserve MSA, LSA and Fauresmith assemblages. In a study aimed at dating the Hutton Sands, Chazan et al. (2013) excavated an undisturbed section of the Hutton Sands at the eastern edge of the area declared as the historic site. A resulting stratigraphy of the Hutton Sands revealed that they occur as at least five horizons of, from bottom to top, compact gray mottled sand, compacted mottled sand, mottled sand, red sand and gray A-horizon bioturbated sand at the top (Chazan et al., 2013). An attempt to date stratum 2 using $^{26}Al/^{10}Be$ burial dating yielded inconclusive results that suggested the deposits had an age of ca. 1 Ma, confirming earlier suggestion by Beaumont and Vogel that the Victoria West assemblage at Canteen Kopje was ca. 0.8 to 1.1 Ma by comparison to the paleomagnetic dating at Doornlaagte site (Li et al., 2017). Dating attempts on the Hutton Sands using OSL were also inconclusive but suggested that the aeolian sand started accumulating atop the gravels ca. 0.3 Ma (Chazan et al., 2013).

3.6 Cornelia Uitzoek

This site is named after the farm (i.e., Uitzoek) in which it is situated. It is located near the town of Cornelia on the eastern part of the Free State Province (Figs. 1 and 2; Brink et al., 2012). The site is well known for its rich faunal assemblage, and it is the type site of the Cornelian Land Mammal Age (Hendey, 1974; Brink, 2005). In addition, the presence of a bonebed which attracted more investigation into the site yielded a hominin specimen (*Homo sp.* molar tooth) as well as Acheulean stone tools that needed to be associated with the stratigraphic sequence (Brink and Rossouw, 2000; Brink, 2004). The basement lithology of this sequence is Ecca Group fines (Brink, 2004). The site fill occurs in a carved basin that is confined with alluvial and colluvial fossiliferous sediments as well as laminated fines and clays (Fig. 10A; Brink, 2004). The site was dated by paleomagnetism to between 1.07 and 0.99 Ma.

3.7 Diepkloof Rock Shelter

Diepkloof Rock Shelter is located 150 km north of Cape Town hosted in the remnant butte of the Table Mountain Group (Figs. 1, 2 and 10B; Miller et al., 2013). The site's bedrock of the Table Mountain Group includes lithologies of low-grade metasedimentary rocks such as conglomerates bedded quartzitic sandstones, shales siltstones. This variable lithology

Fig. 10 Photos of some sites outside the Cradle. (A) Cornelia Uitzoek site contains fossiliferous alluvial and colluvial sediments in gullies formed in the sandstones of the Ecca Group. (B) View from beneath the Diepkloof Rock Shelter formed by differential weathering in metasedimentary rocks of the Table Mountain Group. (C) Duinefontein open-air site where wind polished faunal material has been collected. (D and E) Elandsfontein site on the Stranveld coastal plain is characterized by quartz sands, outcrops of granite and shale, and dune systems. (F) View of the Klasies River cave complex where Middle Stone Age to Late Stage Age hominin remains and associated lithic assemblages have been found. *Photo credits: Andy Herries (A, C-E), Joe Eppel (B).*

has facilitated differential weathering, which in turn carved the cave and its overhangs. The site has yielded a large lithic assemblage which spans several cultural sequences along with ostrich eggshells (some of which are engraved) and faunal remains (Texier et al., 2010; Parkington et al., 2013). The sequence of Diepkloof is very abundant containing over fifty stratigraphic units in the form of discontinuous laminae and beds (Miller et al., 2013). Some of the beds are due anthropogenic inputs representing hearth burning, bedding construction and burning as well as trampling (Miller et al., 2013). The site has been dated numerous times in the variable stratigraphic units with dates ranging from 130 ka to about 45 ka (Rigaud et al., 2006; Texier et al., 2010).

3.8 Doornlaagte

The site is located 37 km west of the town of Kimberly (Figs. 1 and 2; Netterberg, 1974). The site was apparent in 1962 in quarry operation in a Doornlaage farm property. It is because of these operations that a lot of the site was destroyed before investigations had started. The early excavations were well documented (Mason 1967). The site included Acheulean stone tool technology including cleavers, hand-axes, flakes, manuports, anvils and other "heavy duty tools". The abundance of flakes and raw materials suggested it as a production site, meaning the stone tools were being produced at the very same site (Mason, 1967). The site is underlain by basement Dwyka Group shales and Ventersdorp andesites as well as some intrusive Karoo dolerites. The site is found within a white calcareous sediment which is referred to as calcrete. The site lies at the northeastern edge of the Rooidam Vloer, a centripetal drainage area towards the central flats where temporary waters will evaporate or percolate into the shallow subsurface. The stratigraphic sequence contains varying lithologies in varying thicknesses depicting the pan conditions.

3.9 Duinefontein

Duinefontein is the farm south of the village of Melkbosstrand and 35 km from Cape Town in the Western Cape Province (Figs. 1, 2 and 10C). The site has been subjected to a lot of investigation of which the earliest after the reporting of wind polished faunal material (Klein et al., 1999). Since then, the site has been a subject of multiple paleontological, archaeological, chronological and paleoenvironmental investigations (e.g., Hendey 1968, 1974; Klein, 1983; Klein et al., 1991). The site comprises a large faunal assemblage and Acheulean artifacts. These are found in reddish sands that were and subsequently altered by water table which ferruginized them (Klein et al., 1991; Feathers, 2002). The fauna and stone tools of the site provided the basis for age estimates at 151 ± 9 ka and 168 ± 20 ka (Klein et al., 1999). The site was later OSL-dated with estimates in two horizons at >270 ka ago and >290 ka ago (Feathers, 2002).

3.10 Elandsfontein

This site is located on the west coast of South Africa in the Western Cape Province (Figs. 1 and 2) and has been a topic of investigation by various authors due to fossil and artifact remains (including hominin material) as well as in paleoenvironmental proxies (e.g., Drennan, 1953; Singer, 1956; Singer and Crawford, 1958; Singer and Heltne, 1966; Singer and Wymer, 1968; Butzer, 1973; Klein 1988; Deacon, 1998; Avery, 1989; Braun et al., 2013). Elandsfontein forms part of the Stranveld coastal plain which is comprised mostly of quartz sands and shell with some granite and shale outcrops, and the area is characterized by extensive dune systems (Fig. 10D and E; Braun et al., 2013). Elandsfontein is one of the dunefields with the dominant dune plume form; a plume-shaped mass of transgressive dune sand (Roberts et al., 2009). Historically, lithological descriptions have been site-specific and interpreted as a combination of aeolian, fluvial and pedogenic processes (Mabbutt, 1956; Singer and Wymer, 1968; Butzer, 1973; Deacon, 1998). More recent geological work, which involved large scale investigation across space, has recognized stratigraphic patterns across space. This has helped constrain the site into three distinct lithological units as obtained from the excavations (Braun et al., 2013). This categorization has provided the various depositional settings in which the faunal and artifact material is found. Dating in the site has proved very difficult but Braun et al. (2013) placed it within a depositional and chronological framework in which biochronological analysis with correlations from other South African sites suggest the fossil and artifact-bearing horizons are between 1.6 Ma and 0.58 Ma.

3.11 Florisbad

Florisbad is an open-air site located in the central part of the Free State Province, 45 km north-northwest of Bloemfontein (Figs. 1 and 2). The site lithology comprises sands that are intercalated with finer clastic sediments and peats (organic layers). This represents aeolian sand units altered by lacustrine water during dry phases and peats formed from marshes during the dry phases (Toffolo et al., 2017). The bedrock of the site is shales of the Ecca Group (Kuman, 1989). Florisbad is formed in a flat endorheic basin, a depression formed from disruption of drainage (causing choking) due to tectonic movement during the Pliocene (Toffolo et al., 2017). In addition, dolerite intrusions of Ecca Group shales have generated a thermal spring which releases water and gasses. This spring was the point of attraction to the site for hominins and animals. The pH of the water is 9 and has been key in the preservation of bone as well as the precipitation of $CaCO_3$ (Toffolo et al., 2017). The site spans a large, continuous sequence from the last 300 ka and is well known for producing some of the earliest archaic modern human remains (in the form of a partial hominin cranium), several lithic artifact from assemblages in MSA and the LSA as well as a large faunal assemblage. This faunal assemblage is an area of research interest as it marks the type site of the Florisian Large Mammal Age and archives a pre-modern stage of southern African mammalian evolution (Klein, 1984). The oldest dated deposits in the site are the Early MSA-bearing deposits dated to 279 ± 46 ka using OSL dating while the youngest sediments are 2 ka based on radiocarbon dating (Grun et al., 1996).

3.12 Kathu Pan

Kathu Pan is situated in the Northern Cape Province in the central interior of South Africa (Figs. 1 and 2). It is a significant site that not only contains a complete sequence from the Mid-Pleistocene to the Late Holocene but also has allowed for paleoenvironmental reconstructions in this area (Lukich et al., 2020). The sites have recovered a rich assemblage of Acheulean through to LSA artifacts with associated faunal remains. Some of the faunal remains include large (mega) herbivore that are usually rare in this region (Klein, 1988). Kathu Pan consists of a complex of archeological site, which include sinkhole exposures in and around the town of Kathu (Beaumont, 1990, 2004; Wilkins and Chazan, 2012; Walker et al., 2014). The most productive and stratigraphically continuous of these is Kathu Pan 1, which preserves a long history of human occupation and behavior from the ESA (represented by the Acheulean and Fauresmith industries) to MSA to the LSA (Wilkins and Chazan, 2012). The sedimentary sequence of the site was designated into five lithostratigraphic units that have been dated by various methods. The oldest unit is Stratum 4b estimated to be ca. 700–400 ka based on biochronology. It is followed by Stratum 4a and Stratum 3 dated using OSL to 464 ± 47 ka and 291 ± 45 ka, respectively. Then there is Stratum 2 and Stratum 1 dated using radiocarbon to 32-6 and 3.6-2.7 thousand years BP, respectively (Porat et al., 2010).

3.13 Klasies River main site

This site is located near Humansdorp, south of the Klasies River mouth in the southern coast of the Eastern Cape Province (Figs. 1, 2 and 10F). Klasies

River Main Site is a complex comprising of five caves that are closely associated with each other overhanging in a quartz sandstone cliff (Morrissey et al., 2022). This site's complex is regarded as one of the most important MSA-LSA bearing sites in South Africa due to finds such as hominin and faunal remains in association with lithic assemblages that span multiple cultural phases (e.g., Brenner et al., 2020; Deacon, 1989, 2001; Henshilwood and Marean, 2003; Klein, 2001; Wadley, 2015; Wurz, 2013). The stratigraphic units are classified into multiple lithological units (Deacon and Geleijnse, 1988); they lithologically span gravelly bioclastic units at the base to being capped by crusts/flowstones (Morrissey et al., 2022). The stratigraphic sequence is particularly complex as it has been subject to a range of lithological inputs which is compounded by many post-depositional eatures (Morrissey et al., 2022). The site has an upper U-Th date of 126.0 ± 1.5 ka (Wurz et al., 2018) and young ages down to 40 ka (Grine et al., 2017).

3.14 Olieboomspoort

Olieboompoort is located in northern Limpopo in the Waterberg Plateau, a few meters away from the Moloko River tributary known as the Riet Spruit (Figs. 1 and 2). The site is a rock shelter hosted in Paleoproterozoic Waterberg Group rocks, which consist mostly of conglomerates, sandstones and shales (Brandl, 1996). These lithologies have formed part of the raw material sourcing seen in the lithic assemblage (Van Der Ryst, 2007). The sheltered area of the site is around 70 m wide and 7 m tall, which was carved out by erosion of the sandstones by the Riet Spruit (Val et al., 2021). This fluvial influence seems to be a prevailing agent in the site formation as well as diagenetic alteration of the site over time (Val et al., 2021). Nevertheless, the site is categorized by two distinct stratigraphic units; a lower MSA-bearing unit that is gravellier and an upper LSA/Iron Age-bearing unit that is sandier (Val et al., 2021). The site has a large lithic assemblage from the MSA into the LSA to the Iron Age. In addition to this material culture, ochre has also been found identified in the site. The site also has a prolific faunal assemblage that includes prates and reptiles (Val et al., 2021). The site has been estimated at 150 ± 14 ka with combined U-series and ESR dating estimates.

3.15 Penhill

The stone tool bearing site of Penhill Farm is located near the town of Addo along the lower Sundays River in the Eastern Cape Province (Figs. 1 and 2).

The site lies in an area of extensive fluvial evolution unconformably overlain by Jurassic-Cretaceous Algoa Basin rocks (Uitenhage Group shales and mudstones) with upstream lithologies from Cape (quartzites) and Karoo (tillites, shales and sandstones) supergroups as well as Suurberg Group basalts (Ruddock, 1948; Hattingh, 1994; Hattingh and Rust, 1999; Lotter, 2020). The site has produced Acheulean stone tools mostly in the form of large cutting tools (Lotter, 2020). The deposits at Penhill are typically overbank alluvium sediments of gravels and fines exposed spectacularly on vertical walls of Terrace 9 (Fig. 11). The fine sediments are massive, lightly colored silty-sands that do not contain any archaeological material and they overlie a gravel horizon with imbricated pebbles and cobbles (Fig. 11B–D; Lotter and Kuman, 2018). The pebbles and cobbles were used for $^{26}Al/^{10}Be$ isochron burial dating, which yielded an age of 1.36 ± 0.36 Ma for the unit (Erlanger et al., 2012).

Fig. 11 Photos of Penhill site. The deposits are typically overbank alluvium sediments of gravels and fines (A and B). The fine sediments are massive, lightly colored silty-sands that do not contain any archaeological material and they overlie a gravel horizon with imbricated pebbles and cobbles (B–D). *Photo credits: Matt Lotter.*

3.16 Pinnacle Point

Pinnacle Point refers the geographic area around a small headland in the coast of the Indian Ocean, about 10 km from the Mossel Bay point in the Western Cape Province (Figs. 1 and 2). This area has a series of caves and rock shelters formed at the eroded fault breccias of the shear zones in the highly structurally deformed Skurweberg Formation of the Table Mountain Group Sandstones (quartzitic sandstones) (Fig. 12A and B; Bar-Matthews et al., 2010). A series of caves and rock shelters on the same cliff face can be grouped together into a complex, and at the tip of the Pinnacle Point there is Site 13 complex (PP13 A-G). Pinnacle Point Cave 13B (PP13B) has preserved the earliest archaeological evidence of modern humans using marine shellfish and pigments, and the manufacture of bladelets around 164 ka (Marean et al., 2010). The stratigraphy of PP13B, described by Marean et al. (2010), differs laterally within the cave and three sequences of the lightly cemented MSA (LC-MSA) deposit, and two eastern and western areas of non-LC-MSA deposits have been studied. They were found to differ markedly, suggesting a complex history of deposition, erosion and alteration. The stratigraphy of Marean et al. (2010) starts with the lowest (LC-MSA Lower) deposited on top of the Table Mountain Sandstone (TMS) bedrock. This layer is the least cemented of all LC-MSA but contains the most archaeological evidence including materials that humans altered using fire. Three OSL dates of LC-MSA Lower yielded a mean age of 162 ± 5 ka. The succeeding layer is LC-MSA Middle, which is mostly ash with multiple lenses of charcoal and *in situ* hearths, and high densities of marine shellfish. It has been dated to 125 ± 5 ka using OSL. LC-MSA Upper overlies the anthropogenic LC-MSA Middle layer as three heavily cemented layers. The lower LC-MSA Upper layer is an indurated sandy-silty layer with multiple lenses of organic material. The middle layer is a sandy horizon with a lens of shellfish and multiple lenses of organic material. It was dated with four OSL samples that yielded a mean age of 126 ± 4 ka. The top layer is a capping cemented dune that sealed the cave to human occupation around 93 ± 4 ka. The LC-MSA Flowstone with a thickness up to 5 cm is the final unit of LC-MSA deposits that started precipitating immediately after the dune sealed the cave. It is a laminated brown-yellow flowstone that was dated using U/Th and six samples yielded ages from 92 ka to 39 ka.

In the eastern area excavation, seven layers were observed starting with the Lower Roof Spall Facies deposited directly on top of the TMS bedrock. This layer is a clast-supported pebbly breccia to the south and pebbly

Fig. 12 Photos of some sites outside the Cradle. The site of Pinnacle Point is a series of caves and rock shelters formed by shear zones in quartzitic sandstones of the Table Mountain Group (A and B). Sibudu rock shelter contains an extensive cultural sequence that spans pre-Still Bay to Iron Age (C and D). Wonderwerk cave is phreatic tube in the dolomitic hills of Kuruman and contains a lithic sequence that spans the Early Stone Age to the Late Stone Age (E). *Photo credits: Matt Caruana (A, B and D), Andy Herries (C and E).*

sandstone to the north dated to 110 ± 4 ka using OSL dating of two samples. Overlying the Lower Roof Spall Facies is three layers of Upper Roof Spall (URS) and Shelly Brown Sand (UBS) Facies. The URS and UBS occur as individual layers in some places and as inseparable layers in some areas. Generally, the URS comes first, followed by a blended URS-UBS before the UBS on its own. OSL ages of the individual URS and UBS layers overlap in the range 98–91 ka, but the age (ca. 115 ka) of the blended URS-UBS does not overlap. The URS and UBS Facies contain well-preserved hearths with distinct bands of ash, charcoal and baked sediments, and can have lithic artifacts and fauna inside and around them. The URS is overlain by the Truncation Fill, which is a dark, rocky and organic-rich horizon dated using radiocarbon on a sample of charcoal from the boundary with the URS. Then there is the Re-Deposited Disturbance layer of disturbed and re-deposited material from modern activities as evidenced by modern bird feathers and modern artifacts such as cigarette butts. Fine-grained surface sediments are the top layer. In the western area excavation, the deposits seem to resemble a midden or dump compared to the eastern area. The deposit starts with a Boulder Facies, which is an unconsolidated sediment with large boulders and a silty loam matrix. Then a series of Laminated Facies sediment ca. 1 m thick overlies the Boulder Facies. This unit is mostly sterile grayish-brown sand and silt sediment dated between 414 and 349 ka. Then follows the Light Brown Silt Facies dated between 349 and 152 ka. The overlying Dark Brown Sand Facies is a series of sandy layers with a low to moderate density of lithic and faunal material. The OSL ages of these layers range from 159 ± 7 ka to 98 ± 4 ka. Surface sediments are also the top layer in the western area.

3.17 Rietputs 15

This site represents an early Acheulean assemblage near Windsorton about 1 km from the Vaal River, Northern Cape Province (Figs. 1 and 2; Gibbon et al., 2009). This site is within the Vaal River alluvial gravel deposits, which contain an older deposit of higher elevations and a younger gravel of lower elevation, which includes the Rietputs Formation where the site is located. The Rietputs Formation contains a lower, coarser gravel and sand unit overlain by a finer laminated and cross-laminated alluvium with paleosol inclusions. It also contains both fossil material as well as stone artifacts which had been originally designated a middle Pleistocene age by biochronological means with the presence *Elephas recki* and *Metridiocheorus andrewsi* (Klein, 2000). Some researchers have questioned the reliance of these species for biochronological dating. However, absolute dating was applied using $^{26}Al/^{10}Be$ burial dating (Gibbon et al., 2009). This produced a date of ca 1.57 ± 0.22 Ma (individual sample ranges: 1.89 ± 0.19 to 1.34 ± 0.22 Ma) for the lower unit and 1.26 ± 0.10 Ma for the upper unit (Gibbon et al., 2009).

3.18 Rooidam

The Rooidam site lies about 2 km west of Kimberly, in the farmhouse of C.J. Cohn (Figs. 1 and 2). The artifacts are typically of the Fauresmith assemblage (i.e., Acheulean), first recognized in a rubble of a well shaft after which they were systematically excavated (Fock, 1968). The site has been very productive, producing a total assemblage of 6925 artifacts (Fock, 1968). The raw material is usually derived from a nearby (1 km) lydianite, a meta-shale, as well as some diabase of unknown origin (Fock, 1968). The sequence of the site has distinguishable sedimentary layers, but the assemblages lack distinct pockets. This problem is also compounded by the induration of the sediments, which creates difficulty in differentiating layers and appear mostly horizontal (Fock, 1968). This site location is underlain mostly by Karoo dolerite in an area that is usually dominated by Dwyka Group shale basement. The stratigraphic sequence represents lithologies of various conditions in the pan. The site was radiometrically dated with U/Th dating to 115 ± 10 ka for Unit C. There have been attempts for cultural and geomorphic correlations as well as correlations with Doornlaagte (e.g., Butzer et al.,1973; Butzer, 1974; Netterberg, 1969).

3.19 Sibudu

Sibudu is a rock shelter located 40 km N of Durban and 15 km from the Indian Ocean overlooking the Tongati River in the KwaZulu-Natal Province (Figs. 1, 2 and 12C and D). The site has yielded an extensive cultural sequence in the form of stone tools (as well as some bone tools) and faunal assemblage. This begins at pre-Still Bay and proceeds to Still Bay, Howiesons Poort, post-Howieson's, late MSA, final MSA and Iron Age (Wadley, 2006; Wadley and Jacobs, 2006). The stratigraphic sequence is complex as it has both natural and anthropogenic sources. This resulted in a lithological succession that is largely immature sediments derived from weathered roof-rock, aeolian sands as well as chemical nodules of gypsum and carbonate from the decalcification of the deposits (Wadley and Jacobs, 2006). The site was dated by OSL with the oldest dates on the Still Bay layers at 70.5 ± 2.0 ka and the final MSA dates at 34.7 ± 1.7 ka.

3.20 Wonderwerk Cave

Wonderwerk Cave is a 10–12 m long, 140 m wide phreatic tube hosted in the dolomitic hills of Kuruman Hills in the Northern Cape Province (Figs. 1, 2 and 12E). These dolomitic hills belong to the Late Archean-Early Proterozoic Ghaap Group (Eriksson et al., 1993). Wonderwerk has been a subject of study since the 1940s focusing on ESA assemblages (Malan and Cooke, 1941; Malan and Wells 1943, Camp, 1948). Further investigations and excavations have taken place over the years (Butzer et al., 1978;

Butzer, 1984; Beaumont, 1982, 1990, 2004; Beaumont and Vogel, 2006; Chazan et al., 2012). Beaumont's work focused on the lithics and faunal assemblages on LSA deposits (Humphreys and Thackeray, 1983, Thackeray, 1984). The site is particularly unique for preserving an extensive, mostly complete *in situ* sequence from the ESA to LSA lithic sequence as well as with faunal remains (Chazan et al., 2008). It also contains some of the earliest evidence for controlled use of fire (Berna et al., 2012). Considering the host rock, the site contains both clastic and chemical sedimentary deposits; partly calcified silty sands, ash lenses and debitage. These deposits were accumulated primarily by low energy aeolian processes, roof decomposition and minor reworking from water transport (Chazan et al., 2008). In addition, water was actively involved in the formation and diagenesis of the depositional sequence as seen in the erosional contacts, sheet flow and flowstones. The ESA deposits of Wonderwerk have been dated to ca. 0.78–1.96 Ma through multiple dating programs utilizing magnetostratigraphy, $^{26}Al/^{10}Be$ simple burial dating and OSL on clastic sediments and U-Pb on speleothems (Chazan et al., 2008; Matmon et al., 2012; Pickering, 2015; Shaar et al., 2021). The MSA deposits have been dated to ca. 94–240 Ma based on U/Th on speleothems and OSL on clastic sediments (Beaumont and Vogel, 2006; Chazan et al., 2020).

4. Concluding remarks

South Africa has an abundance of sites that contain Late Pliocene and Quaternary deposits, which have played the crucial role of preserving paleoanthropological and archeological remains. The deposits in many sites generally have complex stratigraphic sequences due to a combination of factors such as erosional events during hiatuses or post-deposition and the evolution of the site. For example, in the Cradle some chambers were deroofed and most of the deposits were eroded away or the chambers contain less deposits after their deposits were eroded into deeper and yet inaccessible or unexplored chambers of the cave. In the latter case, the units are hardly preserved and occur as small remnants hanging from high up on the walls, indicating the previous levels once reached by the deposits in those chambers, but the units can be difficult to characterize adequately. Adding to the complications of the natural processes is the fact that geological investigations at most sites are limited and focused to hotspots in the deposits that yield fossils or artifacts. As a result, few sites have comprehensive geology reports of all the deposits in the site. Despite this challenge, though, it would still be impossible to produce a condensed stratigraphic sequence that represents the Pliocene and Quaternary timeline of all the sites even in the Cradle alone. The diverse, complex and incomplete deposits in South African sites allow us, at best, to highlight similarities and differences between sites, and to compare the sequence of paleoanthropological and archeological remains.

Acknowledgments

We are grateful to Andy Herries, Matt Lotter, Matt Caruana, Paloma de la Peña, Dirk van Rooyen, Joe Eppel and Kelita Shadrach for photos and information about some sites. We are grateful to all the researchers who have worked at the sites we have reported here and apologize to those whose work we did not unintentionally include. We thank the National Research Foundation and the South African Department of Higher Education's new Generation of Academics Programme for the varied research funding that made it possible us to become academics and researchers.

References

Avery, G., 1989. Some features distinguishing various types of occurrence at Elandsfontein, Cape Province, South Africa. Palaeontol. Africana 19, 213–220.

Backwell, L.R., D'Errico, F., Banks, W.E., De La Pena, P., Sievers, C., Stratford, D., Lennox, S.J., Wojcieszak, M., Bordy, E.M., Bradfield, J., Wadley, L., 2018. New excavations at Border Cave, KwaZulu-Natal, South Africa. J. Field Archaeol. 43, 417–436.

Balco, G., Rovey, C., 2008. An isochron method for cosmogenic-nuclide dating of buried soils. Am. J. Sci. 308, 1083–1114.

Balter, V., Blichert-Toft, J., Braga, J., Telouk, P., Thackeray, F., Albarède, F., 2008. U–Pb dating of fossil enamel from the Swartkrans Pleistocene hominid site, South Africa. Earth Planet. Sci. Lett. 267 (1-2), 236–246.

Beaumont, P.B., 1982. Aspects of the Northern Cape Pleistocene project. In: Coetzee, J.A., Van Zinderen Bakker, E.M. (Eds.), Palaeoecology of Africa and the Surrounding Islands 15. A.A. Balkema, Cape Town, pp. 41–44.

Beaumont, P.B., 1990. Wonderwerk cave. In: Beaumont, P.B., Morris, D. (Eds.), Guide to the Archaeological Sites in the Northern Cape 101–134. McGregor Museum, Kimberly.

Beaumont, P.B., 2004. Wonderwerk cave. In: Morris, D., Beaumont, P.B. (Eds.), Archaeology in the Northern Cape: Some Key Sites 31–36. McGregor Museum, Kimberly.

Beaumont, P.B., Vogel, J.C., 2006. On a timescale for the past million years of human history in central South Africa. South Afr. J. Sci. 102, 217–228.

Beaumont, P.B., de Villiers, H., Vogel, J., 1978. Modern man in sub-Saharan Africa prior to 49 000 years BP: a review and evaluation with particular reference to Border Cave. South Afr. J. Sci. 74 (11), 409.

Berger, L.R., Tobias, P.V., 1994. New discoveries at the early hominid site of Gladysvale, South Africa. South Afr. J. Sci. 90 (4), 223–226.

Berger, L.R., de Ruiter, D.J., Churchill, S.E., Schmid, P., Carlson, K.J., Dirks, P.H.G.M., Kibii, J.M., 2010. Australopithecus sediba: a new species of Homo like australopith from South Africa. Science 328, 195–204. https://doi.org/10.1126/science.1184944.

Berna, F., Goldberg, P., Horwitz, L.K., Brink, J., Holt, S., Bamford, M., Chazan, M., 2012. Microstratigraphic evidence of in situ fire in the Acheulean strata of Wonderwerk Cave, Northern Cape Province, South Africa. Proc. Natl. Acad. Sci. U. S. A. 9 (20), 1215–1220.

Braga, J., Thackeray, J.F., 2003. Early Homo at Kromdraai B: probabilistic and morphological analysis of the lower dentition. C. R. Palevol 2 (4), 269–279.

Braga, J., Thackeray, J.F., Bruxelles, L., Dumoncel, J., Fourvel, J.B., 2017. Stretching the time span of hominin evolution at Kromdraai (Gauteng, South Africa): recent discoveries. C. R. Palevol 16 (1), 58–70.

Brain, C.K., 1958. The Transvaal Ape-Man-Bearing Cave Deposits. 11 Transvaal Museum.

Brain, C.K., 1970. New finds at the Swartkrans australopithecine site. Nature 225 (5238), 1112–1119.

Brain, C.K., 1981. The Hunters or the Hunted?: An Introduction to African Cave Taphonomy. University of Chicago Press.

Brain, C.K., 1988. New information from the Swartkrans cave of relevance to "robust" australopithecines. In: Grine, F.E. (Ed.), Evolutionary History of the "Robust" Australopithecines. Taylor & Francis, New York, pp. 311–316.

Brain, C.K., 1993. Structure and Stratigraphy of the Swartkrans Cave in the Light of the New Excavations. Swartkrans: A Cave's Chronicle of Early Man. Transvaal Museum, Pretoria, pp. 23–34.

Brain, C.K., 1994. The Swartkrans palaeontological research project in perspective: results and conclusions. South Afr. J. Sci. 90 (4), 220–223.

Brain, C.K., Sillent, A., 1988. Evidence from the Swartkrans cave for the earliest use of fire. Nature 336 (6198), 464–466.

Brandl, G., 1996. The geology of the Ellisras area. In: Explanation of Sheet 2326 (1:250 000): 49. Council for Geoscience, Pretoria.

Braun, D.R., Levin, N.E., Stynder, D., Herries, A.I.R., Archer, W., Forrest, F., Roberts, D.L., Bishop, L.C., Matthews, T., Lehmann, S.B., Pickering, R., Fitzsimmons, K.E., 2013. Mid-pleistocene hominin occupation at Elandsfontein, Western Cape, South Africa. Quat. Sci. Rev. 82, 145–166.

Brenner, M.J., Ryano, K.P., Wurz, S., 2020. Coastal adaptation at Klasies River main site during MIS 5c-d (93,000–110,000 years ago) from a southern Cape perspective. J. Island Coast. Archaeol. 218-245.

Brink, J.S., 2004. The taphonomy of an early/middle pleistocene hyaena burrow at Cornelia-Uitzoek, South Africa. Rev. Paléobiol. 23, 731–740.

Brink, J.S., 2005. The Evolution of the Black Wildebeest, (Connochaetes gnou), and Modern Large Mammal Faunas of Central Southern Africa. PhD Dissertation, University of Stellenbosch, Stellenbosch.

Brink, J.S., Rossouw, L., 2000. New trial excavations at the Cornelia-Uitzoek type locality. Navors. Nas. Mus. Bloemfontein 16, 141–156.

Brink, J.S., Herries, A.I.R., Moggi-Cecchi, J., Gowlett, J.A.J., Bousman, C.B., Hancox, J.P., Grun, R., Eisenmann, V., Adams, J.W., Rossouw, L., 2012. First hominine remains from a ~1.0 million year old bone bed at Cornelia-Uitzoek, Free State Province, South Africa. J. Hum. Evol. 63, 527–535.

Broom, R., 1936. New fossil anthropoid skull from South Africa. Nature 138 (3490), 486–488.

Broom, R., 1938a. Pleistocene anthropoid apes of South Africa. Nature 142, 377–379.

Broom, R., 1938b. Further evidence on the structure of the South African Pleistocene anthropoids. Nature 142 (3603), 897–899.

Broom, R., 1942. The hand of the ape-man, Paranthropus robustus. Nature 149 (3784), 513–514.

Broom, R., 1943. An ankle-bone of the ape-man, Paranthropus robustus. Nature 152 (3867), 689–690.

Broom, R., 1947. Discovery of a new skull of the South African ape-man, Plesianthropus. Nature 159 (4046), 672.

Broom, R., Robinson, J.T., 1949. A new type of fossil man. Nature 164 (4164), 322–323.

Broom, R., Robinson, J.T., 1950. Man contemporaneous with the Swartkrans ape-man. Am. J. Phys. Anthropol. 8 (2), 151–156.

Broom, R., Robinson, J.T., 1952. Swartkrans Ape-Man, Paranthropus crassidens. Transvaal Museum Memoir No. 6, Pretoria.

Bruxelles, L., Clarke, R.J., Maire, R., Ortega, R., Stratford, D., 2014. Stratigraphic analysis of the Sterkfontein StW 573 Australopithecus skeleton and implications for its age. J. Hum. Evol. 70, 36–48.

Bruxelles, L., Maire, R., Couzens, R., Thackeray, J.F., Braga, J., 2016. A revised stratigraphy of Kromdraai. Kromdraai, a birthplace of Paranthropus in the Cradle of Humankind. SUN MeDIA MeTRO, Stellenbosch, South Africa, pp. 31–47.

Bruxelles, L., Stratford, D.J., Maire, R., Pickering, T.R., Heaton, J.L., Beaudet, A., Kuman, K., Crompton, R., Carlson, K.J., Jashashvili, T., McClymont, J., 2019. A multiscale stratigraphic investigation of the context of StW 573 'Little Foot'and Member 2, Sterkfontein Caves, South Africa. J. Hum. Evol. 133, 78–98.

Butzer, K.W., 1973. Re-evaluation of the geology of the Elandsfontein (Hopefield) site, south-western Cape, South Africa. South Afr. J. Sci. 69, 234–238.

Butzer, K.W., 1974. Geo-archeological interpretation of Acheulian calc-pan sites at Doornlaagte and Rooidam (Kimberley, South Africa). J. Archaeol. Sci. 1 (1), 1–25.

Butzer, K.W., Helgren, D.M., Fock, G.J., Stuckenrath, R., 1973. Alluvial terraces of the lower Vaal River, South Africa: a reappraisal and reinvestigation. J. Geol. 81 (3), 341–362.

Butzer, K.W., 1984. Archaeogeology and quaternary environment in the interior of southern Africa. In: Klein, R.G. (Ed.), Southern African Prehistory and Paleoenvironments 1-64. Balkema, Rotterdam.

Butzer, K.W., Stuckenrath, R., Bruzewicz, A.J., Helgren, D.M., 1978. Late Cenozoic paleoclimates of the Ghaap Escarpment, Kalahari margin, South Africa. Quat. Res. 10, 310–339.

Camp, C.L., 1948. University of California African expedition-southern section. Science 10, 550–552.

Channell, J.E.T., Mazaud, A., Sullivan, P., Turner, S., Raymo, M.E., 2002. Geomagnetic excursions and paleointensities in the Matuyama Chron at Ocean Drilling Program sites 983 and 984 (Iceland Basin). J. Geophys. Res. Solid Earth 107 (B6), EPM-1.

Chazan, M., Ron, H., Matmon, A., Porat, N., Goldberg, P., Yates, R., Avery, M., Sumner, A., Horwitz, L.K., 2008. First radiometric dates for the earlier stone age sequence in Wonderwerk Cave, South Africa: preliminary results. J. Hum. Evol. 55, 1–11.

Chazan, M., Avery, D., Bamford, M.K., Berna, F., Brink, J., Fernandez-Jalvo, Y., Goldberg, P., Holt, S., Matmon, A., Porat, N., Ron, H., Rossouw, L., Scott, L., Horwitz, L., 2012. The Oldowan horizon in Wonderwerk Cave, South Africa: archaeological, geological, paleontological and paleoclimatic evidence. J. Hum. Evol. 63 (6), 859–866.

Chazan, M., Porat, N., Sumner, T.A., Horwitz, L.K., 2013. The use of OSL dating in unstructured sands: the archaeology and chronology of the Hutton Sands at Canteen Kopje (Northern Cape Province, South Africa). Archaeol. Anthropol. Sci. 5 (4), 351–363

Chazan, M., Berna, F., Brink, J., Ecker, M., Holt, S., Porat, N., Horwitz, L.K., 2020. Archeology, environment, and chronology of the early middle stone age component of wonderwerk cave. J. Paleolithic Archaeol. 3 (3), 302–335.

Clarke, R.J., 1998. First ever discovery of a well-preserved skull and associated skeleton of Australopithecus. South Afr. J. Sci. 94 (10), 460–463.

Clarke, R.J., 2006. A deeper understanding of the stratigraphy of Sterkfontein fossil hominid site. Trans. R. Soc. South Afr. 61 (2), 111–120.

Clarke, R.J., Kuman, K., 2019. The skull of StW 573, a 3.67 ma Australopithecus prometheus skeleton from Sterkfontein Caves, South Africa. J. Hum. Evol. 134, 102634.

Cooke, H.B.S., 1938. The Sterkfoktein bone breccia: a geological note. South Afr. J. Sci. 35 (12), 204–208.

d'Errico, F., Backwell, L., Villa, P., Degano, I., Lucejko, J.J., Bamford, M.K., Higham, T.F., Colombini, M.P., Beaumont, P.B., 2012. Early evidence of San material culture represented by organic artifacts from Border Cave, South Africa. Proc. Natl. Acad. Sci. U. S. A. 109 (33), 13214–13219.

Dart, R.A., 1925. Australopithecus africanus: the man-ape of South Africa. Nature 115, 195–199.

de Ruiter, D.J., 2003. Revised faunal lists for members 1-3 of Swartkrans, South Africa. Ann. Trans. Mus. 40 (1), 29–41.

Deacon, H.J., 1989. Late Pleistocene palaeoecology and archaeology in the southern Cape, South Africa. In: Mellars, P., Stringer, C.B. (Eds.), The Human Revolution 547–564. Edinburgh University Press, Edinburgh.

Deacon, H.J., 1998. Elandsfontein and Klasies River Revisited. In: Ashton, N., Healy, F., Pettitt, P.B. (Eds.), Stone Age Archaeology: Essays in Honour of John Wymer 23–28. Oxbow Books, Oxford.

Deacon, H.J., 2001. Modern human emergence: an African archaeological perspective. In: Tobias, P.V., Raath, M.A., Moggi-Cecchi, J., Doyle, G.A. (Eds.), Humanity from African Naissance to Coming Millennia—Colloquia in Human Biology and Palaeoanthropology. Florence University Press, Florence, pp. 217–226.

Deacon, H.J., Geleijnse, V.B., 1988. The stratigraphy and sedimentology of the main site sequence, Klasies River, South Africa. South Afr. Archaeol. Bull. 43, 5–14.

Deino, A.L., 2013. Geochronology, 244–264. In: Begun, D.R. (Ed.), A Companion to Paleoanthropology. Blackwell Publishing, Oxford, pp. 244–264.

d'Errico, F., Henshilwood, C., Vanhaeren, M., Van Niekerk, K., 2005. Nassarius kraussianus shell beads from Blombos Cave: evidence for symbolic behaviour in the Middle Stone Age. J. Hum. Evol. 48 (1), 3–24.

Dietl, H., Kandel, A.W., Conard, N.J., 2005. Middle Stone Age settlement and land use at the open-air sites of Geelbek and Anyskop, South Africa. J. Afr. Archaeol. 3, 233–244.

Dirks, P.H.G.M., Berger, L.R., 2013. Hominin-bearing caves and landscape dynamics in the Cradle of Humankind, South Africa. J. Afr. Earth Sci. 78, 109–131.

Dirks, P.H.G.M., Kibii, J.M., Kuhn, B.F., Steininger, C., Churchill, S.E., Kramers, J.D., Pickering, R., et al., 2010. Geological setting and age of Australopithecus sediba from southern Africa. Science 328, 205–208.

Dirks, P.H.G.M., Berger, L.R., Roberts, E.M., Kramers, J.D., Hawks, J., Randolph-Quinney, P.S., Elliott, M., Musiba, C.M., Churchill, S.E., de Ruiter, D.J., Schmid, P., Backwell, L.R., Belyanin, G.A., Boshof, P., Eaves, K.L., Feuerriegel, E.M., Gurtov, A., Harrison, J.d.G., Hunter, R., Kruger, A., Morris, H., Makhubela, T.V., Peixotto, B., Tucker, S., 2015. Geological and taphonomic evidence for deliberate body disposal by the primitive hominin species Homo naledi from the Dinaledi Chamber, South Africa. eLife 2015 (4), e09561. https://doi.org/10.7554/eLife.09561.

Dirks, P.H.G.M., Roberts, E.M., Hilbert-Wolf, H., Kramers, J.D., Hawks, J., Dosseto, A., Duval, M., Elliott, M., Evans, M., Grün, R., Hellstrom, J., Herries, A.I.R., Joannes-Boyau, R., Makhubela, T.V., Placzek, C.J., Robbins, J., Spandler, C., Wiersma, J., Woodhead, J., Berger, L.R., 2017. The age of Homo naledi and associated sediments in the Rising Star Cave, South Africa. eLife 6, e24231. https://doi.org/10.7554/eLife.24231.

Drennan, M.R., 1953. The saldanha skull and its associations. Nature 172, 791–793.

Duller, G.A.T., 2012. Improving the accuracy and precision of equivalent doses determined using the optically stimulated luminescence signal from single grains of quartz. Radiat. Meas. 47 (9), 770–777.

Elliott, M., Makhubela, T., Brophy, J., Churchill, S., Peixotto, B., Feuerriegel, E., Morris, H., Van Rooyen, D., Ramalepa, M., Tsikoane, M., Kruger, A., 2021.

Expanded explorations of the dinaledi subsystem, rising star cave system, South Africa. Paleoanthropology 2021 (1), 15–22.
Eriksson, P.G., Schweitzer, J.K., Bosch, P.J.A., Schereiber, U.M., Van Deventer, J.L., Hatton, C.J., 1993. The Transvaal sequence: an overview. J. Afr. Earth Sci. 16, 25–51.
Erlanger, E.D., Granger, D.E., Gibbon, R.J., 2012. Rock uplift rates in South Africa from isochron burial dating of fluvial and marine terraces. Geology 40 (11), 1019–1022.
Feathers, J.K., 2002. Luminescence dating in less than ideal conditions: case studies from Klasies River Mouth and Duinefontein, South Africa. J. Archaeol. Sci. 29, 177–194.
Fock, G.J., 1968. Rooidam, a sealed site of the First Intermediate. S. Afr. J. Sci. 64 (3), 153–158.
Gibbon, R.J., Granger, D.E., Kuman, K., Partridge, T.C., 2009. Early Acheulean technology in the Rietputs Formation, South Africa, dated with cosmogenic nuclides. J. Hum. Evol. 56 (2), 152–160.
Gibbon, R.J., Pickering, T.R., Sutton, M.B., Heaton, J.L., Kuman, K., Clarke, R.J., Brain, C.K., Granger, D.E., 2014. Cosmogenic nuclide burial dating of hominin-bearing Pleistocene cave deposits at Swartkrans, South Africa. Quat. Geochronol. 24, 10–15.
Granger, D.E., Gibbon, R.J., Kuman, K., Clarke, R.J., Bruxelles, L., Caffee, M.W., 2015. New cosmogenic burial ages for Sterkfontein Member 2 Australopithecus and Member 5 Oldowan. Nature 522, 85–88.
Granger, D.E., Stratford, D., Bruxelles, L., Gibbon, R.J., Clarke, R.J., Kuman, K., 2022. Cosmogenic nuclide dating of Australopithecus at Sterkfontein, South Africa. Proc. Natl. Acad. Sci. U. S. A. 119 (27), 2123516119.
Grine, F.E., Wurz, S., Marean, C.W., 2017. The middle stone age human fossil record from Klasies River main site. J. Hum. Evol. 103, 53–78.
Grun, R., Brink, J.S., Spooner, N.A., Taylor, L., Stringer, C.B., Franciscus, R.G., Murray, A.S., 1996. Direct dating of Florisbad hominid. Nature 382, 500–501.
Hattingh, J., 1994. Depositional environment of some gravel terraces in the Sundays River Valley, Eastern Cape. South Afr. J. Geol. 97, 156–166.
Hattingh, J., Rust, I.C., 1999. Drainage evolution of the Sundays River, South Africa. In: Miller, A., Gupta, A. (Eds.), Varieties in Fluvial Form 145–166. John Wiley and Sons, Chichester.
Hawks, J., Elliott, M., Schmid, P., Churchill, S.E., de Ruiter, D.J., Roberts, E.M., Hilbert-Wolf, H., Garvin, H.M., Williams, S.A., Delezene, L.K., Feuerriegel, E.M., Randolph-Quinney, P., Kivell, T.L., Laird, M.F., Tawane, G., De Silva, J.M., Bailey, S.E., Brophy, J.K., Meyer, M.R., Skinner, M.M., et al., 2017. New fossil remains of Homo naledi from the Lesedi Chamber, South Africa. eLife 6, e24232. https://doi.org/10.7554/eLife.24232.
Hendey, Q.B., 1968. The Melkbos site: an upper Pleistocene fossil occurrence in the South-Western Cape Province. Ann. South Afr. Mus. 52, 89–119.
Hendey, Q.B., 1974. The late Cenozoic Carnivora of the southwestern Cape Province. Ann. South African Mus. 63, 1–369.
Henshilwood, C.S., Marean, C.W., 2003. The origin of modern human behavior: critique of the models and their test implications. Curr. Anthropol. 44, 627–651.
Henshilwood, C.S., Sealy, J.C., Yates, R., Cruz-Uribe, K., Goldberg, P., Grine, F.E., Klein, R.G., Poggenpoel, C., Van Niekerk, K., Watts, I., 2001. Blombos cave, southern Cape, South Africa: preliminary report on the 1992-1999 excavations of the middle stone age levels. J. Archaeol. Sci. 28, 421–448.
Hendey, Q.B., 1982. Langebaanweg. A Record of Past Life. South African Museum, Cape Town.
Henshilwood, C.S., d'Errico, F., Van Niekerk, K.L., Coquinot, Y., Jacobs, Z., Lauritzen, S.E., Menu, M., García-Moreno, R., 2011. A 100,000-year-old ochre-processing workshop at Blombos Cave, South Africa. Science 334 (6053), 219–222.

Henshilwood, C.S., d'Errico, F., Van Niekerk, K.L., Dayet, L., Queffelec, A., Pollarolo, L., 2018. An abstract drawing from the 73,000-year-old levels at Blombos Cave, South Africa. Nature 562 (7725), 115–118.

Hopley, P.J., Herries, A.I., Baker, S.E., Kuhn, B.F., Menter, C.G., 2013. Brief communication: beyond the South African cave paradigm—Australopithecus africanus from Plio–Pleistocene paleosol deposits at Taung. Am. J. Phys. Anthropol. 151 (2), 316–324.

Humphreys, A.J.B., Thackeray, A.I., 1983. Ghaap and Gariep: later stone age studies in the northern Cape. South African Archaeol. Soc. Monograph Ser. Cape Town 2, 1–328.

Hutson, J.M., 2018. The faunal remains from Bundu Farm and Pniel 6, examining the problematic Middle Stone Age archaeological record within the southern African interior. Quat. Int. 466, 178–193.

Jacobs, Z., Jones, B.G., Cawthra, H.C., Henshilwood, C.S., Roberts, R.G., 2020. The chronological, sedimentary and environmental context for the archaeological deposits at Blombos Cave, South Africa. Quat. Sci. Rev. 235, 105850.

Kandel, A.W., Conard, N.J., 2012. Settlement patterns during the earlier and middle stone age around Langebaan Lagoon, Western Cape, South Africa. Quat. Int. 270, 15–29.

Kandel, A.W., Walker, S.J., Conard, N.J., 2006. Near-coastal settlement dynamics at the Anyskop Blowout, an archaeological locality at Langebaanweg, South Africa. Afr. Nat. Hist. 2, 186–187.

Kiberd, P., 2006. Bundu Farm: a report on archaeological and palaeoenvironmental assemblages from a pan site in Bushmanland, Northern Cape, South Africa. South Afr. Archaeol. Bull. 61, 189–201.

Klein, R.G., 1983. Palaeoenvironmental implications of quaternary large mammals in the Fynbos Biome. South Afr. Natl. Sci. Prog. Rep. 75, 116–138.

Klein, R.G., 1984. The large mammals of southern Africa: Late Pliocene to Recent. In: Klein, R.G. (Ed.), Southern African Prehistory and Paleoenvironments. A.A. Balkema, Rotterdam/Boston, pp. 107–146.

Klein, R.G., 1988. The archaeological significance of animal bones from Acheulean sites in southern Africa. Afr. Archaeol. Rev. 6, 3–25.

Klein, R.G., 2000. The earlier stone age of southern Africa. South Afr. Archaeol. Bull. 55, 107–122.

Klein, R.G., 2001. Southern Africa and modern human origins. J. Anthropol. Res. 57, 1–16.

Klein, R.G., Avery, G., Cruz-Uribe, K., Halkett, D., Hart, T., Milo, R.G., Volman, T.P., 1999. Duinefontein 2: an Acheulean site in the western Cape province of South Africa. J. Hum. Evol. 37 (2), 153–190.

Klein, R.G., Cruz-Uribe, K., Beaumont, P.B., 1991. Environmental, ecological, and paleoanthropological implications of the late Pleistocene mammalian fauna from Equus Cave, northern Cape Province, South Africa. Quat. Res. 36, 94–110.

Kramers, J.D., Dirks, P.H.G.M., 2017. The age of fossil StW573 ("Little Foot"): an alternative interpretation of 26Al/10Be burial data. South Afr. J. Sci. 113 (3/4), 45–53.

Kuhn, B.F., Herries, A.I., Pickering, R., Price, G., Baker, S.E., Hopley, P.J., Menter, C., Caruana, M.V., 2016. Renewed investigations at Taung; 90 years after Australopithecus africanus. Palaeontol. Africana 51, 10–26.

Kuman, K., 1989. Florisbad and ≠Gi: The contribution of open-air sites to the study of Middle Stone Age in southern Africa. Unpublished PhD thesis, University of Pennsylvania, Philadelphia.

Kuman, K., Clarke, R.J., 2000. Stratigraphy, artefact industries and hominid associations for Sterkfontein, Member 5. J. Hum. Evol. 38 (6), 827–847.

Kuman, K., Granger, D.E., Gibbon, R.J., Pickering, T.R., Caruana, M.V., Bruxelles, L., Clarke, R.J., Heaton, J.L., Stratford, D., Brain, C.K., 2021. A new absolute date from Swartkrans Cave for the oldest occurrences of Paranthropus robustus and Oldowan stone tools in South Africa. J. Hum. Evol. 156, 103000.

Lacruz, R.S., 2002. Analysis of Stratified Fossil Deposits From the Gladysvale Cave, M.Sc. Dissertation, University of the Witwatersrand.

Lacruz, R.S., Brink, J.S., Hancox, P.J., Skinner, A.R., Herries, A., Schmid, P., Berger, L.R., 2002. Palaeontology and geological context of a Middle Pleistocene faunal assemblage from the Gladysvale Cave, South Africa. Palaeontol. Africana 38 (99), 99–114.

Lacruz, R., Ungar, P., Hancox, P.J., Brink, J.S., Berger, L.R., 2003. Gladysvale: fossils, strata and GIS analysis: preliminary research reports: human origins research in South Africa. South African J. Sci. 99 (5), 283–285.

Li, H., Kuman, K., Lotter, M.G., Leader, G.M., Gibbon, R.J., 2017. The Victoria West: earliest prepared core technology in the Acheulean at Canteen Kopje and implications for the cognitive evolution of early hominids. R. Soc. Open Sci. 4 (6), 170288. https://doi.org/10.1098/rsos.170288.

Lotter, M.G., 2020. A preliminary assessment of large cutting tool production at the Acheulean site of Penhill farm, lower Sundays River valley, Eastern Cape Province, South Africa. South Afr. Archaeol. Bull. 75 (212), 58–69.

Lotter, M.G., Kuman, K., 2018. Atmar and Bernol farms: New Acheulean sites in the lower Sundays River Valley, Eastern Cape province, South Africa. South Afr. Archaeol. Bull. 73 (207), 64–74.

Lotter, M.G., Gibbon, R.J., Kuman, K., Leader, G.M., Forssman, T., Granger, D.E., 2016. A geoarchaeological study of the middle and upper pleistocene levels at Canteen Kopje, Northern Cape Province, South Africa. Geoarchaeol Int. J. 31, 304–323.

Lukich, V., Cowling, S., Chazan, M., 2020. Palaeoenvironmental reconstruction of Kathu Pan, South Africa, based on sedimentological data. Quat. Sci. Rev. 230, 106153.

Mabbutt, J.A., 1956. The physiography and surface geology of the Hopefield fossil site. Trans. R. Soc. South Afr. 35, 21–58.

Makhubela, T.V., Kramers, J.D., 2022. Testing a new combined (U, Th)–He and U/Th dating approach on Plio-Pleistocene calcite speleothems. Quat. Geochronol. 67, 101234.

Malan, B.D., Cooke, H.B.S., 1941. A preliminary account of the Wonderwerk Cave, Kuruman District, South Africa. South Afr. J. Sci. 37, 300–312.

Malan, B.D., Wells, L.H., 1943. A further report on the Wonderwerk Cave, Kuruman. South Afr. J. Sci. 40, 258–270.

Marean, C.W., Bar-Matthews, M., Fisher, E., Goldberg, P., Herries, A., Karkanas, P., Nilssen, P.J., Thompson, E., 2010. The stratigraphy of the Middle Stone Age sediments at Pinnacle Point Cave 13B (Mossel Bay, Western Cape Province, South Africa). Journal of Human Evolution 59 (3–4), 234–255.

Martini, J.E.J., 2006. Karsts and caves. In: Johnson, M.R., Anhaeusser, C.R., Thomas, R.J. (Eds.), The Geology of South Africa. Geological Society of South Africa and Council for Geoscience Publication, Pretoria, pp. 661–668.

Martini, J., Kavalieris, I., 1976. The karst of Transvaal, South Africa. Int. J. Speleol. 8, 229–251.

Martini, J.E.J., Wipplinger, P.E., Moen, H.F.G., Keyser, A., 2003. Contribution to the speleology of Sterkfontein Cave, Gauteng Province, South Africa. Int. J. Speleol. 32, 43–69.

Mason, R.J., 1967. Prehistory as a Science of Change: New Research in the South African Interior. Occusional Papers 1, 1 Archaeological Research Unit, University of the Witwatersrand, pp. 1–19.

Matmon, A., Ron, H., Chazan, M., Porat, N., Horwitz, L.K., 2012. Reconstructing the history of sediment deposition in caves. A case study from Wonderwerk Cave, South Africa. GSA Bull. 124 (3–4), 611–625.

McNabb, J., Sinclair, A., 1996. An Overview of the Archaeological History of the Cave of Hearths and its lithic Assemblages. The Cave of Hearths: Makapan Middle Pleistocene

Research Project: Field Research by Anthony Sinclair and Patrick Quinney, 2001. pp. 12–28.

Miller, C.E., Goldberg, P., Berna, F., 2013. Geoarchaeological investigations at Diepkloof Rock Shelter, Western Cape, South Africa. J. Archaeol. Sci. 40, 3432–3452.

Morrissey, P., Mentzer, S.M., Wurz, S., 2022. A critical review of the stratigraphic context of the MSA I and II at Klasies River Main Site, South Africa. J. Paleolith. Archaeol. 5, 5.

Netterberg, F., 1969. Ages of calcretes in southern Africa. S. Afr. Archaeol. Bull., 88–92.

Netterberg, F., 1974. Observations on calcretes at some archaeological and palaeontological sites in South Africa. Goodwin Series, Progress in Later Cenozoic Studies in South Africa 2. pp. 20–24.

Parkington, J., Rigaud, J.P., Poggenpoel, C., Porraz, G., Texier, P.J., 2013. Introduction to the project and excavation of Diepkloof Rock Shelter (Western Cape, South Africa): a view on the Middle Stone Age. J. Archaeol. Sci. 40, 3369–3375.

Partridge, T.C., 1978. Re-appraisal of lithostratigraphy of Sterkfontein hominid site. Nature 275 (5678), 282–287.

Partridge, T.C., 1982. Some preliminary observations on the stratigraphy and sedimentology of the Kromdraai B hominid site. In: Coetzee, J.A., Van Zinderen Bakker, E.M. (Eds.), Palaeoecology of Africa and the Sur-rounding Islands. vol. 15. CRC Press, Taylor.

Partridge, T.C., Watt, I.B., 1991. The stratigraphy of the Sterkfontein hominid deposit and its relationship to the underground cave system. Palaeontol. Africana 28, 35–40.

Partridge, T.C., Granger, D.E., Caffee, M.W., Clarke, R.J., 2003. Lower pliocene hominid remains from Sterkfontein. Science 300, 607–612.

Peabody, F., 1948. Unpublished Field Notes. Museum of Paleontology. University of California, Berkeley.

Pickering, R., 2015. U–Pb dating small buried stalagmites from Wonderwerk Cave, South Africa: a new chronometer for earlier stone age cave deposits. Afr. Archaeol. Rev. 32 (4), 645–668.

Pickering, R., Kramers, J.D., 2010. Re-appraisal of the stratigraphy and determination of new U-Pb dates for the Sterkfontein hominin site, South Africa. J. Hum. Evol. 59, 70–86.

Pickering, R., Hancox, P.J., Lee-Thorp, J.A., Grün, R., Mortimer, G.E., McCulloch, M., Berger, L.R., 2007. Stratigraphy, U-Th chronology, and paleoenvironments at Gladysvale Cave: insights into the climatic control of South African hominin-bearing cave deposits. J. Hum. Evol. 53, 602–619.

Pickering, R., Kramers, J.D., Hancox, P.J., De Ruiter, D.J., Woodhead, J.D., 2011a. Contemporary flowstone development links early hominin bearing cave deposits in South Africa. Earth Planet. Sci. Lett. 306, 23–32.

Pickering, R., Dirks, P.H.G.M., Jinnah, Z., de Ruiter, D.J., Churchil, S.E., Herries, A.I.R., Woodhead, J.D., et al., 2011b. Australopithecus Sediba at 1.977 Ma and implications for the origins of the genus Homo. Science 333, 1421–1423.

Porat, N., Chazan, M., Grun, R., Aubert, M., Eisenmann, V., Horwitz, L.K., 2010. New radiometric ages for the Fauresmith industry from Kathu Pan, southern Africa: implications for the earlier to middle stone age transition. J. Archaeol. Sci. 37, 269–283.

Reed, K., Kuykendall, K., Herries, A., Hopley, P., Sponheimer, M., Werdelin, L., 2022. Geology, Fauna, and Paleoenvironmental Reconstructions of the Makapansgat Limeworks Australopithecus africanus-Bearing Paleo-Cave. In: Reynolds, S., Bobe, R. (Eds.), African Paleoecology and Human Evolution 66–81. Cambridge University Press, Cambridge, https://doi.org/10.1017/9781139696470.007.

Rigaud, J.P., Texier, P.J., Parkington, J., Poggenpoel, C., 2006. Le mobilier Stillbay et Howiesons Poort de l'abri Diepkloof. La chronologie du Middle Stone Age sud-africain et ses implications. C. R. Palevol 5, 839–849.

Robbins, J.L., Dirks, P.H., Roberts, E.M., Kramers, J.D., Makhubela, T.V., Hilbert-Wolf, H.L., Elliott, M., Wiersma, J.P., Placzek, C.J., Evans, M., Berger, L.R., 2021. Providing context to the Homo naledi fossils: constraints from flowstones on the age of sediment deposits in Rising Star Cave. South Africa. Chem. Geol. 567, 120108.

Roberts, D.L., Bateman, M.D., Murray-Wallace, C.V., Carr, A.S., Holmes, P.J., 2009. West coast dune plumes: climate driven contrasts in dunefield morphogenesis along the western and southern South African coasts. Palaeogeogr. Palaeoclimatol. Palaeoecol. 271, 24–38.

Roberts, D.L., Matthews, T., Herries, A.I.R., Boulter, C., Scott, L., Dondo, C., Mtembi, P., Browning, C., Smith, R.M.H., Haarhoff, P., Bateman, M.D., 2011. Regional and global context of the Late Cenozoic Langebaanweg (LBW) palaeontological site: West Coast of South Africa. Earth Sci. Rev. 106, 191–214.

Ruddock, A., 1948. Terraces in the lower part of the Sundays River Valley, Cape Province. Trans. R. Soc. South Africa 31, 347–370.

Shaar, R., Matmon, A., Horwitz, L.K., Ebert, Y., Chazan, M., Arnold, M., Aumaître, G., Bourlès, D., Keddadouche, K., 2021. Magnetostratigraphy and cosmogenic dating of Wonderwerk Cave: new constraints for the chronology of the South African Earlier Stone Age. Quat. Sci. Rev. 259, 106907.

Singer, R., 1956. The "bone tools" from Hopefield. Am. Anthropol. 58, 1127–1134.

Singer, R., Crawford, J.R., 1958. The significance of the archaeological discoveries at Hopefield, South Africa. J. R. Anthropol. Inst. Great Brit. Ireland 88, 11–19.

Singer, R., Heltne, P.G., 1966. Further notes on a bone assemblage from hopefield, South Africa. Mus. Arqueol. 261-264.

Singer, R., Wymer, J., 1968. Archaeological investigations at the Saldanha skull site in South Africa. South Afr. Archaeol. Bull. 23, 63–74.

Skinner, M.M., Kivell, T.L., Potze, S., Hublin, J.J., 2013. Microtomographic archive of fossil hominin specimens from Kromdraai B, South Africa. J. Hum. Evol. 64 (5), 434–447.

Stratford, D., 2020. The geological setting, cave formation, and stratigraphy of the fossil-bearing deposits at Sterkfontein Caves. Hominin Postcranial Remains from Sterkfontein, South Africa, 1936-1995. In: Zipfel, B., Richmond, B.G., Ward, C.V. (Eds.), Hominin Postcranial Remains From Sterkfontein, South Africa, 1936-1995. Oxford University Press, New York, pp. 8–20.

Sutton, M.B., Pickering, T.R., Pickering, R., Brain, C.K., Clarke, R.J., Heaton, J.L., Kuman, K., 2009. Newly discovered fossil-and artifact-bearing deposits, uranium-series ages, and Plio-Pleistocene hominids at Swartkrans Cave, South Africa. J. Hum. Evol. 57 (6), 688–696.

Texier, P.J., Porraz, G., Parkington, J.E., Rigaud, J.P., Poggenpoel, C., Miller, C.E., Tribolo, C., Cartwright, C., Coudenneau, A., Klein, R.G., Steele, T., Verna, C., 2010. Howiesons Poort tradition of engraving ostrich eggshell containers dated to 60,000 years ago at Diepkloof Rock Shelter, South Africa. Proc. Natl. Acad. Sci. U. S. A. 107, 6180–6185.

Thackeray, J.F., 1984. Man, Animals, and Extinctions: The Analysis of Holocene Faunal Remains From Wonderwerk Cave, South Africa. Unpublished PhD Thesis, Yale University, New Haven.

Thackeray, J.F., 2020. A summary of the history of exploration at the Sterkfontein Caves in the Cradle of Humankind World Heritage Site. In: Zipfel, B., Richmond, B.G., Ward, C.V. (Eds.), Hominin Postcranial Remains From Sterkfontein, South Africa, 1936-1995. Oxford University Press, New York, pp. 3–7.

Toffolo, M.B., Brink, J.S., Van Huyssteen, C., Berna, F., 2017. A microstratigraphic reevaluation of the Florisbad spring site, Free State Province, South Africa: formation processes and paleoenvironment. Geoarchaeology 32, 456–478.

Val, A., De la Peña, P., Duval, M., Bansal, S., Colino, F., Culey, J., Hodgskiss, T., Morrissey, P., Murray, A., Murungi, M., Neumann, F.H., 2021. The place beyond the trees: renewed excavations of the Middle Stone Age deposits at Olieboomspoort in the Waterberg Mountains of the South African Savanna Biome. Archaeol. Anthropol. Sci. 13 (7), 1–32.

Van Der Ryst, M.M., 2007. Seeking Shelter: Later Stone Age Hunters, Gatherers and Fishers of Olieboomspoort in the Western Waterberg, South of the Limpopo. PhD Thesis, University of the Witwatersrand, Johannesburg.

Villa, P., Soriano, S., Tsanova, T., Degano, I., Higham, T.F., d'Errico, F., Backwell, L., Lucejko, J.J., Colombini, M.P., Beaumont, P.B., 2012. Border cave and the beginning of the later stone age in South Africa. Proc. Natl. Acad. Sci. U.S.A. 109 (33), 13208–13213.

Vrba, E.S., 1981. The Kromdraai australopithecine site revisited in 1980; recent investigations and results. Ann. Transvaal Mus. 33 (3), 17–60.

Wadley, L., 2006. Partners in grime: results of multi-disciplinary archaeology at Sibudu Cave. South. Afr. Hum. 18 (1), 315–341.

Wadley, L., 2015. Those marvellous millennia: the Middle Stone Age of Southern Africa. Azania. Archaeol. Res. Afr. 50, 155–226.

Wadley, L., Jacobs, Z., 2006. Sibudu Cave: background to the excavations, stratigraphy and dating. South. Afr. Hum. 18, 1–26.

Walker, J., Cliff, R.A., Latham, A.G., 2006. U-Pb isotopic age of the StW 573 hominid from Sterkfontein, South Africa. Science 314, 1592–1594.

Walker, S.J.H., Lukich, V., Chazan, M., 2014. Kathu townlands: a high density earlier stone age locality in the interior of South Africa. PLoS One 9 (7), e103436.

Wiersma, J.P., Roberts, E.M., Dirks, P.H., 2020. Formation of mud clast breccias and the process of sedimentary autobrecciation in the hominin-bearing (Homo naledi) rising star cave system, South Africa. Sedimentology 67 (2), 897–919.

Wilkins, J., Chazan, M., 2012. Blade production ~500 thousand years ago at Kathu Pan 1, South Africa: support for a multiple origins hypothesis for early Middle Pleistocene blade technologies. J. Archaeol. Sci. 39 (6), 1883–1900.

Wintle, G.A., 2013. Introduction—History of dating methods. In: Scott, A.E. (Ed.), Encyclopedia of Quaternary Science, second ed. Elsevier, pp. 18–27.

Wood, R., 2015. From revolution to convention: the past, present and future of radiocarbon dating. J. Archaeol. Sci. 56, 61–72.

Wurz, S., 2013. Technological trends in the Middle Stone Age of South Africa between MIS 7 and MIS 3. Curr. Anthropol. 54 (8), 305–319.

Wurz, S., Bentsen, S., Reynard, J., Van Pletzen-Vos, L., Brenner, M., Mentzer, S., Pickering, R., Green, H.E., 2018. Connections, culture and environments 100,000 years ago at Klasies River main site. Quat. Int. 495, 102–115.

CHAPTER SEVEN

New lithostratigraphic nomenclature of Neogene rock units in As-Sahabi area, Sirt Basin, northeast Libya

Moftah Hasan El-Shawaihdi[a,*], Noel Thomas Boaz[b,c], Ahmed Muftah[a], and Fathi Salloum[a]

[a]Earth Science Department, Faculty of Sciences, University of Benghazi, Benghazi, Libya
[b]Emory & Henry College School of Health and Science, Marion, VA, United States
[c]International Institute for Human and Evolutionary Research, Integrative Centers for Science and Medicine, Martinsville, VA, United States
*Corresponding author: e-mail address: moftah.elshawaihdi@uob.edu.ly

Contents

Abstract

This paper reviews the lithostratigraphic nomenclature of Neogene formal and informal rock units in the As-Sahabi area of northeast Libya. Formally-named lithostratigraphic units are modified whereas informal units are re-designated and given new names. These modifications are proposed on the basis of regional and local correlation, the

Stratigraphy & Timescales, Volume 7
ISSN 2468-5178
https://doi.org/10.1016/bs.sats.2022.09.005

new identification of an unconformity, and the designation of a new type locality. The lowermost unit is a limestone of Miocene (Langhian-Tortonian) age. This limestone was previously designated as Formation "M" but the term Formation "M" is to be abandoned and this limestone should be named the Benghazi Formation of Ar Rajmah Group. This limestone is capped by a regional unconformity, which is overlain by a 10–40 m thick unit of Miocene (Messinian) sand, calcareous sand, gypsiferous clay, silty clay, and sandy clay. This unit was previously designated as Formation "P" and the lower part of Member "T" of the Sahabi Formation, but these designations (Formation "P" and Member "T" of the Sahabi Formation) are to be abandoned and this unit is named as the Sabkhat Al Qunnayyin Member of the Sahabi Formation. The 10–40 m thick unit of Miocene (Messinian) sand, calcareous sand, gypsiferous clay, silty clay, and sandy clay (Sabkhat Al Qunnayyin Member) is overlain by the following non-gypsiferous units (from base to top) of Miocene (Messinian) age: a 20 m thick unit of white to yellow sand (with balls and lenses of clay); a 10 m thick unit of white sand, green to gray clay, pink to purple clay, and cross-bedded sandstone; and a 4 m thick unit of green clay and sand (with two beds of brown sandstone). Collectively these strata form a 34 m thick interval of non-gypsiferous sand and clay that was previously designated as Members "U1," "UD," and "U2," but these designations (Members "U1," "UD," and "U2") are to be abandoned and these non-gypsiferous strata are collectively named the Qarat Makada Member of the Sahabi Formation. This 34 m thick interval of non-gypsiferous sand and clay (Qarat Makada Member) is capped by a regional unconformity, which is overlain by a 10–40 m thick unit of Pliocene white to green sand, pebbly sand, and sandy clay. This unit was previously designated as Member "V" of the Sahabi Formation, but this designation is to be replaced by the name Qarat Weddah Formation. The 10–40 m thick unit of Pliocene white to green sand, pebbly sand, and sandy clay (Qarat Weddah Formation) is overlain by a 1.5–2 m thick unit of carbonate soil/calcrete. This carbonate soil/calcrete was previously designated as Member "Z" of the Sahabi Formation, but this designation (Member "Z") is to be replaced by the name Shagig Formation. These new formal names should be used onward and all informal names should be discarded. This new lithostratigraphic nomenclature of the Neogene Sahabi units are designated according to the **N**orth **A**merican **C**ommission on **S**tratigraphic **N**omenclature (NACSN, 1983) and the International Stratigraphic Guide (Salvador, 1994).

1. Introduction

As-Sahabi area is a well-known Neogene vertebrate site that covers an area of about $400\,km^2$ in the northeastern part of Sirt Basin in Libya (Fig. 1). The site is bounded by longitudes 20° 48′ 08″ to 20° 54′ 45″ E and latitudes 30° 10′ 58″ to 30° 17′ 36″ N. Neogene strata in As-Sahabi area are exposed along the western edge of the recent Sabkhat Al Qunnayyin, in a general NE-SW trend of elongated intermittent hills (Fig. 2).

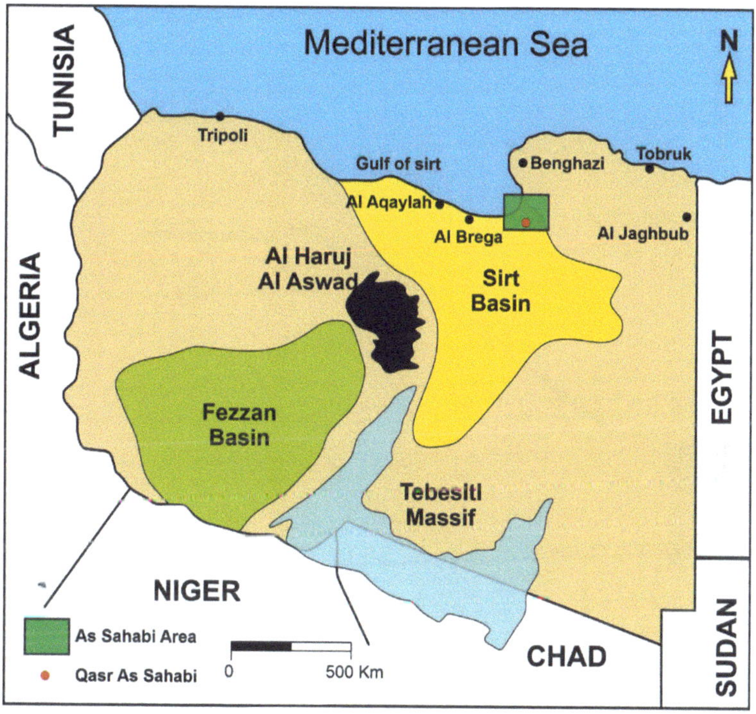

Fig. 1 Map of Libya showing the location of major regions, including As-Sahabi area and Sirt Basin.

Geological research in As-Sahabi area began a century ago with reconnaissance by Italian geologist Ardito Desio of the low-lying sedimentary outcrops near the ancient Roman centenarium, gypsum quarry, and subsequent fort known as Qasr As-Sahabi (Fig. 1). Strata in As-Sahabi area became well-known for their abundant Neogene vertebrate fossils, particularly the type specimen of *Stegotetrabelodon syrticus*, a four-tusked early elephantiid, for which the Libyan Museum of Natural History (now part of the National Museum of Libya) was built and which was featured on two Libyan postage stamps. Much of the detailed stratigraphic research since 1975 has been tied to intensive paleontological investigations undertaken by international teams (International Sahabi Research Project, headed by N. T. Boaz, A. W. Gaziry, and A. El-Arnauti, 1975–1994, subsequently re-organized to include the region of Jabal Zaltan as the East Libyan

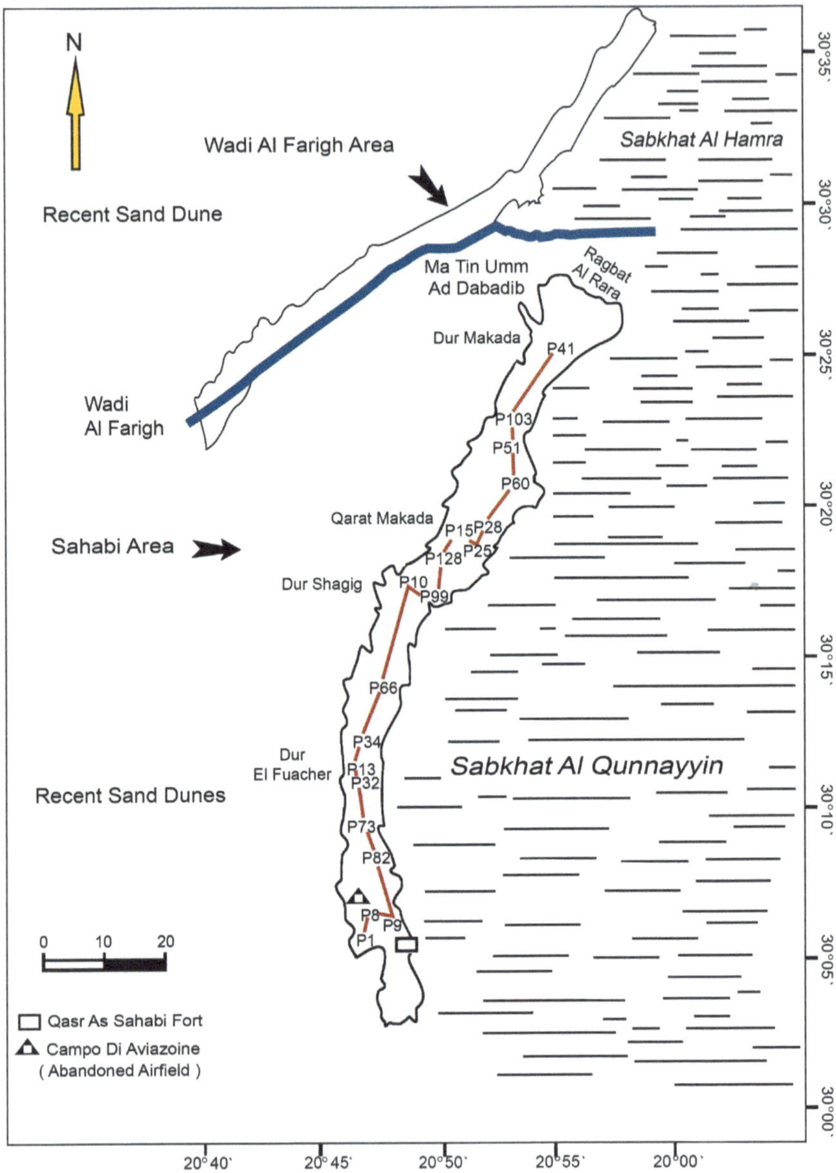

Fig. 2 Map of As-Sahabi area (Libya) showing the location of the localities (Profiles) of the studied sections. The area of this figure is shown by the green box in Fig. 1. The red line is the line of the geologic cross section shown in Fig. 3. The blue line is the course of Wadi Al Farigh. *Modified from El-Shawaihdi, M.H, Mozley, P.S., Boaz, N.T., Salloum, F., Pavlakis, P., Muftah, A., Triantaphyllou, M. 2016. Stratigraphy of the Neogene Sahabi units in the Sirt Basin, Northeast Libya, J. Afr. Earth Sci., 118: 87–106.).*

Neogene Research Project, 1994—present, headed by N. T. Boaz, A. Muftah, and M. El-Shawaihdi). The stratigraphic framework through which the extensive fossil assemblages have been interpreted was based on work by de Heinzelin et al. (1980) and de Heinzelin and El-Arnauti (1982, 1983, 1987). Mineralogical composition of the sediments did not allow absolute geochronological control and the unconsolidated nature of the fossiliferous sediments at Sahabi rendered paleomagnetic studies problematical.

Barr and Walker (1973) first reported two subsurface channels of over 400 m in depth detected by seismic studies near the Qasr As-Sahabi fort in northern Libya. de Heinzelin and El-Arnauti (1987) estimated the location of these subsurface features to be some 20 km south of Qasr As-Sahabi, near the southern extent of surface outcrops of the Sahabi Formation. Griffin (2002, 2006, 2011) related surface erosional features detected through remote sensing studies to an extensive hydrographic drainage network extending southerly past the Tibesti Massif and into northern Chad (Fig. 1). He related these features to the subsurface Sahabi channels discovered by Barr and Walker (1973) and referred to an Eosahabi paleo-river that cut down through Miocene strata during the Late Miocene (Messinian) drawdown of the Mediterranean Sea. Nicolai (2008) used 2D and 3D seismic and aeromagnetic data to trace in detail the last 200 km of the course of the Messinian Eosahabi river northward to its debouchment into the Mediterranean Sea at present Marsa Al Brega (Fig. 1). The mapped course of this paleo-river corresponds in large part to the present depression of the Wadi Al Farigh (Fig. 2), north of Qasr As-Sahabi, flanked by sedimentary outcrops of cross-bedded, peloidal oomicrite and oosparite, alternating with shale, siltstone, and clay that were mapped as the Wadi Farigh Member of the Sahabi Formation (see El-Shawaihdi et al., 2016). Nicolai (2008) documented a riverbed gradient of 1.25 m/km, a channel width of more than 5 km, and a channel depth of 250 m where this paleo-river entered the paleo-Gulf of Sirt. The origin of the Eosahabi River was likely Mega-Lake Chad (Griffin, 2002, 2006, 2011). Muftah et al. (2013) reported geochemical studies indicating that sediments deposited by the Eosahabi River in Unit U1 of the Sahabi Formation originated from upland felsic source rocks of Precambrian age exposed in northeastern Chad, consistent with a river source in Mega-Lake Chad. These results argue against the hypothesis advanced by Carmignani et al. (1990, 2009) that the Sahabi channels were formed by a westerly diversion of the Nile River into the Gulf of Sirt during the Messinian. This suggestion is also vitiated by Chumakov (1973), who identified evidence of deep down-cutting of the Nile River in its present location in Egypt during the Messinian Salinity Crisis (Roveri et al., 2014). Distributions of fossil fish taxa, which have been used to ascertain past hydrographic connections, provide supporting evidence for a Chadian origin of the Eosahabi River (Argyriou et al., 2012; Drake et al., 2008) although not necessarily to the exclusion of other river basins, particularly the Nile and the Niger Rivers. Many fish species are shared throughout a late Neogene Nilo-Sudanian ichthyoprovince (Argyriou et al., 2012).

Drake et al. (2008) proposed a hypothesis to explain the large size of the Eosahabi River, suggesting that two river systems drained Libya during the Late Miocene: (1) an eastern "Sahabi River System" draining Mega-Lake Chad, the Tibesti Mountains, and the Al Kufrah Basin (see also Ghoneim et al., 2012; Paillou et al., 2009, 2012, 2020) into the eastern Gulf of Sirt; and (2) a western "Nashu River System" that drained western Libya and the mountains surrounding the Fezzan Basin (Fig. 1) into the western Gulf of Sirt. They hypothesized that at the end of the Miocene Epoch uplift of the volcanic mountains of Al Haruj Al Aswad in central Libya (Fig. 1) blocked outflow of the Nashu paleo-river to the north and allowed the Eosahabi River to capture much of the Nashu River headwaters. The enlarged catchment area of the Eosahabi River was estimated by these authors to have been more than 4 million km^2, exceeding the modern 3.349 million km^2 catchment area of the Nile River. Another factor that may account for the large size of the Late Miocene Eosahabi River was a migration of the Inter-Tropical Convergence Zone during the Messinian Period that would have greatly increased monsoonal rainfall over Mega-Lake Chad (Gladstone et al., 2007; Maley, 2010; Roveri et al., 2014). The down cutting of the Eosahabi River channels is represented by the top of Member "U2" of the Sahabi Formation, overlain unconformably by the Pliocene Qarat Weddah Formation (El-Shawaihdi et al., 2016). This datum coincides with the onset of evaporite deposition synchronously throughout the basin at 5.96 Ma (Roveri et al., 2014). Re-flooding of the Mediterranean Sea, and the Miocene-Pliocene boundary, occurred at 5.33 Ma (Roveri et al., 2014). These authors, citing Nicolai (2008), note that this regional unconformity recognized at Sahabi is a Margin Erosion Surface (MES) and one of a very few demonstrable offshore-onshore stratigraphic continuities resulting from the Messinian Salinity Crisis in the Mediterranean Basin. They note that "a comprehensive stratigraphic scenario of Messinian events is still unavailable because of the difficulties in combining onshore and offshore observations (Roveri et al., 2014).

Several lithostratigraphic classifications have been proposed for the Neogene rock units in As-Sahabi area; however, the formal lithostratigraphic classification and nomenclature of the area have been in dispute, leading to much confusion for geologists and paleontologists who work in the area (see review in El-Shawaihdi, 2015; El-Shawaihdi et al., 2016). The aim of this publication is to unify the stratigraphy of As-Sahabi area and to eliminate the usage of informal names, following the rules of stratigraphic nomenclature as described in the North American Commission on Stratigraphic Nomenclature (NACSN, 1983) and the International Stratigraphic Guide (Salvador, 1994).

2. Description of lithostratigraphic units of As-Sahabi area

In this study, major localities such as P1, P8, P9, P10, P13, P25, P28, P32, P34, P41, P51, P53, P60, P66, P70, P73, P82, P99, P103, P128 (Figs. 2, 5, 6) were sampled and described in conjunction with extensive revision of previous works as shown in Fig. 7. A geologic cross section of the Neogene strata that outcrop in As-Sahabi area is shown in Fig. 3. These strata are described as follows from base to top:

3. Previous stratigraphic nomenclature

Previous studies by de Heinzelin et al. (1980) and de Heinzelin and El-Arnauti (1982, 1983, 1987) recognized one formal unit named the Sahabi Formation, which they divided into several members (Fig.7) and two informal units named Formations "M" and "P". De Heinzelin et al. (1980) and de Heinzelin and El-Arnauti (1982, 1983, 1987) divided the area into 138 geologic transects (localities) named according to the fossils found (Fig. 7). Giglia (1984), however, recognized two formal units named the Sahabi Formation and the Qarat Weddah Formation (Fig. 7) More recent studies (El-Shawaihdi et al., 2016) have suggested substantial revisions, following the rules of the North American Stratigraphic Code

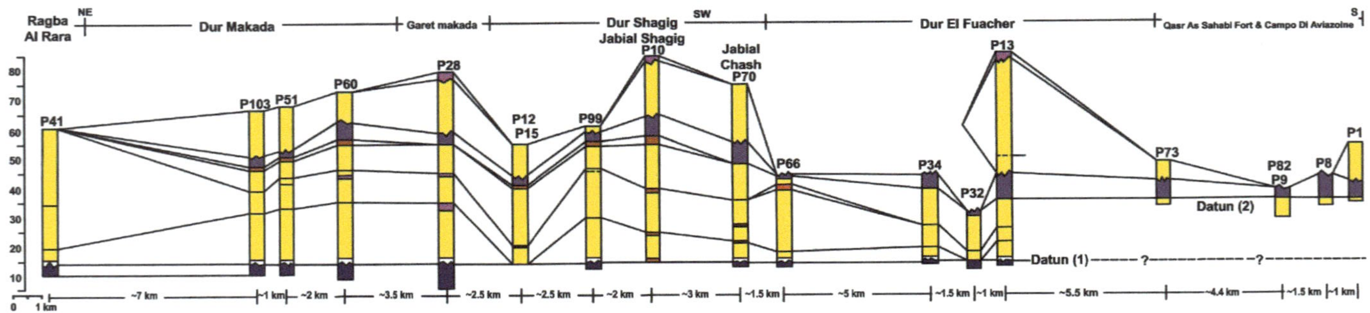

Fig. 3 Geologic cross section of Neogene strata in As-Sahabi area, Libya. Location of the line of section is shown in Fig. 2. Explanation of colors is shown in Fig. 4.

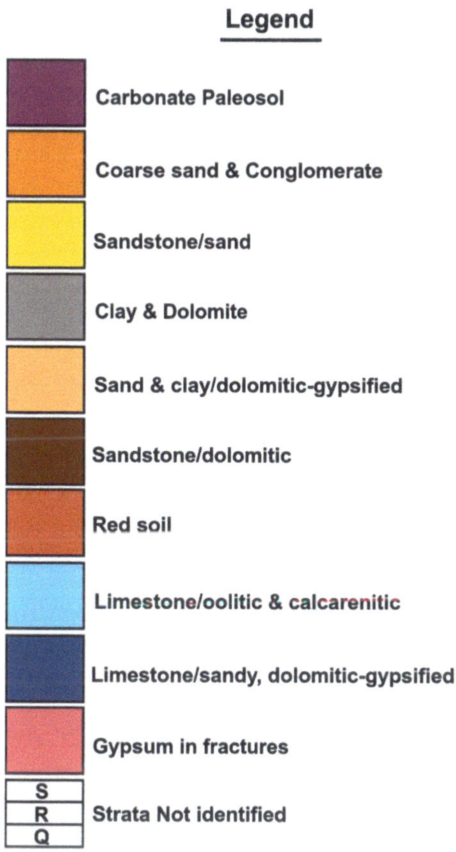

Fig. 4 Legend of key colors used in Figs. 3, 7 and 9.

(North American Commission on Stratigraphic Nomenclature, 1983) and the International Stratigraphic Guide (Salvador, 1994).

4. Methodology

The methods used for revising the stratigraphic nomenclature of Neogene rocks units of As-Salabi area follow the rules of the North American Stratigraphic Code (North American Commission on Stratigraphic Nomenclature, 1983) and the International Stratigraphic Guide (Salvador, 1994). These revisions require an understanding of the distinction between lithostratigraphy and biostratigraphy. Lithostratigraphic units are distinguished on the basis of lithological characteristics (not fossil content), and the fundamental unit of lithostratigraphic nomenclature is the formation. In contrast, biostratigraphic units are distinguished on the basis of fossil content, and the basic unit of biostratigraphy is the biozone. Furthermore, on the basis of Article 3 on the North American Stratigraphic Code (NASC), the procedures to name, establish, revise, redefine, and (or) abandon formal geologic units require publication in a recognized scientific medium issued for the purpose of scientific, public, or permanent record and readily obtainable by purchase or free distribution (inadequate publications include open-file releases, theses, dissertations, scientific meeting abstracts, and map explanations). The publication should consist of a comprehensive statement, which includes the following items: (i) Intent to designate or modify a formal unit; (ii) Designation of category and rank of unit; (iii) Selection and derivation of name; (iv) Specification of stratotype (where applicable); (v) Description of unit; (vi) Definition of boundaries; (vii) Historical background; (viii) Dimensions, shape, and other regional aspects; (ix) Geologic age; (x) Correlations; and (xi) Genesis (where applicable).

5. Revisions of stratigraphic nomenclature for As-Sahabi area of Libya

To produce a unified and formal classification, an attempt to revise and modify the above-mentioned units and establish new units has become an inevitable study. This result is based on recent field work and comprehensive reviews of previous literature. Formations and members with their proposed ages that are planned to be modified and (or) revised are shown in Fig. 7. These units are the Sahabi Formation with its two Members (Sabkhat Al Hamra and Sabkhat Al Qunnayyin) and the Qarat Weddah Formation as designated by Giglia (1984). Formation "M," Formation "P," and the Sahabi Formation with its six Members ("T," "U1," "UD," "U2," "V," "Z") as designated by de Heinzelin et al. (1980) and de Heinzelin and El-Arnauti (1982, 1983, 1987). These revisions to stratigraphic nomenclature are described below for each rock unit (from base to top of the stratigraphic column). For each lithologic unit, there is a description of the lithology, a description of diagnostic fossils, a discussion of the previous stratigraphic nomenclature(s) assigned to this rock unit, a discussion of previous age estimates (biostratigraphy) assigned to this rock unit, and the new stratigraphic nomenclature proposed in this paper for the rock unit. Although the lithologies of the above rock units have been fully described by de Heinzelin et al. (1980) and de Heinzelin and El-Arnauti (1982, 1983, 1987), these units are still being confused with the Sabkhat Al Hamra

Member as designated by Giglia (1984). However, Formation "M" of de Heinzelin et al. (1980) and de Heinzelin and El-Arnauti (1982, 1983, 1987) and the Sabkhat Al Hamra Member of Giglia (1984) are the same unit of Miocene carbonate, but de Heinzelin and El-Arnauti estimated the age to be Middle Miocene whereas Giglia estimated the age to be Upper Miocene, although they used different biostratigraphic schemes. This Miocene carbonate unit is present across much of Libya.

5.1 Unit (1) limestone to sandy limestone

In As-Sahabi area, this unit is approximately 12 m thick and is composed of semi-consolidated limestone (specifically wackestone to packstone), sandy limestone, yellow to dark green, gypsiferous and dolomitic (Fig. 8). The sand consists of medium to fine sand composed of subangular quartz grains and glauconite. This unit contains no obvious sedimentary structures. The lower contact is not exposed, whereas the upper contact is placed immediately below a bed of white to yellow sand and dark green clay with gypsum crystals. This upper contact is not very well exposed but it might be an unconformity. This upper contact is widely exposed in As-Sahabi area. A particularly good exposure in As-Sahabi area is located at Locality P53 (Figs. 2). This unit extends north from As-Sahabi area to the city of Benghazi (Fig. 1) where the unit was described by Klen (1974).

Fossils are predominantly of echinoids (*Clypeaster* sp. and *Echinolampas* sp.), mollusks (*Ostrea* sp., *Strombus* sp., and *Glycymeris* sp.), bryozoans, corals (*Madreporaria* sp.), foraminifera and ostracods (Table 1). Vertebrate remains present are mainly few fossils of sirenian and cetaceans and bones shark teeth (*Oxyrhina* sp.).

In previous studies, Giglia (1984) did not designate these strata as a separate unit (Fig. 7), but included them in the lower part of the Sabkhat El Hamra Member of the Sahabi Formation. de Heinzelin and El-Arnauti (1982, 1983, 1987), however, designated these strata as a separate unit (Fig. 7), which they named Formation "M." This correlation was made on the basis of similar content of selenitic gypsum and similar fossils of gastropods (e.g., *Strombus*, *Conus*, and *Turritella*), pelecypods (e.g., *Cardium*, *Glycymeris*, *Ostrea*, and *Dosinia*), and echinoids (*Clypeaster* and *Echinolampas*). De Heinzelin and El-Arnauti (1987) previously used the similarity of echinoid and mollusk fossils to correlate their Formation "M" with the Sabkhat Al Hamra Member of Giglia (1984). Although formation names should be assigned on the basis of lithologic correlations with type sections rather than

Table 1 Foraminifera and ostracods present in the limestone to sandy limestone unit.

Foraminifera	Ostracods
Globorotalia humerosa,	*Aurila diecii*, *A. albicans*, *Aurila.* sp.,
Ammonia tepida, *Ammonia*	*Cytherella punctatella*, *C. vandenboldi*,
cf. *inflata*,	*Cytherella* cf., *C. vulgate*,
Ammonia beccarii,	*Cytherelliodea* sp.,
Elphidium aculeatum,	*Carinocythereis galilea*, *Chrysocythere cataphracta*,
E. decipiens, *E. crispum*,	*Cyamocytheridea dertonesis*, *Loxoconcha variesculpta*,
Elphidium cf. *sagrum*,	*Loxoconcha gr. ex. Ovulate*,
Elphidium sp.,	*Neomonoceratina mouliana*,
Quinqueloculina spp.,	*Neomonoceratina* cf. *N. laskervi*,
Quinqueloculina oblonga,	*Olimfalunia sicula*, *O. zibinica*,
Operculina sp.,	*Pokornyella italica*, *P. deformis minor*,
Amphistegina sp.,	*Tenedocythereis mediterranea*,
Miogypsinids,	*Neonesidea* cf., *N. mediterranea*, *muricata*,
Borelis melo	*Cytheretta* sp.

fossil content (North American Commission on Stratigraphic Nomenclature, 1983; Salvador, 1994), the fossil assemblages nevertheless support the correlations described in the previous paragraph. For example, the benthic foraminifera *B. melo* is present in Formation "M" and is also present in the lower part of the Sabkhat Al Hamra Member of the Sahabi Formation of Giglia (1984). Likewise, the invertebrate fossils are predominantly mollusks and echinoids, and such fossils have been described in the lower part of the Sabkhat Al Hamra Member of the Sahabi Formation of Giglia (1984) and in Formation "M" of de Heinzelin and El-Arnauti (1983, 1987). Furthermore, de Heinzelin et al. (1980) and de Heinzelin and El-Arnauti (1982, 1983, 1987) mentioned the presence of sirenian (*Metaxytherium*, Domning, 1982; Domning and Thomas, 1987) and cetacean fossils in their Formation "M," and Giglia (1984) mentioned the presence of shark teeth (*Oxyrhina* sp.) in his Sabkhat Al Hamra Member.

This unit does not contain age-diagnostic fossils. The age can be estimated, however, on the basis of the presence of gypsum, and the correlation of this gypsum with the Messinian Salinity Event. Recent work by Krijgsman et al. (1999) dated this event to approximately 5.96–5.32 Ma. As stated above, strata of this unit are correlated with portions of the Sabkhat Al Hamra and Sabkhat Al Qunnayyin Members of Giglia (1984) and with Formation "P" of de Heinzelin and El-Arnauti (1982, 1983). Giglia (1984) postulated that the Sabkhat Al Hamra and Sabkhat Al Qunnayyin Members are of Late Miocene age, specifically Tortonian to early Messinian, on the basis of assemblages of foraminifera (*Globorotalia humerosa, Ammonia tepida, Ammonia* cf. *inflata, Quinqueloculina oblonga, Elphidium aculeatum, E. decipiens*, and *Elphidium* cf. *sagrum*) and ostracods (*Aurila diecii, A. albicans, Cytherella punctatella, C. vandenboldi, Carinocythereis galilea, Chrysocythere cataphracta, Cyamocytheridea dertonesis, Loxoconcha variesculpta, Neomonoceratina mouliana, Olimfalunia sicula, O. zibinica, Pokornyella italica*, and *Tenedocythereis mediterranea*). The most recent age estimate of the ostracods *Chrysocythere cataphracta, Pokornyella italica,* and *Aurila diecii* is Tortonian to Messinian (Guernet, 2005), the most recent age estimate of the ostracod *A. albicans* is Serravallian to Messinian (Guernet, 2005), and the most recent age estimate of the ostracod *Olimfalunia sicula* is Messinian (Guernet, 2005). de Heinzelin et al. (1980) and de Heinzelin and El-Arnauti (1982, 1983) postulated that their Formation "P" is of Messinian age on the basis of the presence of gypsum. Their Formation "P," however, contains no age-diagnostic fossils (Muftah et al., 2008b). The age of this unit may be estimated by the presence of the benthic foraminifera (*B. melo* and *Elphidium* sp.) and by the ostracod *Cytherelliodea* sp. As stated above, the most recent age estimate of *B. melo* is Langhian through middle Messinian (de Graciansky et al., 1998). This unit is also correlated with the Sabkhat Al Hamra Member of the Sahabi Formation of Giglia (1984). Giglia (1984) suggested a Tortonian age for this unit on the basis of assemblages of the foraminifera *B. melo* and the ostracods *Olimfalunia sicula* and *Aurila albicans*. The unit is also correlated with Formation "M" of de Heinzelin and El Arnauti (1983, 1987). de Heinzelin et al. (1980) and de Heinzelin and El-Arnauti (1982, 1983, 1987) postulated that their Formation "M" is of Serravallian age on the basis of its location below the Messinian gypsum of their Formation "P." The presence of sirenian and cetacean fossils in Formation "M" have also been used to suggest a Middle Miocene age for the unit (de Heinzelin and El-Arnauti, 1982; Domning, 1982; Domning and Thomas, 1987). Muftah et al. (2008a,b), however, estimated the age of Formation "M" to be Tortonian on the basis of the presence of certain foraminifera (*Ammonia beccarii, Elphidium crispum, Elphidium* sp., *Quinqueloculina* spp., *Operculina* sp., *Amphistegina* sp., *B. melo, Miogypsinids, Ammonia,* and *Quinqueloculina*) and certain ostracods (*Neonesidea* cf., *N. mediterranea, Chrysocythere cataphracta/muricata, Aurila* sp., *Cytherelliodea* sp., *Cytherella* cf., *C. vulgate, Neomonoceratina* cf. *N. laskervi, Cytheretta* sp., *Pokornyella deformis minor*, and *Loxoconcha gr. ex. Ovulate*). The most recent age estimate of the foraminifera *B. melo* is Langhian through middle Messinian (de Graciansky et al., 1998), whereas the most recent age estimate of the ostracod *Chrysocythere cataphracta* is Tortonian to Messinian (Guernet, 2005) and the most recent age estimate of the ostracod *Aurila albicans* is Serravallian to Messinian (Guernet, 2005). El-Shawaihdi et al. (2014) estimated the age of Formation "M" to be 9.19 Ma (± 0.4 Ma) on the basis of strontium isotope ratios and calcareous nannofossils (*Reticulofenestra pseudoumbilicus, R. haqii, Discoaster loeblichii, D. neorectus, D. bellus, D. berggrenii, Helicosphaera carteri, H. stalis, H. selli, Coccolithus pelagicus, C. miopelagicus, Calcidiscus macintyrei, C. tropicus,* and *Sphenolithus abies*). This assemblage indicates nannofossil biozone NN10b-NN11a of Martini (1971). The biozone NN10b-NN11a has an age range from 9.40 to 7.20 Ma, according to de Graciansky et al. (1998). Both the strontium isotope age estimate of 9.19 Ma and nannofossil biozone age estimate indicate a Tortonian age, according to the geologic time scale of Gradstein et al. (2012).

On the basis of lithologic similarities, this limestone to sandy limestone unit is correlated with the lower An Naggazah Member of El Khums Formation as designated by Salem and Spreng (1978) that was later given a new name as the Wadi Yunis Member of El Khums Formation by Innocenti and Pertusati (1984) and Kodym (1985). El Khums Formation is exposed in the vicinity of Al Aqaylah area and extends to Tripoli (Fig. 1). This limestone to sandy limestone unit can also be correlated with Al Jaghbub Formation. This formation is exposed near Al Jaghbub town and in small outcrops near the city of Darnah, and along the road to Tubroq city (Fig. 1).

In this section, there is an intent to modify the informal names that have been given to this unit. This modification will eliminate the multi-contradictory names and unify the nomenclature usage of this unit. The rules of stratigraphic nomenclature indicate the name that was first published should be retained, and the other names should be discarded. On the basis of regional correlations of lithology, this unit in As-Sahabi area is correlated with a limestone at the city of Benghazi that is named the Benghazi Formation (Desio, 1935; Klen, 1974). In As-Sahabi area, it is herein recommended to retain the name as Benghazi with the rank as a Formation on the basis of regional correlations of lithology. All previous works by de Heinzelin and El-Arnauti (1982, 1983, 1987) have assigned the same locality P53 as the stratotype of their Formation "M," and Giglia (1984) has assigned the type locality for Sabkhat Al Hamra Member near Qarat Al Ush. It is highly recommended to keep the type locality of this unit at the city of Benghazi as specified by Klen (1974).

5.2 Unit (2) gypsiferous sand and clay

In As-Sahabi area, this unit is approximately 20 m thick and is composed predominantly of white to yellow sand, dark green clay, and gypsum (Fig. 8). The sand is fine to very fine sand composed of angular to subangular quartz grains. A few feldspar grains of fine to very fine sand size are also present, and these grains are partially/totally dissolved. This unit is characterized by a large network of fractures (~5 m deep) filled with selenitic gypsum at the bottom of the unit. Another network of fractures (~3.5 m deep) filled with selenitic gypsum is present at the top of this unit (Fig. 8). The unit contains no other obvious sedimentary structures. The upper contact of this unit is placed at the top of a bed of sand and clay with gypsum, above which lies a bed of non-gypsiferous white sand with clay balls and lenses that is described later. A particularly good exposure in As-Sahabi area is located at Locality P51 (Fig. 2).

The fossils in this unit are restricted to a few mollusk and echinoids fragments. These fossils are either partially or completely transformed into gypsum.

In previous studies, Giglia (1984) did not designate these sediments as a separate unit but rather included them as the upper part of the Sabkhat Al Hamra Member of the Sahabi Formation (Fig. 7). de Heinzelin and El-Arnauti (1982, 1983, 1987), however, designated these sediments as a separate unit (Fig. 7), which they named Formation "P." El-Shawaihdi et al. (2016) included this gypsiferous sand and clay unit within the lower part of his newly named "lower member" of the Sahabi Formation.

This gypsiferous sand and clay unit does not contain age-diagnostic fossils. The age can be estimated, however, on the basis of the presence of gypsum, and the correlation of this gypsum with the Messinian Salinity Event. Recent work by Krijgsman et al. (1999) dated this event to approximately 5.96 5.32 Ma. The age of this unit can also be estimated on the basis of correlation of this unit with portions of the Sabkhat Al Hamra and Sabkhat Al Qunnayyin Members (Giglia, 1984) and with Formation "P" (de Heinzelin and El-Arnauti, 1982, 1983). Giglia (1984) postulated that the Sabkhat Al Hamra and Sabkhat Al Qunnayyin Members are of Late Miocene age, specifically Tortonian to early Messinian, on the basis of assemblages of foraminifera (*Globorotalia humerosa*, *Ammonia tepida*, *Ammonia* cf. *inflata*, *Quinqueloculina oblonga*, *Elphidium aculeatum*, *E. decipiens*, and *Elphidium* cf. *sagrum*) and ostracods (*Aurila diecii*, *A. albicans*, *Cytherella punctatella*, *C. vandenboldi*, *Carinocythereis galilea*, *Chrysocythere cataphracta*, *Cyamocytheridea dertonesis*, *Loxoconcha variesculpta*, *Neomonoceratina mouliana*, *Olimfalunia sicula*, *O. zibinica*, *Pokornyella italica*, and *Tenedocythereis mediterranea*). The most recent age estimate of the ostracods *Chrysocythere cataphracta*, *Pokornyella italica,* and *Aurila diecii* is Tortonian to Messinian (Guernet, 2005), the most recent age estimate of the ostracod *A. albicans* is Serravallian to Messinian (Guernet, 2005), and the most recent age estimate of the ostracod *Olimfalunia sicula* is Messinian (Guernet, 2005). de Heinzelin et al. (1980) and de Heinzelin and El-Arnauti (1982, 1983) postulated that their Formation "P" is of Messinian age on the basis of the presence of gypsum. Their Formation "P", however, contains no age-diagnostic fossils (El-Shawaihdi, 2015; El-Shawaihdi et al., 2016).

On the basis of similar lithology, this gypsiferous sand and clay unit is considered to be the lateral equivalent of a unit of cross-bedded pelmicrite, oomicrite, and oosparite, alternating with restricted deposits of siltstone, shale, clay, gypsiferous clay, and gypsum that was mapped as the Wadi Al Qattarah Member of the Al Rajmah Formation as designated by Klen (1974) or Wadi Al Qattarah Formation as designated by El-Hawat and Abdulsamad (2004). In turn, the same units are correlated with the Sabkhat Al Qunnayyin Member as designated by Giglia (1984) and with the Wadi Al Farigh Member as designated by Giglia (1984). On the basis of lithologic similarities, this gypsiferous sand and clay unit can be correlated with the An Naggazah Member of El Khums Formation as designated by Salem and Spreng (1978). This An Naggazah Member, however, was recently updated and given a new name as the Wadi Yunis Member of El Khums Formation by Innocenti and Pertusati (1984) and Kodym (1985). Similarly, this unit is correlated with Quwayrat Al Jibs Member of El Khums Formation as designated by Innocenti and Pertusati (1984) and Kodym (1985).

In this section, there is an intent to modify the informal names that have been given to this unit. This modification will eliminate the multi-contradictory names and will unify the nomenclature usage of this unit. The lower gypsiferous sand and clay part of this unit collectively with the gypsiferous sand and clay of Formation "P" of de Heinzelin and El-Arnauti (1982, 1983) was named as the "lower member" of the Sahabi Formation by El-Shawaihdi et al. (2016). It is highly recommended to name these gypsiferous sand and clay units of the "lower member" (collectively Formation "P" and lower part of Member "T") as the Sabkhat Al Qunnayyin with a rank as a Member belonging to the Sahabi Formation on the basis of regional correlations. Previous works by de Heinzelin and El-Arnauti (1982, 1983) have assigned Locality P51 (Fig. 2) on the western

Fig. 5 General field view of Locality P28 showing the upper studied strata in the As-Sahabi area.

side of Sabkhat Al Qunnayyin as the type section for Formation "P" and Locality 116 on the western side of Sabkhat Al Qunnayyin (Fig. 2) as the type locality for Member "T". The authors strongly recommend keeping the type section of this newly named Sabkhat Al Qunnayyin Member as Locality P51 (Figs. 2, 3, 5), which is located at Longitude 30° 16′ 48″ E and Latitude 20° 53′ 37″ N as designated by de Heinzelin and El-Arnauti (1982, 1983, 1987).

5.3 Unit (3) non-gypsiferous sand/sandstone and clay

In As-Sahabi area, this unit is 10–40 m thick and is composed of white sand, calcareous sandstone, green-brown non-gypsiferous clay, silty clay, and sandy clay (Fig. 8). The sand and sandstone beds consist of medium to fine sand composed of rounded quartz grains. The sand and sandstone display cross-bedding with dip angles ranging from 25 to 30° toward the south-southwest. The lower part of this unit is characterized by fractures (~1 m deep) that are filled with selenitic gypsum and this lower part is now included with Formation "P" and both make the gypsiferous "lower member" of the Sahabi Formation. The lower contact is difficult to recognize due to the lithologic similarities with the underlying gypsiferous sand and clay unit. The upper contact is placed immediately below the first appearance of a bed of medium sand with clay balls, which is described as part of the overlying non-gypsiferous sand and clay unit. A particularly good exposure in As-Sahabi area is located Locality P116 (Fig. 2).

Fossils in this unit are sparse and limited to pelecypods (predominantly *Gryphaea*) and vertebrate remains. According to Boaz et al. (1982, 1987), vertebrate fossils mostly belong to Sirenia, Anthracotheriidae, Bovidae, and Carnivore (*Ursidae, Viverridae, hyaenidae, Felidae,* and *Phocidae*; Howell, 1982, 1987). This unit is extensively bioturbated with *Ophiomorpha* burrows (Table 2).

In previous studies, Giglia (1984) designated these strata as a separate unit, naming it the Sabkhat Al Qunnayyin Member of the Sahabi Formation (Fig. 7). de Heinzelin et al. (1980) and de Heinzelin and El-Arnauti (1982, 1983, 1987) also designated these strata as a separate unit, which they named Member "T" of the Sahabi Formation (Fig. 7). El-Shawaihdi et al. (2016) separated the upper non-gypsiferous sand and clay part of Member "T" and included it with the newly designated "upper member" of the Sahabi Formation.

The age of this non-gypsiferous sand/sandstone and clay unit is estimated to be Late Miocene to Early Pliocene on the basis of correlation with Member "T" of de Heinzelin and El-Arnauti (1987). de Heinzelin and El-Arnauti (1987) estimated that their Member "T" is of Early Pliocene age because they correlated their underlying Formation "P" with the Messinian Salinity event. Domning (1982) also estimated the age of Member "T" as post-Messinian on the basis of the presence of Sirenia such as *Metaxytherium medium, M. serresii,* and *M. forestii* and their occurrence overlying the evaporitic Formation "P."

Table 2 Marine and terrestrial mammal fossils found in the non gypsiferous sand and clay unit as exposed in the As Sahabi area.

Vertebrates	Species	Researcher(s)
Sirenia	*Metaxytherium serresii*;	Domning (1982) and Domning and Thomas (1987)
Anthracotheriidae	*Merycopotamus petrocchii, Libycosaurus petrocchii*;	Bonarelli (1947), Gaziry (1982a,b, 1987a,b,c), and Pavlakis and Boaz (2008)
Bovidae	*Gazella* sp., *Prostrepsiceros libycus, Raphicerus* sp., *Hippotragus* sp., *Miotragocerus cyrenaicus*;	Lehmann (1982)

On the basis of lithologic similarities, this non-gypsiferous sand and clay unit is correlative with the An Naggazah Member of El Khums Formation as designated by Salem and Spreng (1978). Also, this non-gypsiferous sand and clay unit may be correlative with Quwayrat Al Jibs Member of El Khums Formation as designated by Innocenti and Pertusati (1984) and Kodym (1985).

In this section, there is an intent to modify the informal names that have been given to this unit. It is highly recommended to include this non-gypsiferous sand and clay unit of the "upper member" collectively with Members "U1," "UD," and "U2" and named it Qaret Makada with a rank as a Member belonging to the Sahabi Formation. The authors strongly recommend assigning Locality P28 (Figs. 2, 3, 5), which is located at Longitude 30° 13′ 55.11″ E and Latitude 20° 51′ 31.98″ N, as the type section of this newly named member at As-Sahabi area. Locality P28 is located at Qaret Makada, which is an area of small hills located at a geographic place known as Dur Makada along the northern part of As-Sahabi area (Figs. 2, 3, 5).

5.4 Unit (4) non-gypsiferous sand and clay

In As-Sahabi area, this unit is ~20 m thick and is composed of white to yellow sand with small balls and lenses of clay (Fig. 8). The sand is medium to fine sand composed of quartz grains. The lower part is transitional from white sand with clay to yellow-orange sandstone cemented by hematite in upper part of the unit. The sand and sandstone display cross-bedding with dip angles ranging from 30 to 35° toward the southwest. Large concretions (2–3 m long) composed of sandstone are present within the sand and sandstone beds. The upper contact of this unit is placed at the top of a bed of yellow sand with dark green clay. This bed of yellow sand is overlain by a bed of white, brown, and yellow sand, which is described as part of the overlying unit. A particularly good exposure in As-Sahabi area is located at Locality P16 (Fig. 2) as designated by de Heinzelin and El-Arnauti (1982, 1983, 1987).

Fossils in this unit include silicified wood fragments, as well as many well-preserved vertebrate remains (Boaz et al., 1982, 1987). Vertebrate fossils from this unit mostly belong to Bovidae, Anthracotheriidae several carnivore taxa of *Ursidae, Viverridae, Hyaenidae, Felidae*, and *Phocidae,* Proboscidea, Equidae, Suidae, and Hippopotamidae (Table 3).

Table 3 Marine and terrestrial mammal fossils found in the non gypsiferous sand and clay of Unit (4) as exposed in the As Sahabi area.

Vertebrates	Species	Researcher(s)
Bovidae	*Dyticodorcas libycus*, *Kobus subdolos*, *Miotragocerus cyrenaicus*, *Bovini Gen. indet. *"Leptobos" syrticus*	Gentry (2008)
Anthracotheriidae	*Libycosaurus petrocchii*	Pavlakis and Boaz (2008)
Carnivore	*Agriotherium* cf. *africanum*, *Vivera howelli*, *Dinocrocuta senyureki*, *Adcrocuta eximia*, *Hyaenictitheriun namaquensis*, *Amphimachairodus* aff. *kabir*	Rook and Sardella (2008) and Werdelin and Peigné (2010)
Proboscidea	*Stegotetrabelodon syrticus*, *Amebelodon cyrenaicus*, **Anancus petrocchii*	Sanders (2008)
Equidae	*Eurygnathohippus feibeli*, *Sivalhippus* sp. *Cremohipparion* aff. *matthewi*	Bernor et al. (2012, 2020, 2008)
Suidae	**Nyanzachoerus kanamensis*, *Nyanzachoerus syrticus*, *Nyanzachoerus* cf. *devauxi*	Gallai et al. (2008)
Sirenia	+*Metaxytherium serresii*	Domning (1982), Domning and Thomas (1987), and Bianucci et al. (2008)
Hippopotamidae	*Hexaprotodon sahabiensis*	Pavlakis (2008)
Rodentia	*Atlantoxerus getulus*, *Abudhabia yardangi*, *Progonomys* cf. *mauritanicus*	Agustì (2008)
Cetacea	*Lagenorhynchus* sp., Iniidae indet. +Balaenopteridae sp. indet.	Whitmore (1987)

All taxa listed are not necessarily represented at this locality but are present at contiguous localities in the same stratigraphic unit. *Note*: The plus sign indicates that this unit as exposed at Locality P28 contain only these mammalian taxa. Asterisks indicate that taxa are limited to described at Localities P28 and P10.

In previous studies, Giglia (1984) did not designate these sediments as a separate unit but included them within the Qarat Weddah Formation (Garet Uedda Formation of Di Cesare et al., 1963) (Fig. 7). de Heinzelin et al. (1980) and de Heinzelin and El-Arnauti (1982, 1983, 1987), however, designated these sediments as a separate unit and named it Member "U1" of the Sahabi Formation (Fig. 7). On the basis of similar lithology, El-Shawaihdi et al. (2016) included this non-gypsiferous sand and clay unit as the lower unit of "upper member" of the Sahabi Formation.

On the basis of Suidae fossils in this unit, Cooke (1982) estimated the age to be close to the Miocene/Pliocene boundary. Monthe (1987) subsequently identified certain rodent fossils (mostly gerbils and squirrels) in this unit, and tentatively estimated these fossils to be of Late Miocene to Late Pliocene age. Agustì et al. (2000) and Agustì (2008) later determined that the rodent fossils described by Monthe (1987) predominantly belong to *Abudhabia yardangi*, which suggests a Late Miocene (pre Messinian) age.

This unit might be correlative with the lower part of Qarat Weddah Formation (Garet Uedda Formation of Di Cesare et al., 1963). It is also could be correlated with the lower part of Sabkhat Al Qunnayyin Member of Giglia (1984).

In this section, there is an intent to modify this informal names that have been given to this unit. As stated in the previous section, El-Shawaihdi et al. (2016) included this non-gypsiferous sand and clay unit within the "upper member" of the Sahabi Formation. It is highly recommended to name this "upper member" as Qaret Makada with a rank as a Member belonging to the Sahabi Formation. The authors strongly recommend assigning Locality P28 (Figs. 2, 3, 5), which is located at Longitude 30° 13′ 55.11″ E and Latitude 20° 51′ 31.98″ N, as the type section of this newly named member at As-Sahabi area. Locality P28 is located at Qaret Makade, which is an area of small hills located at a geographic place known as Dur Makada along in the northern part of As-Sahabi area (Figs. 2, 3, 5).

5.5 Unit (5) non-gypsiferous sand and clay (sandstone-cemented by dolomite)

In As-Sahabi area, this unit is ~3 m thick and composed of yellow-brown sandstone cemented by dolomite (Fig. 8). The sandstone consists of fine to very fine sand composed of angular to sub-angular quartz grains. The sandstone displays low-angle cross-bedding with dip angles of less than 5° toward the northeast. The upper contact of this unit is placed at the top of a bed of yellow-brown cross-bedded dolomite-cemented sandstone, above which lies a bed of highly bioturbated dolomite-cemented sandstone that is considered to be the lowest bed of this unit (described below). This unit is considered as a marker bed in limited areas in the northern part of As-Sahabi area. A particularly good exposure in As-Sahabi area is located at Locality P14 (Fig. 2) as designated de Heinzelin et al. (1980) and de Heinzelin and El-Arnauti, 1982, 1983, 1987.

No vertebrate remains have been found in this unit. However, the bed of dolomite-cemented sandstone contains a few benthic foraminifera, such as *Ammonia beccarii*, *Amphistegina* sp., and *Elphidium* spp.

In previous studies, Giglia (1984) did not designate the strata of this unit as a separate unit, but included them within the Qarat Weddah Formation (Garet Uedda Formation of Di Cesare et al., 1963) (Fig. 7). de Heinzelin et al. (1980) and de Heinzelin and El-Arnauti (1982, 1983, 1987), however, designated these strata as a separate unit, which they named Member "UD" of the Sahabi Formation (Fig. 7). El-Shawaihdi et al. (2016) included this this unit/Member "UD" within their newly named "upper member" of the Sahabi Formation.

The age of this unit may be estimated by the presence of the benthic foraminifera *Ammonia beccarii*, *Amphistegina* sp., and *Elphidium* spp. The most recent age estimate of *A. beccarii* and *Elphidium* spp. is Late Pliocene to Early Pleistocene (Jenkins and Murray, 1989; Pèrez-Asensio and Aguirre, 2010). As stated above, Unit 4 at Locality P10 is correlated with Member "UD" of de Heinzelin and El-Arnauti (1987). de Heinzelin and El-Arnauti (1987) estimated Member "UD" to be of Early Pliocene age on the basis of stratigraphic position. Muftah et al. (2008a) also estimated an Early Pliocene age for Member "UD" on the basis of the presence of the foraminifera *Ammonia beccarii*, *Amphistegina* sp., and *Elphidium* spp., and the stratigraphic position.

In this section, there is an intent to modify this informal name that has been given to this unit. As stated above, El-Shawaihdi et al. (2016) included this sandstone cemented by dolomite with the non-gypsiferous sand and clay unit that they named as the "upper member" of the Sahabi Formation. It is highly recommended to name this "upper member" as Qaret Makada with a rank as a Member belong to the Sahabi Formation. The authors strongly recommend assigning Locality P28 (Figs. 2, 3, 5), which is located at Longitude 30° 13′ 55.11″ E and Latitude 20° 51′ 31.98″ N, as the type section of this newly named member at As-Sahabi area. Locality P28 is located at Qaret Makade, which is an area of small hills located at a geographic place known as Dur Makada along in the northern part of As-Sahabi area (Figs. 2, 3, 5).

5.6 Unit (6) non-gypsiferous sand and clay that is capped by an unconformity

In As-Sahabi area, this unit is 4–10 m thick and is composed of beds of brown to pale orange sandstone cemented by dolomite, yellow-brown to white sandy clay, dark green clay, and dark green shale (Fig. 8). The sand in this unit consists of medium to fine sand composed predominantly of subangular quartz grains. Both the lowermost and uppermost beds of this unit are composed of dolomite-cemented sandstone. The bed of dolomite-cemented sandstone at the base of the unit contains mollusk shells, is greatly bioturbated, and displays cross-bedding with dip angles up to 25° toward the northeast. The dolomite-cemented sandstone beds contain fragments of ostracods and gastropods that are greatly dolomitized, whereas the clay and shale beds contain fossils of mostly vertebrates. The upper contact of this unit is an unconformity at the top of a bed of dolomite-cemented sandstone that contains many dolomitized gastropods. This unconformity is overlain by a bed of white sand with dolomite and gypsum crystals that is described below. A particularly good exposure in As-Sahabi area is located at Locality P71 (Fig. 2) as designated by de Heinzelin et al. (1980) and de Heinzelin and El-Arnauti (1982, 1983, 1987).

According to Boaz et al. (1982, 1987), vertebrate fossils from this unit mostly belong to Bovidae, Anthracotheriidae, several carnivore taxa of *Ursidae, Viverridae, Hyaenidae, Felidae,* and *Phocidae,* Proboscidea, Equidae, Suidae, and Hippopotamidae (Table 4).

In previous studies, Giglia (1984) did not designate these sediments as a separate unit, but rather included them with the Qarat Weddah Formation (Garet Uedda Formation) (Fig. 7). de Heinzelin et al. (1980) and de Heinzelin and El Arnauti (1982, 1983, 1987), however, designated these sediments as a separate unit, which they named Member "U2" of the Sahabi Formation (Fig. 7).

The age of this unit may be estimated by the presence of different vertebrate fossils such as Bovidae (*Gazella* sp., *Prostrepsiceros libycus*, *Raphicerus* sp., *Hippotragus* sp., and *Miotragocerus cyrenaicus)*, Anthracotheriidae (*Merycopotamus petrocchii* and *Libycosaurus petrocchii)*, Proboscidea (*Stegotetrabelodon lybicus*, *Amebelodon* sp., and *Stegotetrabelodon sahabianus*), Equidae (*Hipparion* sp. cf.

Table 4 Marine and terrestrial mammal fossils found in the non gypsiferous sand and clay of Unit (6) as exposed in the As Sahabi area

Vertebrates	Species	Researcher(s)
Bovidae	*Gazella* sp., *Prostrepsiceros libycus*, *Raphicerus* sp., *Hippotragus* sp., *Miotragocerus cyrenaicus*	Lehmann (1982)
Anthracotheriidae	*Merycopotamus petrocchii*, *Libycosaurus petrocchii*	Bonarelli (1947), Gaziry (1982a,b, 1987a,b,c), and Pavlakis and Boaz, 2008
Proboscidea	*Stegotetrabelodon lybicus*, *Amebelodon* sp., *Stegotetrabelodon sahabianus*	Gaziry (1982a,b, 1987a,b,c)
Equidae	*Hipparion* sp. cf. *sitifense*, *Hipparion* sp. cf. *africanum*, *Cremohipparion* aff. *matthewi*, *Hipparion* sp.	Tobien (1982), Bernor et al. (1987), and Bernor and Scott (2003)
Suidae	*Nyanzachoerus kanamensis*, *Nyanzachoerus* cf. *syrticus*, *Nyanzachoerus* cf. *devauxi*	Cooke (1982, 1987)
Hippopotamidae	*Hexaprotodon sahabiensis*	Gaziry (1982a,b, 1987a, b,c) and Pavlakis, 2008

sitifense, *Hipparion* sp. cf. *africanum*, *Cremohipparion* aff. *matthewi*, and *Hipparion* sp.), Suidae (*Nyanzachoerus kanamensis*, *Nyanzachoerus* cf. *syrticus*, and *Nyanzachoerus* cf. *devauxi*), and Hippopotamidae (*Hexaprotodon sahabiensis*). According to (Boaz et al., 2008, 1982, 1987, and references therein), these vertebrate fossils suggest an age of Late Miocene to Early Pliocene. Member "U2" is estimated to be of Early Pliocene age on the basis of stratigraphic position (de Heinzelin and El-Arnauti (1982, 1983, 1987). However, Whitmore (1982, 1987) estimated the age of Member "U2" to be Late Miocene on the basis of the presence of Delphinidae referred to cf. *Lagenorhynchus*. Furthermore, Gentry (2008) estimated Member "U2" to be of Late Miocene to earliest Pliocene age on the basis of the presence of Bovidae (*Hippotragus* sp. and *Miotragocerus cyrenaicus*).

This unit may be correlative with the lower part of the Qarat Weddah Formation (Garet Uedda Formation of Di Cesare et al., 1963). It is also could be correlated with the middle part of Sabkhat Al Qunnayyin Member of Giglia (1984).

In this section, there is an intent to modify the informal name that has been given to this unit. As stated above, El-Shawaihdi et al. (2016) included this non-gypsiferous sand and clay unit within the "upper member" of the Sahabi Formation. It is highly recommended to name this "upper member" as Qaret Makada with a rank as a Member belonging to the Sahabi Formation. The authors strongly recommend assigning Locality P28 (Figs. 2, 3, 5), which is located at Longitude 30° 13′ 55.11″ E and Latitude 20° 51′ 31.98″ N, as the type section of this newly named member at As-Sahabi area. Locality P28 is located at Qaret Makade, which is an area of small hills located at a geographic place known as Dur Makada along the western part of the recent Sabkhat Al Qunnayyin within As-Sahabi area (Figs. 2, 3, 5).

5.7 Unit (7) sand and pebbly sand that is capped by an unconformity

In As-Sahabi area, this unit is 10–40 m thick, and is composed of white and green sand, pebbly sand to conglomeratic sandstone, sandy clay, and clay balls, with lenses of gray dolomite and gypsum crystals. The sand in this unit consists of coarse to medium sand composed predominantly of rounded to subrounded quartz grains. The sand also contains a few pink to light brown feldspar grains that are up to 2 mm in diameter. A 10–15 cm thick bed of pebbles (up to 4 cm in diameter) and sand is present approximately 10 m above the base of the unit. A 2 m thick bed of pebbles (up to 6 cm in diameter) and sand is present at the top of the unit, approximately 38 m above the base of the unit. The sand, pebbly sand, and conglomeratic sandstone display cross-bedding with dip angles up to 30° toward the southwest. The lower 10 m of this unit contain abundant vertebrate fossils. The upper contact of this unit is an unconformity, above which lies a unit weathered carbonate soil/calcrete that is described below. A particularly good exposure in As-Sahabi area is located at Locality P13 (Fig. 2) as designated by de Heinzelin et al. (1980) and de Heinzelin and El-Arnauti (1982, 1983, 1987).

According to Boaz et al. (1982, 1987), vertebrate fossils from this unit mostly belong to Hippopotamidae, Anthracotheriidae, Bovidae, Proboscidea, Equidae, and Suidae (Table 5). Silicified wood fragments are present near the base of this unit and in some beds located approximately 15–20 m above the base of the unit. No microfossils have been identified in this unit.

Table 5 Marine and terrestrial mammal fossils found in the non gypsiferous sand and clay of Unit (7) as exposed in the As Sahabi area.

Vertebrates	Species	Researcher(s)
Bovidae	*Gazella* sp., *Prostrepsiceros libycus*, *Raphicerus* sp., *Hippotragus* sp., *Miotragocerus cyrenaicus*	Lehmann (1982)
Anthracotheriidae	*Merycopotamus petrocchii*, *Libycosaurus petrocchii*	Bonarelli (1947), Gaziry (1982a,b, 1987a,b,c), and Pavlakis and Boaz (2008)
Proboscidea	*Stegotetrabelodon lybicus*, *Amebelodon* sp., *Stegotetrabelodon sahabianus*	Gaziry (1982a,b, 1987a,b,c)
Equidae	*Hipparion* sp. cf. *sitifense*, *Hipparion* sp. cf. *africanum*, *Cremohipparion* aff. *matthewi*, *Hipparion* sp.	Tobien (1982), Bernor et al. (1987), and Bernor and Scott (2003)
Suidae	*Nyanzachoerus kanamensis*, *Nyanzachoerus* cf. *syrticus*, *Nyanzachoerus* cf. *devauxi*	Cooke (1982, 1987)
Hippopotamidae	*Hexaprotodon sahabiensis*	Gaziry (1982a,b, 1987a,b,c) and Pavlakis (2008)

In previous studies, Giglia (1984) did not designate these sediments as a separate unit but included them with the Qarat Weddah Formation (Garet Uedda Formation) (Fig. 7). de Heinzelin et al. (1980) and de Heinzelin and El-Arnauti (1982, 1983, 1987), however, designated these sediments as a separate unit, which they named Member "V" of the Sahabi Formation (Fig. 7).

The age of this unit may be estimated by the presence of similar vertebrate fossils that are similar to those found in the underlying unit of non-gypsiferous sand and clay (Hippopotamidae, Bovidae, Anthracotheriidae, Equidae, Suidae, and Proboscidea). According to (Boaz et al., 2008, 1982, 1987 and references therein), these vertebrate fossils suggest an age of Late Miocene to Early Pliocene. As stated above, this unit is correlated with Member "V" of de Heinzelin and El-Arnauti (1982, 1983, 1987). de Heinzelin and El-Arnauti (1982, 1983, 1987) estimated the age of Member "V" to be of Early Pliocene on the basis of stratigraphic position. Gentry (2008) estimated the age of Member "V" to be Late Miocene to earliest Pliocene on the basis of the presence of Bovidae (*Hippotragus* sp. and *Miotragocerus cyrenaicus*).

This unit may be correlative with the upper part of Qarat Weddah Formation (Garet Uedda Formation of Di Cesare et al., 1963). It is also could be correlated with the upper part of Sabkhat Al Qunnayyin Member of Giglia (1984).

In this section, there is an intent to modify the informal name that has been given to this unit. El-Shawaihdi et al. (2016) correlated this sand and pebbly sand unit with the Qarat Weddah Formation of Giglia (1984) or Garet Uedda Formation of Di Cesare et al. (1963). The authors strongly recommend keeping the name Qarat Weddah Formation on the basis of the strong lithologic similarities between both units. This designation is recommended on the basis of the change in lithology from sand and clay to coarse sand to pebbly sand that mark the presence of an unconformity separating the overlying Qarat Weddah Formation from the underlying newly named Qaret Makada Member of the Sahabi Formation. The authors strongly recommend assigning locality P10 (Figs, 2, 3, 6), which is located at Longitude 30° 11′ 19.00″ E and Latitude 20° 48′ 34.00″ N, as the type section of this formation at As-Sahabi area. An alternative reference section is locality P70 (Fig. 2), which is located at Longitude 30° 09′ 59.00″ E and Latitude 20° 48′ 37.00″ N, because this locality also contains a thick section of this formation.

5.8 Unit (8) carbonate soil/calcrete

In As-Sahabi area, this unit is 1.5 m thick, and is composed of yellow to light brown carbonate. The unit contains several concretions (3–4 cm diameter). The unit contains no obvious sedimentary structures. A particularly good exposure in As-Sahabi area is located at Locality P130 (Fig. 2) as designated by de Heinzelin and El Arnauti (1982, 1983, 1987). This unit contain some shells of terrestrial gastropods.

In previous studies, Giglia (1984) did not designate these sediments as a separate unit, but rather described them as a calcareous paleosol (caliche or calcrete) covering the top of the Qarat Weddah Formation (Garet Uedda Formation) (Fig. 7). de Heinzelin et al. (1980) and de Heinzelin and El-Arnauti (1982, 1983, 1987) designated these sediments as a separate unit, which they named Member "Z" of the Sahabi Formation (Fig. 7). Muftah (2013) separated this member from the Sahabi Formation and raised it to formation rank using the name Formation "Z." This designation of Formation "Z" was done on the basis of the recognition of an unconformity between the unit of carbonate soil/calcrete (Formation "Z") and the underlying unit sand and pebbly sand (named Member "V").

This unit does not contain age-diagnostic fossils. However, the age of this unit is estimated to be Late Pliocene to Pleistocene on the basis of lithostratigraphic correlation with the top part of the Qarat Weddah Formation (Garet Uedda Formation of Di Cesare et al., 1963).

This unit may be correlative with the upper part of Qarat Weddah Formation (Garet Uedda Formation of Di Cesare et al., 1963). It is also could be correlated with the Qarat Weddah Formation of Giglia (1984).

In this section, there is an intent to modify the informal name that has been given to this unit. The authors strongly suggest naming this unit as Shagig and support the formation rank given by Muftah (2013). The Shagig name comes from the big mountain named as Jabail Shagig (Locality P10; Figs. 2, 3, 6) at the northern part of As-Sahabi area. This designation of Shagig Formation is also suggested on the basis of the presence of an unconformity between the unit of carbonate soil/calcrete and the underlying unit of sand and pebbly sand that is mapped as the Qarat Weddah Formation (Garet Uedda Formation of Di Cesare et al., 1963).

Fig. 6 General field view of Locality P10 showing most of the studied units in the As-Sahabi area.

The authors strongly recommend assigning locality P10 (Figs. 2, 3, 6), which is located at Longitude 30° 11′ 19.00″ E and Latitude 20° 48′ 34.00″ N, as the type section of this formation at As-Sahabi area. An alternative reference section is locality P70 (Fig. 2), which is located at Longitude 30° 09′ 59.00″ E and Latitude 20° 48′ 37.00″ N, because this locality also contains a thick section of this formation.

6. Discussion

The Miocene carbonate unit is equivalent to Formation "M" of de Heinzelin and El-Arnauti (1982, 1983, 1987), the Regima Limestone of Desio (1935), the Sabkhat Al Hamra Member of the Sahabi Formation of Giglia (1984), the Wadi Al Farigh Member of the Sahabi Formation of Giglia (1984), and the Benghazi Member of Ar Rajmah Formation of Klen (1974). This unit of carbonate strata is capped by a local unconformity. The rules of stratigraphic nomenclature indicate that the name that was first published (Regima Limestone) should be retained, and the other stratigraphic names should be discarded. In this case, however, the name Regima Limestone was formally changed to Ar Rajmah Formation by Klen (1974) and later to Ar Rajmah Group by El-Hawat and Abdulsamad (2004). Therefore, the name Ar Rajmah Group should be used (Fig. 9).

The Miocene gypsiferous sand and mud unit is equivalent to Formation "P" of de Heinzelin and El-Arnauti (1982, 1983, 1987), and the Sabkhat Al Qunnayyin Member of the Sahabi Formation of Giglia (1984). An unconformity separates these strata from the underlying carbonate strata (Ar Rajmah Group). The name for these strata that was first published in accordance to the rules of stratigraphic nomenclature is the Sabkhat Al Qunnayyin Member of the Sahabi Formation (Giglia, 1984), and this is the name that should be used. This gypsiferous sand and mud should be considered as the lower member of the Sahabi Formation. The name Sahabi Formation should be restricted to the post-carbonate strata (i.e., above the Ar Rajmah Group) that are below the regional unconformity at the base of the Lower Pliocene sand and pebbly sand mapped as the Qarat Weddah Formation (Fig. 9).

The Upper Miocene units of non-gypsiferous sand and clay are equivalent to upper part of Member "T" and Members "U1," "UD," and "U2" of de Heinzelin and El-Arnauti (1982, 1983, 1987) These units that are capped by a regional unconformity (Fig. 9). These units should be referred to collectively as the Qaret Makada Member of the Sahabi Formation. This designation of Qaret Makada Member should be restricted to the strata that lie above the gypsiferous sand and mud of Sabkhat Al Qunnayyin Member of the Sahabi Formation and below the regional unconformity at the base of Lower Pliocene sand and pebbly sand that is designated as the Qarat Weddah Formation (Fig. 9). The Qaret Makada Member is capped by a major unconformity of late Messinian age that is correlated with the regression and desiccation of the Mediterranean Sea and with Eosahabi Channel cutting during the Messinian Salinity Crisis. As noted above, the name for the Sahabi Formation is retained but with modification and renaming of its informal members and its boundaries. In this regard the name Sahabi Formation should be restricted to the post-carbonate strata (i.e., above the Benghazi Formation of Ar Rajmah Group) that are below the regional unconformity at the base of the Lower Pliocene sand and pebbly sand mapped as the Qarat Weddah Formation of Giglia (1984) (Fig. 9).

The Lower Pliocene sand and pebbly is equivalent to the Member "V" of de Heinzelin and El-Arnauti (1982, 1983, 1987) and to the Garet Uedda Formation of Di Cesare et al. (1963). It appears to be that all of these names have been applied to correlative strata of identical lithology (Fig. 9). Thus, the rules of stratigraphic nomenclature indicate that the name that was first published (Garet Uedda Formation) should be retained, and the other stratigraphic names should be discarded (Figs. 7,9). In this case, however, Carmignani (1984) formally changed the name Garet Uedda Formation to Qarat Weddah Formation on the basis of the new Arabic Transcription (the new transcription of Garet Uedda is Qarat Weddah), and therefore the Qarat Weddah Formation is now the new name of this rock unit and this revised spelling should be used (Fig. 9).

The Pleistocene pedogenically altered carbonate soil/calcrete is equivalent to the Member "Z" of de Heinzelin and El-Arnauti (1982, 1983, 1987) and to the top part of the Qarat Weddah Formation as mapped by Giglia (1984). The authors support the designation by Muftah (2013) that this carbonate soil/calcrete unit should be raised to formation rank and be given the name Shagig. This carbonate soil/calcrete (Shagig Formation) should be considered as the uppermost and youngest rock unit in the As-Sahabi area (Fig. 9).

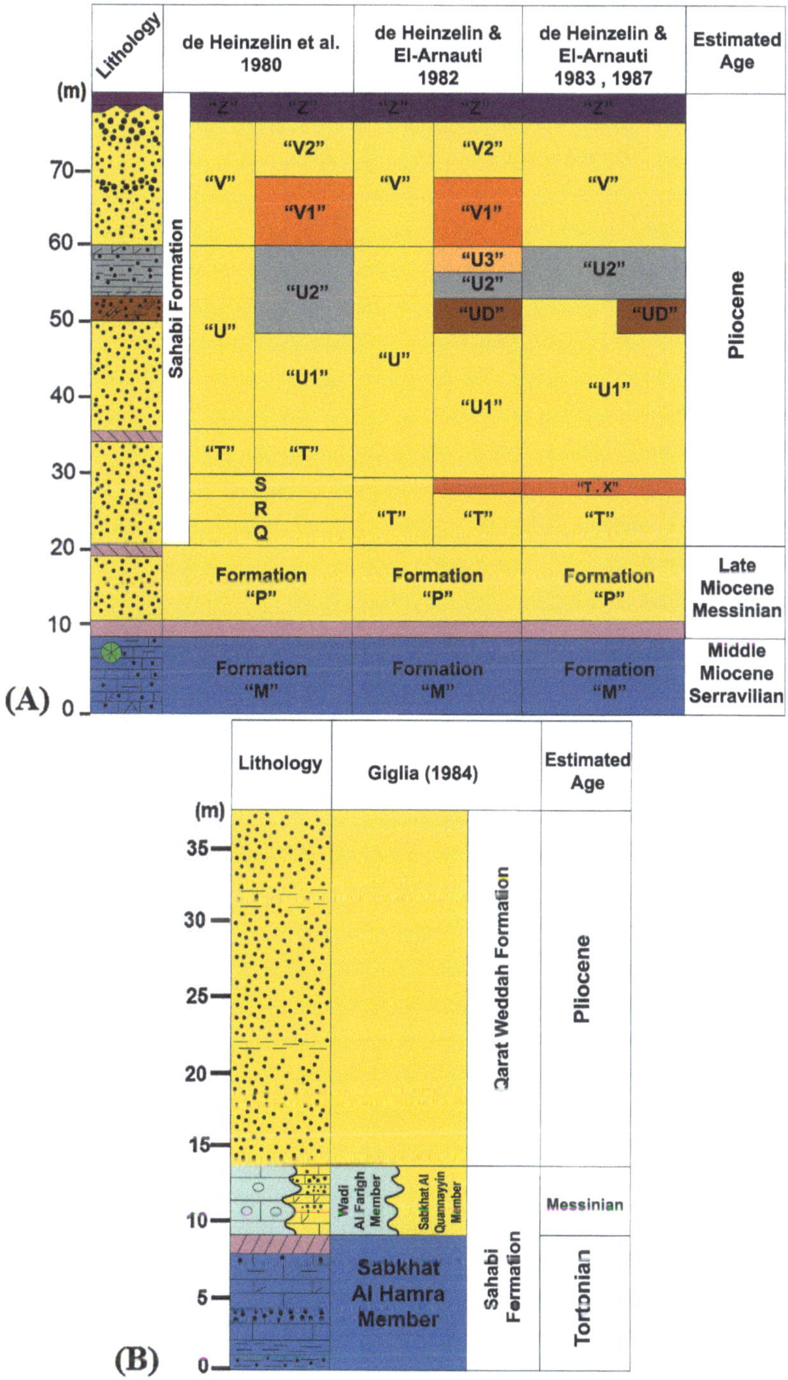

Fig. 7 Previous lithostratigraphic classifications and nomenclatures of Neogene strata in As-Sahabi area of northeast Libya. (A) The lithostratigraphy designated by de Heinzelin et al. (1980) and de Heinzelin and El-Arnauti (1982, 1983, 1987). (B) The lithostratigraphy designated by Giglia (1984). Explanation of colors is shown in Fig. 4.

Fig. 8 Field close-up views showing the dominant lithology of Units 1 to 7 at the As-Sahabi area.

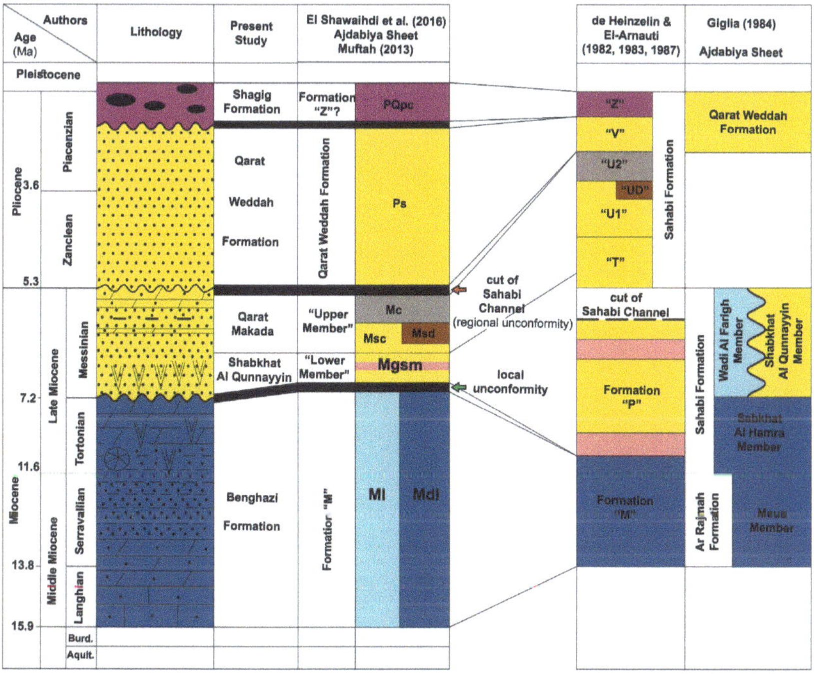

Fig. 9 An updated chronostratigraphic chart summarizing the new nomenclatures and lithologic descriptions for the Neogene stratigraphic units in As-Sahabi area, Libya. Aquit. = Aquitanian; Burd. = Burdigalian. The time scale is from (Gradstein et al., 2012). *Modified from El-Shawaihdi, M.H. 2015. Geochronology, Petrology, and Stratigraphy of Neogene Units in the Sahabi Area, Northeastern Sirt Basin, Libya, New Mexico Institute of Mining and Technology (N. M. Tech.), PhD Dissertation, 152 p., El-Shawaihdi, M.H, Mozley, P.S., Boaz, N.T., Salloum, F., Pavlakis, P., Muftah, A., Triantaphyllou, M. 2016. Stratigraphy of the Neogene Sahabi units in the Sirt Basin, Northeast Libya, J. Afr. Earth Sci., 118: 87–106.)*

7. Conclusions

Recent field work combined with a review of the previous geological literature from the As-Sahabi and Wadi Al Farigh areas of northeast Libya have resulted in a regional correlation of Neogene strata in As-Sahabi area with equivalent units in nearby areas. As a result of this work, it is suggested that the use of certain informal stratigraphic names in As-Sahabi area should be discontinued. These suggestions are listed as follows:

- Formation "M" of de Heinzelin and El-Arnauti (1982, 1983, 1987) is a Miocene carbonate unit that is equivalent to the Regima Limestone of Desio (1935), the Sabkhat Al Hamra Member of the Sahabi Formation of Giglia (1984), and the Wadi Al Farigh Member of the Sahabi Formation of Giglia (1984). This unit of carbonate strata is capped by a regional unconformity. The rules of stratigraphic nomenclature indicate that the name that was first published (Regima Limestone) should be retained, and the other stratigraphic names should be discarded. In this case, however, the name Regima Limestone was formally changed to Ar Rajmah Formation by Klen (1974) and later to Ar Rajmah Group by El-Hawat and Abdulsamad (2004).
- Formation "P" and lower part of Member "T" of de Heinzelin and El-Arnauti (1982, 1983, 1987) is a unit of Miocene gypsiferous sand and mud that is equivalent to the Sabkhat Al Qunnayyin Member of the Sahabi Formation of Giglia (1984). An unconformity separates these strata from the underlying carbonate strata (Ar Rajmah Group). The name for these strata that was first published in accordance to the rules of stratigraphic nomenclature is the Sabkhat Al Qunnayyin Member of the Sahabi Formation (Giglia, 1984), and this is the name that should be used. This gypsiferous sand and mud should be considered as the lower gypsiferous member of the Sahabi Formation, and the name Sahabi Formation should be restricted to the post-carbonate strata (i.e., above the Ar Rajmah Group) that are below the regional unconformity at the base of the Lower Pliocene sand and pebbly sand mapped as the Qarat Weddah Formation.
- The upper part of Members "T," "U1," "UD," and "U2" of de Heinzelin and El-Arnauti (1982, 1983, 1987) consist of Upper Miocene (Messinian) sand and mud that are capped by a regional unconformity represented by the cut in the Sahabi Channel. These units should be referred to collectively as the upper member and named as the Qaret Makada Member of the Sahabi Formation. This designation of an upper member should be restricted to the strata that lie above the gypsiferous sand and mud designated as the Sabkhat Al Qunnayyin Member of the Sahabi Formation and below the regional unconformity at the base of Lower Pliocene sand and pebbly sand designated as the Qaret Weddah Formation.
- Member "V" of de Heinzelin and El-Arnauti (1982, 1983, 1987) is a unit of Lower Pliocene sand and pebbly sand that is equivalent to the Qarat Weddah Formation (Garet Uedda Formation) Formation of Di Cesare

et al. (1963). The name that was first published following the rules of stratigraphic nomenclature (Garet Uedda Formation) should be used. In this case, however, the name Garet Uedda Formation was formally changed to Qarat Weddah Formation by Carmignani (1984) on the basis of new Arabic transcription, and this new transcription (Qarat Weddah Formation) should be used for this stratigraphic unit.

- Member "Z" of de Heinzelin and El-Arnauti (1982, 1983, 1987) is a unit of pedogenically altered carbonate that is equivalent to the top part of the Qarat Weddah Formation (Garet Uedda Formation) as mapped by Giglia (1984). This unit has been named Shagig and ranked as a formation. Therefore, the stratigraphic term Shagig Formation should replace both previous names of Member "Z" and Formation "Z" (Muftah, 2013) and should be used accordingly onward.

Acknowledgments

The authors are very grateful to the USGS stratigrapher, Professor Christopher Swezey for his support, encourage and reading the manuscript several times. His comments and edits have extremely improved the manuscript and put it in this shape. We also would like to thank the faculty and staff members of the Department of Earth Sciences at the Faculty of Science, University of Benghazi, Libya. The senior author (MHE) would like to send special thanks to my supervisors Prof. Ahmed El Hawat and Prof. Ali El-Arnauti for their support and help during the field work since 1994. Special thanks to engineers Seraj El Shawaihdi and Mohamed El Ebaidi for redrawing the figures with high resolution.

Declarations

None of the authors have any direct or indirect conflicts of interest or competing financial interests to report.

References

Agustì, J., 2008. New data on the rodent fauna from As-Sahabi, Libya. In: Boaz, N.T., El-Arnauti, A., Pavlakis, P., Salem, M.J. (Eds.), Circum Mediterranean Geology and Biotic Evolution During the Neogene Period: The Perspective From Libya. 5. Garyounis Scientific Bulletin, pp. 139–143.

Agustì, J., El-Arnauti, A., Moyà Solà, S., Köhler, M., Galobart, A., Gaete, R., Llenas, M., 2000. Results of a field-campaign in the Late Miocene of the Sahabi Formation (Libya). In: *XIth Regional Committee on Mediterranean Neogene Stratigraphy (RCMNS) Congress*, Fez (Morocco). Abstracts. 1, p. 59.

Argyriou, T., Otero, O., Pavlakis, P., Noel, T.B., et al., 2012. Description and paleobiogeographical implications of new Semlikiichthys (Teleostei, Perciformes) fish material from the Late Miocene deposits of Sahabi, Libya. Geobios 45, 429–436. https://doi.org/10.1016/j.geobios.2012.01.002.

Barr, F.T., Walker, B.R., 1973. Late Tertiary channel system in northern Libya and its implications on Mediterranean sea-level changes. In: *Initial Reports, Deep Sea Drilling Project*, XIII. 2, pp. 1244–1250.

Bernor, R.L., Scott, R.S., 2003. New interpretations of the systematics, biogeography and paleoecology of the Sahabi Hipparions (latest Miocene) (Libya). Geodiversitas 25 (2), 297–319.

Bernor, R.L., Heissig, K., Tobien, H., 1987. Early Pliocene Perissodactyla from Sahabi, Libya. In: Boaz, N.T., El-Arnauti, A., Gaziry, A.W., de Heinzelin, J., Boaz, D.D. (Eds.), Neogene Paleontology and Geology of Sahabi. Alan R. Liss, New York, pp. 233–254.

Bernor, R.L., Kaiser, T.M., Wolf, D., et al., 2008. Revisiting Sahabi Equid Species Diversity, Biogeographic Patterns and Diet Preferences. In: Garyounis Scientific Bulletin, pp. 159–167 (Special Issue, 5).

Bernor, R.L., Boaz, N.T., Rook, L., 2012. *Eurygnathohippus feibeli* (Perissodactyla: Mammalia) from the Late Miocene of As-Sahabi (Libya) and its Evolutionary and Biogeographic Significance. Boll. Soc. Paleontol. Ital. 51 (1), 39–48.

Bernor, R.L., Boaz, N.T., Cirlli, O., El-Shawaihdi, M.H., Rook, L., et al., 2020. Sahabi *Eurygnathohippus feibeli*: Its systemic, stratigraphic, chronologic and biogeoraphic contexts. Rivista Italiana di Paleontologia e Stratigrafia 126 (2), 561–581. https://doi.org/10.13130/2039-4942/13937.

Bianucci, G., Carone, G., Domning, D.P., Landini, W., Rook, L., Sorbi, S., et al., 2008. Peri-Messinian dwarfing in Mediterranean Metaxytherium (Mammlia, Sirenia):: evidence of habitat degradation related to the Messinian salinity crisis. In: Boaz, N.T., El-Arnauti, A., Pavlakis, P., Salem, M.J. (Eds.), Circum Mediterranean Geology and Biotic Evolution during the Neogene Period: The Perspective from Libya, 5. Grayounis Scientific Bulletin, pp. 145–157. Special Issue.

Boaz, N.T., Gaziry, A.W., De Heinzelin, J., El-Arnauti, A., 1982. Results from the International Sahabi Research Project (Geology and Paleontology). In: Garyounis Scientific Bulletin. Special Issue 4.

Boaz, N.T., El-Arnauti, A., Gaziry, A.W., De Heinzelin, J., Boaz, D.D., 1987. *Neogene Paleontology and Geology of Sahabi*. Alan R. Liss, New York, pp. 401–pp.

Boaz, N.T., El-Arnauti, A., Agusti, J., Bernor, R., Pavlakis, P., Rook, L., 2008. Temporal, lithostratigraphic, and biochronologic setting of the Sahabi Formation, North-Central Libya. In: Salem, M.J., El-Arnauti, A., El-Sogher, A. (Eds.), The Geology of East Libya, Sedimentary Basins of Libya. *Earth Science Society of Libya*, Tripoli, pp. 233–246.

Bonarelli, G., 1947. Dinosaurio Fossile del Sahara Cirenaico. Rivista di Biologia Coloniale. Roma 8, 23–33.

Carmignani, L., 1984. Geological map of Libya 1:250.000, Sheet: Wadi Al Hamim (NH 34-7). *Industrial Research Center* (IRC), Tripoli, Libya. 1 Sheet and Explanatory Booklet, 98 pp.

Carmignani, L., Giglia, G., Pertusati, P.C., et al., 1990. The Qasr As-Sahabi succession and the Neogene evolution of the Sirte Basin (Libya). J. African Earth Sci. 10, 753–769.

Carmignani, L., Salvini, R., Bonciani, F., 2009. Did the Nile flow to the Gulf of Sirt during the late Miocene? Italian J. Geosci. (Boll. Soc. Geol. It.) 128, 2, 403–408.

Chumakov, I.S., 1973. Geological history of the Mediterranean at the end of the Miocene—the beginning of the Pliocene according to new data. In: Deep Sea Drilling Project (DSDP). 13. Reports and Publications, p. 144.

Cooke, H.B.S., 1982. A preliminary appraisal of fossil Suidae from Sahabi, Libya. In: Boaz, N.T., Gaziry, A.W., de Heinzelin, J., El-Arnauti, A. (Eds.), Results from the International Sahabi Research Project (Geology and Paleontology). Garyounis Scientific Bulletin, pp. 71–82. Special Issue 4.

Cooke, H.B.S., 1987. Fossil Suidae from Sahabi, Libya. In: Boaz, N.T., El-Arnauti, A., Gaziry, A.W., De Heinzelin, J., Boaz, D.D. (Eds.), *Neogene Paleontology and Geology of Sahabi*. Alan R. Liss, Inc., New York, pp. 255–266.

de Graciansky, P.-C., Hardenbol, J., Jacquin, T., Vail, P.R. (Eds.), 1998. Mesozoic and Cenozoic Sequence Stratigraphy of European Basins. 60. SEPM *Special Publication*. 786p.

de Heinzelin, J., El-Arnauti, A., 1982. Stratigraphy and geological history of the Sahabi and related formations. In: Boaz, N.T., Gaziry, A.W., de Heinzelin, J., El-Arnauti, A. (Eds.), Results from the International Sahabi Research Project (Geology and Paleontology). Garyounis Scientific Bulletin, pp. 5–12. Special Issue 4.

de Heinzelin, J., El-Arnauti, A. 1983. Geology of the Sahabi Area with a Geological Map at Scale 1:25 000. (Eds., the Section of Cartography and Photo-Interpretation of the Royal Museum for Central Africa, Tervuren, Belgium) in commission of the Research Center, Garyounis University, Benghazi, Libya, 59p with Annexes.

de Heinzelin, J., El-Arnauti, A., 1987. The Sahabi formation and related deposits. In: Boaz, N.T., El-Arnauti, A., Gaziry, A.W., de Heinzelin, J., Boaz, D.D. (Eds.), Neogene Paleontology and Geology of Sahabi. Alan R. Liss, New York, pp. 1–21.

de Heinzelin, J., El-Arnauti, A., Gaziry, W.A., 1980. A preliminary revision of the Sahabi Formation. In: Salem, M.J., Busrewil, M.T. (Eds.), The Geology of Libya. **I**, pp. 127–133.

Desio, A., 1935. Appunti geologici sui dintorni di Sahabi (Sirtica). Rend. R. Ist. Lomb. Sci. Lett. 68, 137–144. (ser. 2).

Di Cesare, F., Franchino, A., Sommaruga, C., 1963. The Pliocene-Quaternary of Giarabub Erg Region. Rev. Inst. Francais Pétrole 18 (10), 1344–1362.

Domning, D.P., 1982. Fossil Sirenia from the Sahabi Formation. In: Boaz, N.T., Gaziry, A.W., de Heinzelin, J., El-Arnauti, A. (Eds.), Results from the International Sahabi Research Project (Geology and Paleontology). Garyounis Scientific Bulletin, pp. 29–32. Special Issue 4.

Domning, D.P., Thomas, H., 1987. Metaxytherium serresii (Mammalia: Sirenia) from the early Pliocene of Libya and France: a reevaluation of its morphology, phyletic position, and biostratigraphic and Paleoecological significance. In: Boaz, N.T., El-Arnauti, A., Gaziry, A.W., de Heinzelin, J., Boaz, D.D. (Eds.), Neogene Paleontology and Geology of Sahabi. Alan R. Liss, New York, pp. 205–232.

Drake, N.A., El-Hawat, A.S., Salem, M.S., 2008. The development, decline, and demise of the As-Sahabi River system over the last seven million years. In: Boaz, N.T., El-Arnauti, A., Pavlakis, P., Salem, M.J. (Eds.), Circum Mediterranean Geology and Biotic Evolution during the Neogene Period: The Perspective from Libya. Garyounis Scientific Bulletin, pp. 95–109. Special Issue 5.

El-Hawat, A.S., Abdulsamad, E.O., 2004. Geology and archeology of Cyrenaica north eastern Libya. In: Guerrieri, L., Rischia, I., Serva, L. (Eds.), Field Trip Guidebooks, 32nd International Geological Congress (Florence, Italy; 20–28 August 2004): Memorie Descriptive Della Carta Geologica d'Italia, **62** (1), from PR01 to B15. Agenzia per la Protezione dell'Ambiente e per I Servizi Tecnici (APAT), Rome, Italy. 29 pp.

El-Shawaihdi, M.H., 2015. Geochronology, Petrology, and Stratigraphy of Neogene Units in the Sahabi Area, Northeastern Sirt Basin, Libya. New Mexico Institute of Mining and Technology (N. M. Tech.). PhD Dissertation, 152 p.

El-Shawaihdi, M.H., Muftah, A.M., Mozley, P.S., Boaz, N.T., 2014. New age constraints for Neogene sediments of the Sahabi area, Libya (Sirt Basin) using strontium isotope ($^{87}Sr/^{86}Sr$) geochronology and calcareous nannofossils. J. Afr. Earth Sci. 89, 42–49.

El-Shawaihdi, M.H., Mozley, P.S., Boaz, N.T., Salloum, F., Pavlakis, P., Muftah, A., Triantaphyllou, M., 2016. Stratigraphy of the Neogene Sahabi units in the Sirt Basin, Northeast Libya. J. Afr. Earth Sci. 118, 87–106.

Gallai, G., Bernor, R.L., Rook, L., et al., 2008. Systemitics and biogeographic relationships of the Sahabi Suidae. In: Boaz, N.T., El-Arnauti, A., Pavlakis, P., Salem, M.J. (Eds.), *Circum Mediterranean Geology and Biotic Evolution during the Neogene Period: The Perspective from Libya*. 5. Garyounis Scientific Bulletin (Special Issue, pp. 169–178.

Gaziry, A.W., 1982a. Anthracotheriidae and Hippopotamidae from the Sahabi formation. In: Boaz, N.T., Gaziry, A.W., de Heinzelin, J., El-Arnauti, A. (Eds.), Results from the International Sahabi Research Project (Geology and Paleontology). Garyounis Scientific Bulletin, pp. 63–70. Special Issue 4.

Gaziry, A.W., 1982b. Proboscidea from the Sahabi formation. In: Boaz, N.T., Gaziry, A.W., de Heinzelin, J., El-Arnauti, A. (Eds.), Results from the International Sahabi Research Project (Geology and Paleontology). *Garyounis Scientific Bulletin*, pp. 101–108. Special Issue 4.

Gaziry, A.W., 1987a. Remains of the Proboscidea from the early Pliocene of the Sahabi, Libya. In: Boaz, N.T., El-Arnauti, A., Gaziry, A.W., de Heinzelin, J., Boaz, D.D. (Eds.), Neogene Paleontology and Geology of Sahabi. Alan R. Liss, New York, pp. 183–203.

Gaziry, A.W., 1987b. Merycopotamus petrocchii (Artiodactyla, Mammalia) from Sahabi, Libya. In: Boaz, N.T., El-Arnauti, A., Gaziry, A.W., de Heinzelin, J., Boaz, D.D. (Eds.), Neogene Paleontology and Geology of Sahabi. Alan R. Liss, New York, pp. 287–302.

Gaziry, A.W., 1987c. Hexaprotodon sahabiensis (Artiodactyla, Mammalia): a new Hippopotamus from Libya. In: Boaz, N.T., El-Arnauti, A., Gaziry, A.W., de Heinzelin, J., Boaz, D.D. (Eds.), Neogene Paleontology and Geology of Sahabi. Alan R. Liss, New York, pp. 303–315.

Gentry, A.W., 2008. New records of Bovidae from Sahabi Formation. In: Boaz, N.T., El-Arnauti, A., Pavlakis, P., Salem, M.J. (Eds.), Circum Mediterranean Geology and Biotic Evolution During the Neogene Period: The Perspective from Libya. Garyounis Scientific Bulletin, pp. 205–215. Special Issue 5.

Ghoneim, E., Benedetti, M., El-Baz, F., 2012. An integrated remote sensing and GIS analysis of the Kufrah Paleoriver, eastern Sahara. Geomorphology 139-140, 242–257.

Giglia, G., 1984. Geological map of Libya, 1:250.000 Sheet: Ajdabiya (NH 34–6). *Industrial Research Center* (IRC), Tripoli, Libya. 1 Sheet and Explanatory Booklet, 93 pp.

Gladstone, R., Flecker, R., Valdes, P., Lunt, D., Markwick, P., 2007. The Mediterranean hydrologic budget from a late Miocene global climate simulation. Palaeogeogr. Palaeoclimatol. Palaeoecol. 251, 254–267.

Gradstein, F.M., Ogg, J.G., Schmitz, M.D., Ogg, G.M., 2012. The Geologic Time Scale 2012. 1 (p. 1-436) and 2 (p. 437-1144), Elsevier, Amsterdam.

Griffin, D.L., 2002. Aridity and humidity: two aspects of the late Miocene climate of North Africa. Palaeogeogr. Palaeoclimatol. Palaeoecol. 182, 65–91.

Griffin, D.L., 2006. Late Neogene Sahabi Rivers of the Sahara and their climatic and environmental implications for the Chad Basin. J. Geol. Soc. London 163, 905–921.

Griffin, D.L., 2011. The late Neogene Sahabi Rivers of the Sahara and the Hamadas of the eastern Libya-Chad border area. Palaeogeogr. Palaeoclimatol. Palaeoecol. 309, 176–185.

Guernet, C., 2005. Ostracodes et Stratigraphie du Néogène et du Quaternaire Méditerranéens. Rev. Micropaleontol. 48, 83–121.

Howell, F.C., 1982. Preliminary observations on Carnivora from the Sahabi Formation (Libya). In: Boaz, N.T., Gaziry, A.W., de Heinzelin, J., El-Arnauti, A. (Eds.), Results from the International Sahabi Research Project (Geology and Paleontology). *Garyounis Scientific Bulletin*, pp. 49–61. Special Issue 4.

Howell, F.C., 1987. Preliminary observations on Carnivora from the Sahabi formation (Libya). In: Boaz, N.T., El-Arnauti, A., Gaziry, A.W., de Heinzelin, J., Boaz, D.D. (Eds.), Neogene Paleontology and Geology of Sahabi. Alan R. Liss, New York, pp. 153–181.

Innocenti, F., Pertusati, P., 1984. Geological Map of Libya, 1:250.000, Sheet: Al Aqaylah (NH 34–5). *Industrial Research Center* (IRC), Tripoli, Libya. 1 Sheet and Explanatory Booklet, 105 pp.

Jenkins, D.G., Murray, J.W., 1989. Stratigraphic Atlas of Fossils Foraminifera, 2nd edition. *British Micropaleontology Society Series*. 593 pp.

Klen, L., 1974. Geological Map of Libya, 1:250.000, Sheet: Benghazi (NI 34–14). *Industrial Research Center* (IRC), Tripoli, Libya. 1 Sheet and Explanatory Booklet, 56 pp.

Kodym, O., 1985. Geological Map of Libya, 1:250.000, Sheet: Sabkhat Ghuzayil (NH 34–10). *Industrial Research Center* (IRC), Tripoli, Libya. 1 sheet and Explanatory Booklet (104 pp).

Krijgsman, W., Hilgen, F., Raffi, I., Sierro, F., Wilson, D., 1999. Chronology, causes, and progression of the Messinian salinity crisis. Nature 400, 652–655.

Lehmann, U., 1982. Preliminary report on Sahabi Bovidae. In: Boaz, N.T., Gaziry, A.W., de Heinzelin, J., El-Arnauti, A. (Eds.), Results from the International Sahabi Research Project (Geology and Paleontology). *Garyounis Scientific Bulletin*, Special Issue 4, pp. 87–93.

Maley, J., 2010. Climate and Paleoenvironment evolution in north tropical Africa from the end of the tertiary to the upper quaternary. Palaeoecol. Afr. 30, 227–278.

Martini, E., 1971. Standard tertiary and quaternary calcareous Nannoplankton zonation. In: Farinnacci, A. (Ed.), Proceeding of the Second Conference. **2**. Edizioni Tecnoscienca, Roma, pp. 739–785.

Monthe, J., 1987. Small-mammal fossils from the Pliocene Sahabi formation of Libya. In: Boaz, N.T., El-Arnauti, A., Gaziry, A.W., de Heinzelin, J., Boaz, D.D. (Eds.), Neogene Paleontology and Geology of Sahabi. Alan R. Liss, New York, pp. 135–144.

Muftah, A., 2013. Biostratigraphic and Temporal Relations Between the Neogene Sahabi and Marada Formations, Libya. In: Contribution to the Age Determination of their Contained Mammalian Paleofaunas. University of Athens, Greece, p. 226 (PhD Dissertation).

Muftah, A.M., Salloum, F.M., El-Shawaihdi, M.H., Al-Faitouri, M.S., 2008a. A contribution to the stratigraphy of formations of the As-Sahabi area, Sirt Basin, Libya. In: Boaz, N.T., El-Arnauti, A., Pavlakis, P., Salem, M.J. (Eds.), Circum Mediterranean Geology and Biotic Evolution during the Neogene Period: The Perspective from Libya. Garyounis Scientific Bulletin, Special Issue 5, pp. 33–45.

Muftah, A.M., El-Mehaghag, A.A., Starkie, S., 2008b. Biostratigraphical notes on the As Sahabi stratigraphic boreholes 1 and 2, Sirt Basin, Libya. In: Boaz, N.T., El-Arnauti, A., Pavlakis, P., Salem, M.J. (Eds.), Circum Mediterranean Geology and Biotic Evolution during the Neogene Period: The Perspective from Libya. Garyounis Scientific Bulletin, pp. 47–57. Special Issue 5.

Muftah, A.M., Pavlakis, P., Godelitsas, A., Gamaletsos, P., Boaz, N.T., et al., 2013. Paleogeography of the Eosahabi River in Libya: new insights into the mineralogy, geochemistry, and paleontology of member "U1" of the Sahabi formation, northeastern Libya. J. African Earth Sci. 78, 86–96. https://doi.org/10.1016/j.jafrearsci.2012.09.016.

North American Commission on Stratigraphic Nomenclature, 1983. North American stratigraphic code. Am. Assoc. Pet. Geol. Bull. 67, 841–875.

Paillou, P., Schuster, M., Tooth, S., Farr, T., Rosenqvist, A., Lopez, S., Malezieux, J.-M., 2009. Mapping of a major Paleodrianage system in eastern Libya using orbital imaging radar: the Kufrah River. Earth Planet. Sci. Lett. 277, 327–333.

Paillou, P., Tooth, S., Lopez, S., 2012. The Kufrah Paleodrianage system in Libya: a past connection to the Mediterranean Sea? C. R. Geosci. 344, 406–414.

Paillou, P., Lopez, S., Marais, E., Scipal, K., 2020. Mapping Paleohydrology of the ephemeral Kuiseb River, Namibia, from Radar Remote Sensing Water 12 (5), 1441.

Pavlakis, P., Boaz, N.T., 2008. Newly discovered remains of As-Sahabi Anthracotheriidae. In: Boaz, N.T., El-Arnauti, A., Pavlakis, P., Salem, M.J. (Eds.), Circum Mediterranean Geology and Biotic Evolution during the Neogene Period: The Perspective from Libya. Garyounis Scientific Bulletin, Special Issue 5, pp. 189–204.

Pèrez-Asensio, L.N., Aguirre, J., 2010. Benthic foraminiferal assemblages in temperate coral-bearing deposits from the late Pliocene. J. Foraminiferal Res. 40 (1), 61–78.

Rook, L., Sardella, R., 2008. An overview of the as Sahabi carnivore guild with description of new specimens from E.L.N.R.P. field surveys. In: Garyounis Scientific Bulletin, pp. 257–263. (Special Issue, No. 5).

Roveri, M., Flecker, R., Krijgsman, W., Lofi, J., Lugli, S., Manzi, V., Sierro, F.J., Bertini, A., Camerlenghi, A., De Lange, G., Govers, R., Hilgen, F.J., Hübscher, C., Meijer, P.-T., Stoica, M., 2014. The Messinian salinity crisis: past and future of great challenge for marine sciences. Mar. Geol. 352, 25–58.

Salem, M.J., Spreng, A.C., 1978. Middle Miocene stratigraphy, Al Khums area Northwestern Libya. In: Salem, M.J., Busrewil, M.T. (Eds.), Geology of Libya. 2nd Symposium on Geology of Libya, **I**, Tripoli, pp. 97–116.

Salvador, A. (Ed.), 1994. International Stratigraphic Guide, second edition. *International Union of Geological Sciences and Geological Society of America*, Boulder, Colorado (U.S.A.) (214 p).

Sanders, W.J., 2008. Review of fossil proboscidae from the late miocene-early pliocene site of As Sahabi, Libya. In: Boaz, N.T., El-Arnauti, A., Pavlakis, P., Salem, M.J. (Eds.), *Circum Mediterranean Geology and Biotic Evolution During the Neogene Period: The Perspective From Libya*, Special Issue. 5. Garyounis Scientific Bulletin, pp. 241–256.

Tobien, H., 1982. Preliminary report on the Equidae (Perissodactyla, Mammalia) from the Sahabi Formation (Libya). In: Boaz, N.T., Gaziry, A.W., de Heinzelin, J., El-Arnauti, A. (Eds.), Results from the International Sahabi Research Project (Geology and Paleontology). *Garyounis Scientific Bulletin*, pp. 83–85. Special Issue 4.

Werdelin, L., Peigné, S., 2010. Chapter: 32. In: Werdelin, L., Sanders, W.J. (Eds.), *Cenozoic Mammals of Africa*. University of California Press, Carnivora, pp. 603–657.

Whitmore, F.C., JR., 1982. Remains of Delphinidae from the Sahabi Formation. In: Boaz N.T., Gaziry A.W., de Heinzelin J., El-Arnauti A. (Eds.), *Results from the International Sahabi Research Project (Geology and Paleontology)*. Garyounis Scientific Bulletin 27–28.

Whitmore Jr., F.C., 1987. Cetacae from the Sahabi Formation, Libya. In: Boaz, N.T., El-Arnauti, A., Gaziry, A.W., De Heinzelin, J., Boaz, D.D. (Eds.), *Neogene Paleontology and Geology of Sahabi*. Alan R. Liss, New York, pp. 145–151.

Kolibri 360 GmbH
Ferdinand-Jühlke-Straße 7
99095 Erfurt, Deutschland
produktsicherheit@kolibri360.de